Ar

Terms and Military Symbols

December 2015

United States Government
US Army

Contents

Distribution Restriction: Approved for public release; distribution is unlimited.

This publication supersedes ADRP 1-02, dated 2 February 2015.

Figures

Tables

Preface

Army Doctrine Reference Publication (ADRP) 1-02 constitutes approved Army doctrinal terminology and symbology for general use. It builds on the foundational doctrine established in Army Doctrine Publication (ADP) 1-02.

The principal audience for ADRP 1-02 is all members of the profession of Arms. Commanders and staffs of Army headquarters serving as joint task force or multinational headquarters should also refer to applicable joint or multinational doctrine concerning the range of military operations and joint or multinational forces. Trainers and educators throughout the Army will also use this publication.

Commanders, staffs, and subordinates ensure their decisions and actions comply with applicable U.S., international, and, in some cases, host nation laws and regulations. Commanders at all echelons ensure their Soldiers operate in accordance with the law of war and the rules of engagement. (See Field Manual [FM] 27-10.)

This publication implements the following international agreements:

- AAP 15(2014). *NATO Glossary of Abbreviations Used in NATO Documents and Publications (English and French)*. 5 May 2014.
- STANAG 1059 (ED. 8). *Letter Codes for Geographical Entities*. 19 February 2004.
- STANAG 1241 (ED. 5). *NATO Standard Identity Description Structure for Tactical Use*. 6 April 2005.
- STANAG 2019 (ED. 6)/APP-6 (C). *NATO Joint Military Symbology*. 24 May 2011.
- STANAG 2291 (ED. 4)/APP-19 (D). *NATO Combat Engineer Glossary*. November 2003.
- STANAG 3680 (ED. 5)/AAP-6 (2012) (2). *NATO Glossary of Terms and Definitions (English and French)*. 20 April 2014.

ADRP 1-02 uses joint terms where applicable.

ADRP 1-02 applies to the Active Army, Army National Guard/Army National Guard of the United States, and United States Army Reserve unless otherwise stated.

The proponent of ADRP 1-02 is the United States Army Combined Arms Center. The preparing agency is the Combined Arms Doctrine Directorate, United States Army Combined Arms Center. Send written comments and recommendations on DA Form 2028 (*Recommended Changes to Publications and Blank Forms*) to Commander, U.S. Army Combined Arms Center and Fort Leavenworth, ATTN: ATZL MCK D (ADRP 1-02), 300 McPherson Avenue, Fort Leavenworth, KS 66027 2337; by e-mail to usarmy.leavenworth.mccoe.mbx.cadd-org-mailbox@mail.mil; or submit an electronic DA Form 2028.

Introduction

This revision of Army Doctrine Reference Publication (ADRP) 1-02 compiles definitions of all Army terms approved for use in Army doctrinal publications, including Army doctrine publications (ADPs), Army Doctrine Reference Publications (ADRPs), field manuals (FMs), and Army techniques publications (ATPs). It also includes joint terms appearing in the glossaries of Army doctrinal publications as of September 2015. ADRP 1-02 also lists shortened forms (whether considered acronyms or abbreviations) approved for use in Army doctrinal publications. In addition, unlike the 2013 edition of ADRP 1-02, this revision incorporates North Atlantic Treaty Organization (NATO) terms appearing in the glossaries of Army doctrinal publications as of January 2014.

This publication is augmented by the Army Dictionary online. Changes to terminology occur more frequently than traditional publication media can be updated. The terminology and symbology database, known as the Army Dictionary, is updated monthly to reflect the latest editions of Army publications. (To access the database, go to https://jdeis.js.mil/jdeis/index.jsp?pindex=207, and login with a common access card.) This database is an official Department of Defense (DOD) Web site, maintained by the Combined Arms Doctrine Directorate in collaboration with the Joint Staff Directorate for Joint Force Development. The site is part of the Joint Doctrine, Education, and Training Electronic Information System. It includes all Army doctrinal terms and all military symbols in MIL-STD 2525D, including air, land, maritime, space, activities control measures, and meteorological symbols. While the database includes the same joint terms appearing in ADRP 1-02, readers should consult Joint Publication (JP) 1-02 for up-to-date joint terminology.

ADRP 1-02 also provides a single standard for developing and depicting hand drawn and computer-generated military symbols for situation maps, overlays, and annotated aerial photographs for all types of military operations. It is the Army proponent publication for all military symbols and complies with Department of Defense (DOD) Military Standard (MIL-STD) 2525D. The symbology chapters of this ADRP focus primarily on military symbols applicable to Army land operations. When communicating instructions to subordinate units, commanders and staffs from company through corps echelons should use this publication as a dictionary of operational terms and military symbols.

ADRP 1-02 is organized as follows:

- Chapter 1 presents terms.
- Chapter 2 presents acronyms, abbreviations, and country codes.
- Chapter 3 introduces military symbology basics.
- Chapters 4 through 7 provide icons for units, individuals, organizations, equipment, installations, and activities.
- Chapter 8 introduces control measure symbols.
- Chapter 9 discusses tactical mission tasks.
- Chapter 10 discusses the course of action sketch.

The terminology entries in chapter 1 of this publication fall into three categories:

- Definitions applicable to the Army only.
- Joint (DOD) definitions commonly used in Army publications.
- North Atlantic Treaty Organization (NATO) definitions commonly used in Army publications.

For each term and definition, a proponent publication is cited in parentheses after the definition.

Definitions applicable to the Army only. The Army definition is preceded by "(Army)" if the term also has a joint definition that differs from the Army definition. (See the definition for "attack position" listed below.) If the term has no associated joint definition, the Army definition is not preceded by "(Army)". (See the definition for "situational understanding" listed below.) All Army unique definitions must be followed by the proponent Army publication in parentheses, as in the following examples:

attack position - (Army) The last position an attacking force occupies or passes through before crossing the line of departure. (ADRP 3-90)

situational understanding – The product of applying analysis and judgment to relevant information to determine the relationship among the operational and mission variables to facilitate decisionmaking. (ADP 5-0)

Definitions that are joint (DOD) and appear in the glossaries of Army publications. Each joint definition is preceded by "(DOD)." A cross-reference such as "See ADRP X-YY" follows the definition, signifying the publication discussing Army usage of the term, as in the following example:

airspace control – (DOD) A process used to increase operational effectiveness by promoting the safe, efficient, and flexible use of airspace. (JP 3-52) See ADRP 3-90, ADRP 5-0, FM 3-52, and FM 3-90-1.

Definitions that are NATO and appear in the glossaries of Army publications. Each NATO definition is preceded by "(NATO)." A cross-reference such as "See ADRP X-YY" follows the definition, signifying the publication discussing Army usage of the term, as in the following example:

explosive ordnance disposal incident – (NATO) The suspected or detected presence of unexploded explosive ordnance, or damaged explosive ordnance, which constitutes a hazard to operations, installations, personnel or material. Not included in this definition are the accidental arming or other conditions that develop during the manufacture of high explosive material, technical service assembly operations or the laying of mines and demolition charges. (STANAG 3680) See ATP 4-32.

DOD and NATO terms with multiple definitions. The related definition number will precede the definition to denote the definition(s) applicable to Army doctrine. For example the DOD term "demonstration" has two different definitions (1 and 2) for this term, and the Army only uses definition number 2 in doctrine:

demonstration – (DOD) 2. In military deception, a show of force in an area where a decision is not sought that is made to deceive an adversary. It is similar to a feint but no actual contact with the adversary is intended. (JP 3-13.4) See FM 3-90-1.

In addition, two other descriptors may appear after a definition:

- Also called.
- See also.

Also called. If a term has a shortened form (acronym or abbreviation) approved for doctrinal use, the shortened form appears after the definition, preceded by *also called*, as in the following example:

after action review – A guided analysis of an organization's performance, conducted at appropriate times during and at the conclusion of a training event or operation with the objective of improving future performance. It includes a facilitator, event participants, and other observers. Also called AAR. (ADRP 7-0)

See also. If related terms are defined elsewhere in ADRP 1-02, they are cross-referenced after the definition. The related terms are bolded and preceded by "See also," as in the following example:

area security – A security task conducted to protect friendly forces, installation routes, and actions within a specific area. (ADRP 3-90) See also **area reconnaissance, security operations, rear area security**.

The acronym and abbreviation entries listed in section I of chapter 2 are Army and joint. Shortened forms applicable only to Army doctrine are shown in boldface, to distinguish Army from joint usage. Acronyms are added to ADRP 1-02 and the online "Army Dictionary" when a defined term has an associated acronym or if an acronym has doctrinal cross branch or functional usage. The cross branch usage criterion for considering an acronym for inclusion is that it must appear in two or more doctrinal manuals that are not branch specific or related.

The symbology chapters (chapters 3 through 10) provide detailed requirements for composing and constructing symbols. The rules for building a set of military symbols allow enough flexibility for users to create any symbol to meet their operational needs. Although this publication serves as the Army proponent for military symbols, within DOD, MIL-STD 2525D is the proponent for military symbols. This publication compiles control measure symbols. Readers can find defined terms used for symbology in chapter 1, including cross-references to publications that discuss usage of control measure symbols. All control measure symbols in this publication are linked to doctrine.

Chapter 1

Military Terms

This chapter presents selected military terms.

—A—

access control point – A corridor at the installation entrance through which all vehicles and pedestrians must pass when entering or exiting the installation. (ATP 3-39.32)

acknowledge – A directive from the originator of a communication requiring the addressee(s) to advise the originator that his communication has been received and understood. This term is normally included in the electronic transmission of orders to ensure the receiving station or person confirms the receipt of the order. (FM 6-02.53)

actions on contact – A series of combat actions, often conducted simultaneously, taken upon contact with the enemy to develop the situation. (ADRP 3-90)

active air defense – (DOD) Direct defensive action taken to destroy, nullify, or reduce the effectiveness of hostile air and missile threats against friendly forces and assets. (JP 3-01) See ADRP 3-09, ADRP 3-90, FM 3-90-1, ATP 3-27.5.

activity – (DOD) 1. A unit, organization, or installation performing a function or mission. 2. A function, mission, action, or collection of actions. Also called ACT. (JP 3-0) See ATP 3-55.12.

Adaptive Planning and Execution system – (DOD) A Department of Defense system of joint policies, processes, procedures, and reporting structures, supported by communications and information technology, that is used by the joint planning and execution community to monitor, plan, and execute mobilization, deployment, employment, sustainment, redeployment, and demobilization activities associated with joint operations. Also called APEX system. (JP 5-0) See ATP 3-05.2.

administrative contracting officer – Contracting officers whose duties are limited to contract administration. Also called ACO. (ATTP 4-10)

administrative control – (DOD) Direction or exercise of authority over subordinate or other organizations in respect to administration and support. Also called ADCON. (JP 1) See ADRP 5-0, FM 3-94, FM 6-0, ATP 3-53.1, ATP 4-32.16.

administrative movement – A movement in which troops and vehicles are arranged to expedite their movement and conserve time and energy when no enemy ground interference is anticipated. (FM 3-90-2)

advanced trauma management - Resuscitative and stabilizing medical or surgical treatment provided to patients to save life or limb and to prepare them for further evacuation without jeopardizing their well-being or prolonging the state of their condition. (FM 4-02)

adversary – (DOD) A party acknowledged as potentially hostile to a friendly party and against which the use of force may be envisaged. (JP 3-0) See ADRP 3-0, ADRP 3-37, FM 3-07, FM 3-53, FM 3-98, ATP 3-07.5, ATP 3-53.1, ATP 3-53.2.

aerial port – (DOD) An airfield that has been designated for the sustained air movement of personnel and materiel as well as an authorized port for entrance into or departure from the country where located. Also called APORT. See also port of debarkation, port of embarkation. (JP 3-17) See FM 4-01, ATP 3-17.2.

aeromedical evacuation – (DOD) The movement of patients under medical supervision to and between medical treatment facilities by air transportation. Also called AE. (JP 4-02) See ATP 4-02.2.

after action review – A guided analysis of an organization's performance, conducted at appropriate times during and at the conclusion of a training event or operation with the objective of improving future performance. It includes a facilitator, event participants, and other observers. Also called AAR. (ADRP 7-0)

agility – The ability of friendly forces to react faster than the enemy. (ADRP 3-90)

air and missile defense – (DOD) Direct [active and passive] defensive actions taken to destroy, nullify, or reduce the effectiveness of hostile air and ballistic missile threats against friendly forces and assets. Also called AMD. (JP 3-01) See ATP 3-14.5, ATP 3-27.5. (Army) The direct defensive actions taken to protect friendly forces by destroying or reducing the effectiveness of hostile air and ballistic missile threats against friendly forces and assets in support of joint force commanders' objectives. (ADRP 3-09)

air assault – (DOD) The movement of friendly assault forces by rotary-wing aircraft to engage and destroy enemy forces or to seize and hold key terrain. (JP 3-18) See FM 3-90-1, FM 3-99.

air assault force – (DOD) A force composed primarily of ground and rotary-wing air units organized, equipped, and trained for air assault operations. (JP 3-18) See FM 3-99.

air assault operation – (DOD) An operation in which assault forces, using the mobility of rotary wing assets and the total integration of available firepower, maneuver under the control of a ground or air maneuver commander to engage enemy forces or to seize and hold key terrain. (JP 3-18) See FM 3-99.

airborne assault – (DOD) The use of airborne forces to parachute into an objective area to attack and eliminate armed resistance and secure designated objectives. (JP 3-18) See FM 3-99.

airborne mission coordinator – (DOD) The designated individual that serves as an airborne extension of the component commander or supported commander responsible for the personnel recovery mission. Also called AMC. (JP 3-50) See ATP 3-55.6.

airborne operation – (DOD) An operation involving the air movement into an objective area of combat forces and their logistic support for execution of a tactical, operational, or strategic mission. (JP 3-18) See FM 3-99.

air apportionment – (DOD) The determination and assignment of the total expected effort by percentage and/or by priority that should be devoted to the various air operations for a given period of time. (JP 3-0) See ATP 3-52.2.

air defense – (DOD) Defensive measures designed to destroy attacking enemy aircraft or missiles in the atmosphere, or to nullify or reduce the effectiveness of such attack. Also called AD. (JP 3-01) See FM 3-01, FM 3-01.7, ATP 3-01.18.

air defense artillery – (Army) The defensive measures designated to destroy attacking enemy aircraft or missiles in the atmosphere, or to nullify or reduce the effectiveness of such attack either through surveillance actions or active engagements of aerial threat. (ADRP 3-09)

airdrop – (DOD) The unloading of personnel or materiel from aircraft in flight. (JP 3-17) See ATP 4-48.

airfield – (DOD) An area prepared for the accommodation (including any buildings, installations, and equipment), landing, and takeoff of aircraft. See also departure airfield; landing area; landing site. (JP 3-17) See FM 3-99, ATP 3-17.2.

air-ground operations – The simultaneous or synchronized employment of ground forces with aviation maneuver and fires to seize, retain, and exploit the initiative. Also called AGO. (FM 3-04)

airhead – (DOD) 1. A designated area in a hostile or potentially hostile operational area that, when seized and held, ensures the continuous air landing of troops and materiel and provides the maneuver space necessary for projected operations. Also called a lodgment area. (JP 3-18) See FM 3-99.

airhead line – (DOD) A line denoting the limits of the objective area for an airborne assault. (JP 3-18) See FM 3-99.

air interdiction – (DOD) Air operations conducted to divert, disrupt, delay, or destroy the enemy's military surface capabilities before it can be brought to bear effective against friendly forces, or to otherwise achieve objectives that are conducted at such distances from friendly forces that detailed integration of each air mission with the fire and movement of friendly forces is not required. (JP 3-03) See FM 3-09, ATP 3-04.64, ATP 3-09.34, ATP 3-52.2, ATP 3-55.6.

airland – (DOD) Move by air and disembark, or unload, after the aircraft has landed or while an aircraft is hovering. (JP 3-17) See ATP 4-48.

air liaison officer – (DOD) The senior tactical air control party member attached to a ground unit who functions as the primary advisor to the ground commander on air power. An air liaison officer is usually an aeronautically rated officer. Also called ALO. (JP 3-09.3) See FM 4-01, FM 6-05.

Air Mobility Command – (DOD) The Air Force component command of the United States Transportation Command. Also called AMC. (JP 3-17) See FM 4-01.

air movement – (DOD) Air transport of units, personnel, supplies, and equipment including airdrops and air landings. (JP 3-17) See FM 3-99, ATP 4-48.

air movements – (Army) Operations involving the use of utility and cargo rotary-wing assets for other than air assaults. (FM 3-90-2)

airspace control – (DOD) A process used to increase operational effectiveness by promoting the safe, efficient, and flexible use of airspace. (JP 3-52) See ADRP 3-90, ADRP 5-0, FM 3-52, FM 3-90-1, ATP 3-52.1, ATP 3-52.2.

airspace control area – (DOD) Airspace that is laterally defined by the boundaries of the operational area, and may be subdivided into airspace control sectors. (JP 3-01)

airspace control authority – (DOD) The commander designated to assume overall responsibility for the operation of the airspace control system in the airspace control area. Also called ACA. (JP 3-52) See FM 3-01, FM 3-09, FM 3-52, ATP 3-04.64, ATP 3-52.1, ATP 3-52.2.

airspace control order – (DOD) An order implementing the airspace control plan that provides the details of the approved requests for airspace coordinating measures. It is published either as part of the air tasking order or as a separate document. Also called ACO. (JP 3-52) See FM 3-09, FM 3-52, ATP 2-01, ATP 3-01.15, ATP 3-04.64, ATP 3-06.1, ATP 3-60.2, ATP 3-09.34, ATP 3-52.3.

airspace control plan – (DOD) The document approved by the joint force commander that provides specific planning guidance and procedures for the airspace control system for the joint force operational area. Also called ACP. (JP 3-52) See FM 3-01, ATP 3-04.64, ATP 3-09.34, ATP 3-52.1, ATP 3-52.2, ATP 3-52.3.

airspace control system – (DOD) An arrangement of those organizations, personnel, policies, procedures, and facilities required to perform airspace control functions. Also called ACS. (JP 3-52) See FM 3-52.

airspace coordinating measures – (DOD) Measures employed to facilitate the efficient use of airspace to accomplish missions and simultaneously provide safeguards for friendly forces. Also called ACM. (JP 3-52) See FM 3-09, FM 3-52, FM 3-99, ATP 3-09.34, ATP 3-52.1, ATP 3-52.2, ATP 3-60.2.

airspace coordination area – (DOD) A three-dimensional block of airspace in a target area, established by the appropriate ground commander, in which friendly aircraft are reasonably safe from friendly surface fires. The airspace coordination area may be formal or informal. Also called ACA. (JP 3-09.3) See FM 3-09, FM 3-99, ATP 3-60.2, ATP 3-09.24.

airspace management – (DOD) The coordination, integration, and regulation of the use of airspace of defined dimensions. (JP 3-52) See FM 3-96.

air support operations center – (DOD) The principal air control agency of the theater air control system responsible for the direction and control of air operations directly supporting the ground combat element. It coordinates air missions requiring integration with other supporting arms and ground forces. It normally collocates with the Army tactical headquarters senior fire support coordination center within the ground combat element. Also called ASOC. (JP 3-09.3) See ATP 3-04.64, ATP 3-60.2.

air tasking order – (DOD) A method used to task and disseminate to components, subordinate units, and command and control agencies projected sorties, capabilities and/or forces to targets and specific missions. Also called ATO. (JP 3-30) See FM 3-09, ATP 2-01, ATP 3-01.15, ATP 3-04.64, ATP 3-06.1, ATP 3-60.2.

air terminal – (DOD) A facility on an airfield that functions as an air transportation hub and accommodates the loading and unloading of airlift aircraft and the intransit processing of traffic. (JP 3-17) See ATP 4-13.

alkalinity – The content of carbonates, bicarbonates, hydroxides, and occasionally berates, silicates, and phosphates in water. (ATP 4-44)

alliance – (DOD) The relationship that results from a formal agreement between two or more nations for broad, long-term objectives that further the common interests of the members. (JP 3-0) See ADRP 3-0, FM 3-07, FM 3-16, FM 4-95.

allocation – (DOD) Distribution of limited forces and resources for employment among competing requirements. (JP 5-0) See FM 3-09, ATP 2-01.

all-source intelligence – (DOD) 1. Intelligence products and/or organizations and activities that incorporate all sources of information, most frequently including human intelligence, imagery intelligence, measurement and signature intelligence, signals intelligence, and open-source data in the production of finished intelligence. See FM 3-24. 2. In intelligence collection, a phrase that indicates that in the satisfaction of intelligence requirements, all collection, processing, exploitation, and reporting systems and resources are identified for possible use and those most capable are tasked. See also **intelligence.** (JP 2-0) See ATP 3-05.20. (Army) The integration of intelligence and information from all relevant sources to analyze situations or conditions that impact operations. (ADRP 2-0)

alternate position – A defensive position that the commander assigns to a unit or weapon for occupation when the primary position becomes untenable or unsuitable for carrying out the assigned task. (ADRP 3-90)

alternate supply route – A route or routes designated within an area of operations to provide for the movement of traffic when main supply routes become disabled or congested. Also called ASR. (FM 4-01) See also **area of operations, main supply route**.

altitude – The vertical distance of a level, a point or an object considered as a point, measured from mean sea level. (ATP 3-09.30)

ambulance control point – This consists of a Soldier (from the ambulance company or platoon) stationed at a crossroad or road junction where ambulances may take one of two or more directions to reach loading points. The Soldier, knowing from which location each loaded ambulance has come, directs empty ambulances returning from the rear. (ATP 4-02.2)

ambulance exchange point – A location where a patient is transferred from one ambulance to another en route to a medical treatment facility. This may be an established point in an ambulance shuttle or it may be designated independently. Also called AXP. (ATP 4-02.2)

ambulance loading point – This is the point in the shuttle system where one or more ambulances are stationed ready to receive patients for evacuation. (ATP 4-02.2)

ambulance relay point – This is a point in the shuttle system where one or more empty ambulances are stationed. They are ready to advance to a loading point or to the next relay post to replace an ambulance that has moved. As a control measure, relay points are generally numbers from front to rear. (ATP 4-02.2)

ambulance shuttle system – This is an effective and flexible method of employing ambulances during operations. It consists of one or more ambulance loading points, relay points, and when necessary, ambulance control points, all echeloned forward from the principal group of ambulances, the company location, or basic relay points as tactically required. (ATP 4-02.2)

ambush – An attack by fire or other destructive means from concealed positions on a moving or temporarily halted enemy. (FM 3-90-1)

ammunition load – A support package designed or tailored specifically for munitions operations. (ATP 4-35)

ammunition supply point – An ammunition support activity operated by one or more modular ammunition platoons. (ATP 4-35)

ammunition support activity – Locations that are designated to receive, store, maintain, and provide munitions support to Army forces. (FM 4-30)

ammunition transfer holding point – A designated site operated by a brigade support battalion distribution company where ammunition is received, transferred, or temporarily stored to supported units within a brigade combat team. Also called ATHP. (ATP 4-35)

amphibious operation – (DOD) A military operation launched from the sea by an amphibious force, embarked in ships or craft with the primary purpose of introducing a landing force ashore to accomplish the assigned mission. (JP 3-02) See ATP 3-52.3.

anticipation – The ability to foresee operational requirements and initiate actions that satisfy a response without waiting for an operation order or fragmentary order. (ADP 4-0)

antiterrorism – (DOD) Defensive measures used to reduce the vulnerability of individuals and property to terrorist acts, to include rapid containment by local military and civilian forces. Also called AT. (JP 3-07.2) See ATP 2-01.3.

apportionment – (DOD) In the general sense, distribution of forces and capabilities as the starting point for planning. (JP 5-0) See FM 3-09, ATP 2-01, ATP 3-52.2.

approach march – The advance of a combat unit when direct contact with the enemy is intended. (ADRP 3-90)

area air defense commander – (DOD) The component commander with the preponderance of air defense capability and the required command, control, and communications capabilities who is assigned by the joint force commander to plan and execute integrated air defense operations. Also called AADC. (JP 3-01) See FM 3-01, ATP 3-27.5, ATP 3-52.2.

area command – In unconventional warfare, the irregular organizational structure established within an unconventional warfare operational area to command and control irregular forces advised by Army Special Forces. (ATP 3-05.1)

area defense – A defensive task that concentrates on denying enemy forces access to designated terrain for a specific time rather than destroying the enemy outright. (ADRP 3-90)

area of influence – (DOD) A geographical area wherein a commander is directly capable of influencing operations by maneuver or fire support systems normally under the commander's command or control. (JP 3-0) See ADRP 3-0, ADRP 3-90, FM 3-90-1, FM 3-94, ATP 2-01.3, ATP 2-19.4.

area of interest – (DOD) That area of concern to the commander, including the area of influence, areas adjacent thereto, and extending into enemy territory. This area also includes areas occupied by enemy forces who could jeopardize the accomplishment of the mission. Also called AOI. (JP 3-0) See ADP 3-0, ADRP 3-90, FM 3-24, FM 3-90-1, ATP 2-19.4, ATP 3-55.6, ATP 4-02.55, ATP 5-0.1.

area of operations – (DOD) An operational area defined by the joint force commander for land and maritime forces that should be large enough to accomplish their missions and protect their forces. Also called AO. (JP 3-0) See ADP 1-01, ADRP 3-0, ADRP 3-90, FM 3-07, FM 3-24, FM 3-52, FM 3-90-1, FM 4-40, FM 6-05, ATP 1-06.3, ATP 2-01.3, ATP 3-09.34, ATP 3-52.2, ATP 3-53.2, ATP 3-55.6, ATP 3-60.2, ATP 4-02.2.

area of responsibility – (DOD) The geographical area associated with a combatant command within which a geographic combatant commander has authority to plan and conduct operations. Also called AOR. (JP 1) See ATP 3-52.2, ATP 3-55.6, ATP 4-43.

area reconnaissance – A form of reconnaissance that focuses on obtaining detailed information about the terrain or enemy activity within a prescribed area. (ADRP 3-90)

area security – A security task conducted to protect friendly forces, installation routes, and actions within a specific area. (ADRP 3-90) See also **area reconnaissance, security operations, rear area security**.

area support – Method of logistics, medical support, and personnel services in which support relationships are determined by the location of the units requiring support. Sustainment units provide support to units located in or passing through their assigned areas. (ATP 4-90)

ARFOR – The Army component and senior Army headquarters of all Army forces assigned or attached to a combatant command, subordinate joint force command, joint functional command, or multinational command. (FM 3-94)

Army Civilian Corps – A community within the Army Profession composed of civilians serving in the Department of the Army. (ADRP 1)

Army core competencies – The Army's essential and enduring capabilities that define the Army's fundamental contributions to the Nation's security. (ADP 1-01)

Army design methodology – A methodology for applying critical and creative thinking to understand, visualize, and describe unfamiliar problems and approaches to solving them. (ADP 5-0)

Army doctrine – Fundamental principles, with supporting tactics, techniques, procedures, and terms and symbols, used for the conduct of operations and which the operating force, and elements of the institutional Army that directly support operations, guide their actions in support of national objectives. It is authoritative but requires judgment in application. (ADP 1-01)

Army ethic – The evolving set of laws, values, and beliefs, embedded within the Army culture of trust that motivates and guides the conduct of the Army professionals bound together in common moral purpose. (ADRP 1)

Army Health System – A component of the Military Health System that is responsible for operational management of the health service support and force health protection missions for training, predeployment, deployment, and postdeployment operations. Army Health System includes all mission support services performed, provided, or arranged by the Army Medical Department to support health service support and force health protection mission requirements for the Army and as directed, for joint, intergovernmental agencies, coalition, and multinational forces. (FM 4-02)

Army leader – Anyone who by virtue of assumed role or assigned responsibility inspires and influences people to accomplish organizational goals. Army leaders motivate people both inside and outside the chain of command to pursue actions, focus thinking, and shape decisions for the greater good of the organization. (ADP 6-22)

Army personnel recovery – The military efforts taken to prepare for and execute the recovery and reintegration of isolated personnel. (FM 3-50)

Army Profession – A unique vocation of experts certified in the ethical design, generation, support, and application of landpower, serving under civilian authority and entrusted to defend the Constitution and the rights and interests of the American people. (ADRP 1)

Army professional – A Soldier or Army Civilian who meets the Army Profession's certification criteria in character, competence, and commitment. (ADRP 1)

Army requirements review board – The Army force commander's established board to review, validate, approve, and prioritize selected contract support requests. Also called ARRB. (ATP 4-92)

Army Service component command – (DOD) Command responsible for recommendations to the joint force commander on the allocation and employment of Army forces within a combatant command. Also called ASCC. (JP 3-31) See FM 3-94.

Army special operations aviation – Designated Active Component forces and units organized, trained, and equipped specifically to conduct air mobility, close combat attack, and other special air operations. (ADRP 3-05)

Army special operations forces – (DOD) Those Active and Reserve Component Army forces designated by the Secretary of Defense that are specifically organized, trained, and equipped to conduct and support special operations. Also called ARSOF. (JP 3-05) See FM 3-05, FM 3-18, ATP 3-75.

Army team building – A continuous process of enabling a group of people to reach their goals and improve their effectiveness through leadership and various exercises, activities and techniques. (FM 6-22)

artillery target intelligence zone – An area in enemy territory that the commander wishes to monitor closely. (FM 3-09)

art of command – The creative and skillful exercise of authority through timely decisionmaking and leadership. (ADP 6-0)

art of tactics – This consists of three interrelated aspects: the creative and flexible array of means to accomplish assigned missions, decisionmaking under conditions of uncertainty when faced with a thinking and adaptive enemy, and understanding the effects of combat on Soldiers. (ADRP 3-90)

assailable flank – A flank which is exposed to attack or envelopment. (ADRP 3-90) See also **flank**.

assault echelon – (Army) The element of a force that is scheduled for initial assault on the objective area. (ADRP 1-02)

assault position – A covered and concealed position short of the objective, from which final preparations are made to assault the objective. (ADRP 3-90)

assault time – The moment to attack the initial objective throughout the geographical scope of the operation. (ADRP 3-90)

assembly area – (Army) An area a unit occupies to prepare for an operation. (FM 3-90-1)

assessment – (DOD) 1. A continuous process that measures the overall effectiveness of employing joint force capabilities during military operations. See FM 3-07, FM 3-24. 2. Determination of the progress toward accomplishing a task, creating a condition, or achieving an objective. See ADP 3-37, ADP 5-0, ADRP 3-37, ADRP 5-0, FM 3-13, FM 3-24, FM 6-0, ATP 2-01, ATP 4-13, ATP 5-0.1, ATP 6-01.1. 3. Analysis of the security, effectiveness, and potential of an existing or planned intelligence activity. 4. Judgment of the motives, qualifications, and characteristics of present or prospective employees or "agents." (JP 3-0) See FM 3-07.

asset visibility – (DOD) Provides users with information on the location, movement, status, and identity of units, personnel, equipment, and supplies, which facilitates the capability to act upon that information to improve overall performance of the Department of Defense's logistics practices. Also called AV. (JP 3-35) See FM 4-01, ATP 3-35, ATP 4-0.1.

assign – (DOD) 1. To place units or personnel in an organization where such placement is relatively permanent, and/or where such organization controls and administers the units or personnel for the primary function, or greater portion of the functions, of the unit or personnel. (JP 3-0) See ADRP 5-0, FM 3-09, FM 6-0.

assumption – (DOD) A supposition on the current situation or a presupposition on the future course of events, either or both assumed to be true in the absence of positive proof, necessary to enable the commander in the process of planning to complete an estimate of the situation and make a decision on the course of action. (JP 5-0) See FM 6-0, ATP 2-19.3, ATP 5-0.1.

assured mobility – A framework—of processes, actions, and capabilities—that assures the ability of a force to deploy, move, and maneuver where and when desired, without interruption or delay, to achieve the mission. (ATTP 3-90.4)

attach – (DOD) 1. The placement of units or personnel in an organization where such placement is relatively temporary. (JP 3-0) See ADRP 5-0, FM 3-09, FM 6-0.

attack – An offensive task that destroys or defeats enemy forces, seizes and secures terrain, or both. (ADRP 3-90) See **also defeat, deliberate attack, demonstration, destroy; feint, offensive operations, raid, secure, seize, spoiling attack**.

attack by fire – A tactical mission task in which a commander uses direct fires, supported by indirect fires, to engage an enemy force without closing with the enemy to destroy, suppress, fix, or deceive that enemy. (FM 3-90-1) See also **destroy, fix, frontal attack, support by fire, suppress, tactical mission task**.

attack by fire position – The general position from which a unit conducts the tactical task of attack by fire. (ADRP 3-90) See also **attack by fire**.

attack guidance matrix – A targeting product approved by the commander, which addresses the how and when targets are engaged and the desired effects. Also called AGM. (ATP 3-60)

attack position – (Army) The last position an attacking force occupies or passes through before crossing the line of departure. (ADRP 3-90)

authenticate – (DOD) A challenge given by voice or electrical means to attest to the authenticity of a person, message, or transmission. (JP 3-50) See ATP 3-50.3.

authority – The delegated power to judge, act or command. (ADP 6-0)

auxiliary – For the purpose of unconventional warfare, the support element of the irregular organization whose organization and operations are clandestine in nature and whose members do not openly indicate their sympathy or involvement with the irregular movement. (ADRP 3-05)

available-to-load date – (DOD) A date specified for each unit in a time-phased force and deployment data indicating when that unit will be ready to load at the point of embarkation. Also called ALD. (JP 5-0) See FM 4-01.

avenue of approach – (DOD) An air or ground route of an attacking force of a given size leading to its objective or to key terrain in its path. Also called AA. (JP 2-01.3) See FM 6-0, ATP 2-01.3, ATP 2-19.4. (Army) The air or ground route leading to an objective (or key terrain in its path) that an attacking force can use. (ADRP 3-90)

axis of advance – (Army) The general area through which the bulk of a unit's combat power must move. (ADRP 3-90) See also **attack, movement to contact, offensive operations**.

azimuth – A horizontal angle measured clockwise from a north base line that could be true north, magnetic north, or grid north. (ATP 3-09.50)

azimuth of fire – The direction, expressed in mils, that a firing unit is laid (oriented) on when it occupies a position. (ATP 3-09.50)

azimuth of the orienting line – The direction from the orienting station to a designated end of the orienting line. (ATP 3-09.50)

—B—

back-azimuth – The direction equal to the azimuth plus or minus 3200 mils. (ATP 3-09.50)

backbrief – A briefing by subordinates to the commander to review how subordinates intend to accomplish their mission. (FM 6-0)

ballistic missile – (DOD) Any missile which does not rely upon aerodynamic surfaces to produce lift and consequently follows a ballistic trajectory when thrust is terminated. (JP 3-01) See ATP 3-14.5, ATP 3-27.5.

banking support – The provision of cash, non-cash and electronic commerce mechanisms necessary to support the theater procurement process and host nation banking infrastructure. (FM 1-06)

base – (DOD) A locality from which operations are projected or supported. (JP 4-0) See ADRP 3-0, FM 3-14, FM 4-95, ATP 3-91.

base camp – An evolving military facility that supports that military operations of a deployed unit and provides the necessary support and services for sustained operations. (ATP 3-37.10)

base cluster – (DOD) In base defense operations, a collection of bases, geographically grouped for mutual protection and ease of command and control. (JP 3-10) See ATP 3-91.

base defense – (DOD) The local military measures, both normal and emergency, required to nullify or reduce the effectiveness of enemy attacks on, or sabotage of, a base, to ensure that the maximum capacity of its facilities is available to U.S. forces. (JP 3-10) See ADRP 3-37, ATP 3-91.

base defense zone – (DOD) An air defense zone established around an air base and limited to the engagement envelope of short-range air defense weapons systems defending that base. Base defense zones have specific entry, exit, and identification, friend or foe procedures established. Also called BDZ. (JP 3-52) See ATP 3-52.3.

basic load – (DOD) The quantity of supplies required to be on hand within, and which can be moved by, a unit or formation. It is expressed according to the wartime organization of the unit or formation and maintained at the prescribed levels (JP 4-09). See ATP 4-35.

basic load (ammunition) – (Army) The quantity of nonnuclear ammunition that is authorized and required by each Service to be on hand for a unit to meet combat needs until resupply can be accomplished. It is expressed in rounds, units or unity of weight, as appropriate. (FM 3-01.7)

battalion – A unit consisting of two or more company-, battery-, or troop-size units and a headquarters. (ADRP 3-90) See also **battery, company**.

battalion aid station – The forward-most medically staffed treatment location organic to a maneuver battalion. (ATP 4-02.3)

battalion task force – A maneuver battalion-size unit consisting of a battalion headquarters, at least one assigned company-size element, and at least one attached company-size element from another maneuver or support unit (functional and multifunctional). (ADRP 3-90)

battery – A company-size unit in a fires or air defense artillery battalion. (ADRP 3-90)

battle – A set of related engagements that lasts longer and involves larger forces than an engagement. (ADRP 3-90) See also **campaign, engagement, major operation**.

battlefield coordination detachment – (DOD) An Army liaison located in the air operations center that provides selected operational functions between the Army forces and the air component commander. Also called BCD. (JP 3-03) See FM 6-05, ATP 3-01.15, ATP 3-09.13, ATP 3-60.2.

battle damage assessment – (DOD) The estimate of damage composed of physical and functional damage assessment, as well as target system assessment, resulting from the application of lethal or nonlethal military force. Also called BDA. (JP 3-0) See ATP 3-55.6, ATP 3-60.1, ATP 3-60.2.

battle handover line – A designated phase line on the ground where responsibility transitions from the stationary force to the moving force and vice versa. Also called BHL. (ADRP 3-90) See also **handover line, phase line**.

battle injury – (DOD) Damage or harm sustained by personnel during or as a result of battle conditions. Also called BI. (JP 4-02) See ATP 4-02.55.

battle management – (DOD) The management activities within the operational environment based on the commands, direction, and guidance given by appropriate authority. Also called BM. (JP 3-01) See ATP 3-27.5, ATP 3-52.2.

battle position – 1. A defensive location oriented on a likely enemy avenue of approach. (ADRP 3-90) 2. For attack helicopters, an area designated in which they can maneuver and fire into a designated engagement area or engage targets of opportunity. Also called BP. (ADRP 1-02) See also **airspace coordination area, avenue of approach**.

battle rhythm – A deliberate cycle of command, staff, and unit activities intended to synchronize current and future operations. (FM 6-0)

beach capacity – The per day estimate expressed in terms of measurement tons, weight tons, or cargo unloaded over a designated strip of shore. (ATP 4-13)

begin morning civil twilight – (DOD) The period of time at which the sun is halfway between beginning morning and nautical twilight and sunrise, when there is enough light to see objects clearly with the unaided eye. Also called BMCT. (JP 2-01.3) See ATP 2-01.3.

begin morning nautical twilight – (DOD) The start of that period where, in good conditions and in the absence of other illumination, the sun is 12 degrees below the eastern horizon and enough light is available to identify the general outlines of ground objects and conduct limited military operations. Also called BMNT. (JP 3-09.3) See ATP 2-01.3.

be-prepared mission – A mission assigned to a unit that might be executed. (FM 6-0) See also **on-order mission**.

biological agent – (DOD) A microorganism (or a toxin derived from it) that causes disease in personnel, plants, or animals or causes the deterioration of materiel. (JP 3-11) See ATP 3-05.11.

biological weapon – (DOD) An item of material which projects, disperses, or disseminates a biological agent including arthropod vectors. (JP 3-11) See ATP 4-02.84.

biometrics – (DOD) The process of recognizing an individual based on measurable anatomical, physiological, and behavioral characteristics. (JP 2-0) See FM 3-16, FM 3-24, ATP 3-90.15.

biometrics-enabled intelligence – (Army) The information associated with and or derived from biometric signatures and the associated contextual information that positively identifies a specific person and or matches an unknown identity to a place, activity, device, component, or weapon. Also called BEI. (ADRP 2-0)

blister agent – (DOD) A chemical agent that injures the eyes and lungs, and burns or blisters the skin. Also called vesicant agent. (JP 3-11) See ATP 3-05.11.

block – A tactical mission task that denies the enemy access to an area or prevents his advance in a direction or along an avenue of approach. Block is also an obstacle effect that integrates fire planning and obstacle effort to stop an attacker along a specific avenue of approach or to prevent the attacking force from passing through an engagement area. (FM 3-90-1) See also **avenue of approach, contain, disrupt, fix, tactical mission task, turn**.

blue kill box – A fire support and airspace coordination measure used to facilitate the attack of surface targets with air-to-surface munitions without further coordination with the establishing headquarters. Also called BKB. (ATP 3-09.34)

board – A grouping of predetermined staff representatives with delegated decision authority for a particular purpose or function. (FM 6-0) See also **working group, battle rhythm**.

boost phase – (DOD) That portion of the flight of a ballistic missile or space vehicle during which the booster and sustainer engines operate. (JP 3-01) See ATP 3-27.5.

boundary – (DOD) A line that delineates surface areas for the purpose of facilitating coordination and deconfliction of operations between adjacent units, formations, or areas. (JP 3-0) See FM 3-09, FM 3-90-1, FM 3-99.

bounding overwatch – A movement technique used when contact with enemy forces is expected. The unit moves by bounds. One element is always halted in position to overwatch another element while it moves. The overwatching element is positioned to support the moving unit by fire or fire and movement. (FM 3-90-2)

box formation – A unit formation with subordinate elements arranged in a box or square, or two elements up and two elements back. It is a flexible formation that provides equal firepower in all directions. It is generally used when the enemy location is known. This formation can cause 50 percent of force to be decisively engaged at the same time, therefore limiting the combat power available to maneuver against an enemy. (FM 3-90-1) See also **column formation, echelon formation**.

branch – (DOD) 1. A subdivision of any organization. See FM 3-07. 2. A geographically separate unit of an activity, which performs all or part of the primary functions of the parent activity on a smaller scale. See FM-3-07. 3. An arm or service of the Army. See FM 3-07. 4. The contingency options built into the base plan used for changing the mission, orientation, or direction of movement of a force to aid success of the operation based on anticipated events, opportunities, or disruptions caused by enemy actions and reactions. (JP 5-0) See FM 3-07, FM 6-0, ATP 2-01.

breach – A tactical mission task in which the unit employs all available means to break through or establish a passage through an enemy defense, obstacle, minefield, or fortification. (FM 3-90-1) See also **tactical mission task**.

breach area – The area where a breaching operation occurs. It is established and fully defined by the higher headquarters of the unit conducting breaching operations. (ATTP 3-90.4) See also **breaching operation**.

breaching operation – Operation conducted to allow maneuver despite the presence of obstacles. Breaching is a synchronized combined arms operation under the control of the maneuver commander. Breaching operations begin when friendly forces detect an obstacle and begin to apply the breaching fundamentals, and they end when battle handover has occurred between follow-on forces and a unit conducting the breaching operation. (ATTP 3-90.4) See also **follow-on forces**.

breakbulk ship – (DOD) A ship with conventional holds for stowage of breakbulk cargo and a limited number of containers, below or above deck, and equipped with cargo-handling gear. (JP 4-09) See FM 4-01.

breakout – An operation conducted by an encircled force to regain freedom of movement or contact with friendly units. It differs from other attacks only in that a simultaneous defense in other areas of the perimeter must be maintained. (ADRP 3-90) See also **encirclement, follow and support, main body**.

breakpoint chlorination – The application of chlorine to water containing free ammonia. (ATP 4-44)

breakthrough – A rupturing of the enemy's forward defenses that occurs as a result of a penetration. A breakthrough permits the passage of an exploitation force. (FM 3-90-1) See also **attack, exploitation, penetration**.

bridgehead – In gap crossing operations, an area on the enemy's side of the linear obstacle that is large enough to accommodate the majority of the crossing force, has adequate terrain to permit defense of the crossing sites, provides security of crossing forces from enemy direct fire, and provides a base for continuing the attack. (ATTP 3-90.4) See also **crossing site**.

bridgehead force – A force that assaults across a gap to secure the enemy side (the bridgehead) to allow the buildup and passage of a breakout force during river crossing operations. (ATTP 3-90.4) See also **bridgehead**.

brigade – A unit consisting of two or more battalions and a headquarters company or detachment. (ADRP 3-90) See also **battalion, division**.

brigade combat team – (Army) A combined arms organization consisting of a brigade headquarters, at least two maneuver battalions, and necessary supporting functional capabilities. Also called BCT. (ADRP 3-90)

brigade support area – A designated area in which sustainment elements locate to provide support to a brigade. Also called BSA. (ATP 4-90)

buffer zone – (DOD) 1. A defined area controlled by a peace operations force from which disputing or belligerent forces have been excluded. Also called area of separation in some United Nations operations. Also called BZ. See also line of demarcation; peace operations. (JP 3-07.3) See ATP 3-07.31.

bypass – A tactical mission task in which the commander directs his unit to maneuver around an obstacle, position, or enemy force to maintain the momentum of the operation while deliberately avoiding combat with an enemy force. (FM 3-90-1) See also **tactical mission task**.

bypass criteria – Measures during the conduct of an offensive operation established by higher headquarters that specify the conditions and size under which enemy units and contact may be avoided. (ADRP 3-90) See also **bypass**.

—C—

cache – (DOD) A source of subsistence and supplies, typically containing items such as food, water, medical items, and/or communications equipment, packaged to prevent damage from exposure and hidden in isolated locations by such methods as burial, concealment, and/or submersion, to support isolated personnel. (JP 3-50) See ATP 3-05.1.

call for fire – A request for fire containing data necessary for obtaining the required fire on a target. (FM 3-09)

call for fire zone – A radar search area from which the commander wants to attack hostile firing systems. (FM 3-09)

call forward area – In gap-crossing operations, waiting areas within the crossing area where final preparations are made. (ATTP 3-90.4)

campaign – (DOD) A series of related major operations aimed at achieving strategic and operational objectives within a given time and space. (JP 5-0) See ADRP 3-0, ATP 3-07.5.

campaign plan – (DOD) A joint operation plan for a series of related major operations aimed at achieving strategic or operational objectives within a given time and space. (JP 5-0) See FM 6-0.

canalize – (Army) A tactical mission task in which the commander restricts enemy movement to a narrow zone by exploiting terrain coupled with the use of obstacles, fires, or friendly maneuver. (FM 3-90-1) See also **tactical mission task**.

capacity building – The process of creating an environment that fosters host-nation institutional development, community participation, human resources development, and strengthening of managerial systems. (FM 3-07)

captured enemy documents and media – Any piece of recorded information previously under enemy control regardless of its form—written, printed, engraved, and photographic matter as well as recorded media and media devices—that pertains to the enemy, weather, or terrain that are under the U.S. Government's physical control and are not publicly available. (ATP 2-91.8)

carrier-owned containers – Containers owned or leased by the ocean liner carrier for the movement of intermodal cargo. (ATP 4-12)

casualty – (DOD) Any person who is lost to the organization by having been declared dead, duty status – whereabouts unknown, missing, ill, or injured. (JP 4-02) See ATP 4-02.2, ATP 4-02.55.

casualty evacuation – (DOD) The unregulated movement of casualties that can include movement both to and between medical treatment facilities. Also called CASEVAC. (JP 4-02) See ATP 3-07.31, ATP 3-55.6, ATP 4-01.45. (Army) Nonmedical units use this to refer to the movement of casualties aboard nonmedical vehicles or aircraft without en route medical care. (FM 4-02)

casualty operations – The process of recording, reporting, verifying, and processing casualty information from unit level to Headquarters, Department of the Army; notifying appropriate individuals and agencies; and providing casualty notification and assistance to the primary next of kin. (ATP 1-0.2)

catastrophic event – (DOD) Any natural or man-made incident, including terrorism, which results in extraordinary levels of mass casualties, damage, or disruption severely affecting the population, infrastructure, environment, economy, national morale, and/or government functions. (JP 3-28) See ADRP 3-28.

C-day – (DOD) The unnamed day on which a deployment operation commences or is to commence. (JP 5-0) See FM 4-01, FM 6-0.

center of gravity – (DOD) The source of power that provides moral or physical strength, freedom of action, or will to act. Also called COG. (JP 5-0) See ADRP 3-0, FM 3-24, ATP 3-05.20, ATP 3-53.2, ATP 3-57.60, ATP 3-57.80, ATP 5-0.1.

censor zone – An area from which radar is prohibited from reporting acquisitions. Normally placed around friendly weapon systems to prevent them from being acquired by friendly radars. (FM 3-09)

certification – Verification and validation of an Army professional's character, competence, and commitment to fulfill responsibilities and successfully perform assigned duty with discipline and to standard. (ADRP 1)

character – Dedication and adherence to the Army Ethic, including Army Values, as consistently and faithfully demonstrated in decisions and actions. (ADRP 1)

characteristic – A feature or quality that marks an organization or function as distinctive or is representative of that organization or function. (ADP 1-01)

checkpoint – A predetermined point on the ground used to control movement, tactical maneuver, and orientation. Also called CP. (ADRP 1-02)

chemical agent – (DOD) A chemical substance that is intended for use in military operations to kill, seriously injure, or incapacitate mainly through its physiological effects. (JP 3-11) See ATP 3-05.11.

chemical, biological, radiological, and nuclear environment – (DOD) An operational environment that includes chemical, biological, radiological, and nuclear threats and hazards and their potential resulting effects. Also called CBRN environment. (JP 3-11) See ATP 3-05.11.

chemical, biological, radiological, and nuclear consequence management – Chemical, biological, radiological, and nuclear consequence management consists of actions taken to plan, prepare, respond to, and recover from chemical, biological, radiological, and nuclear incidents that require force and resource allocation beyond passive defense capabilities. (FM 3-11)

chemical, biological, radiological, and nuclear defense – (Army) Chemical, biological, radiological, and nuclear active defense comprises measures taken to defeat an attack with chemical, biological, radiological, and nuclear weapons by employing actions to divert, neutralize, or destroy those weapons or their means of delivery while en route to their target. (FM 3-11) (DOD) Measures taken to minimize or negate the vulnerabilities and/or effects of a chemical, biological, radiological, or nuclear hazard incident. Also called CBRN defense. (JP 3-11) See ATP 4-02.84.

chemical, biological, radiological, or nuclear incident – (DOD) Any occurrence, resulting from the use of chemical, biological radiological and nuclear weapons and devices; the emergence of secondary hazards arising from counterforce targeting; or the release of toxic industrial materials into the environment, involving the emergence of chemical biological, radiological and nuclear hazards. (JP 3-11) See ADP 3-28, ATP 3-55.6, ATP 4-02.3.

chemical, biological, radiological, and nuclear operations – Chemical, biological, radiological, and nuclear operations include the employment of tactical capabilities that counter the entire range of chemical, biological, radiological, and nuclear threats and hazards through weapons of mass destruction proliferation prevention; weapons of mass destruction counterforce; chemical, biological, radiological, and nuclear defense; and chemical, biological, radiological, and nuclear consequence management activities. Chemical, biological, radiological, and nuclear operations support operational and strategic objectives to combat weapons of mass destruction and operate safely in a chemical, biological, radiological, and nuclear environment. (FM 3-11)

chemical, biological, radiological, and nuclear responders – Chemical, biological, radiological, and nuclear responders are Department of Defense military and civilian personnel who are trained to respond to chemical, biological, radiological, and nuclear incidents and certified to operate safely at the awareness, operations, technician, or installation level according to Section 120, Part 1910, Title 29, Code of Federal Regulations and National Fire Protection Association 472. (FM 3-11)

chemical, biological, radiological, or nuclear sample management – Chemical, biological, radiological, or nuclear sample management is the process whereby chemical, biological, radiological, or nuclear samples are collected, packaged, transported, stored, transferred, analyzed, tracked, and disposed. It begins with the decision to collect chemical, biological, radiological, or nuclear samples and continues to the reporting of information produced by the final analysis of that sample. This process includes safeguarding and prioritizing chemical, biological, radiological, or nuclear samples, tracking their movements and analytical status, and reporting the end result of sample analysis. The chemical, biological, radiological, or nuclear sample management process establishes procedures, guidelines, and constraints at staff and unit levels to protect and preserve the integrity of chemical, biological, radiological, or nuclear samples that may have tactical, operational, and/or strategic implications. (ATP 3-11.37)

chemical, biological, radiological, and nuclear threats – Chemical, biological, radiological, and nuclear threats include the intentional employment of, or intent to employ, weapons or improvised devices to produce chemical, biological, radiological, and nuclear hazards. (FM 3-11)

chemical warfare – (DOD) All aspects of military operations involving the employment of lethal and incapacitating munitions/agents and the warning and protective measures associated with such offensive operations. Also called CW. (JP 3-11) See ATP 3-05.11.

chemical weapon – (DOD)Together or separately, (a) a toxic chemical and its precursors, except when intended for a purpose not prohibited under the Chemical Weapons Convention; (b) a munition or device, specifically designed to cause death or other harm through toxic properties of those chemicals specified in (a), above, which would be released as a result of the employment of such munition or device; (c) any equipment specifically designed for use directly in connection with the employment of munitions or devices specified in (b), above. (JP 3-11) See ATP 3-05.11.

chief of fires – The senior fires officer at division and higher headquarters level who is responsible for advising the commander on the best use of available fire support resources, providing input to necessary orders, and developing and implementing the fires support plan. (ADRP 3-09)

chief of mission – (DOD) The principal officer (the ambassador) in charge of a diplomatic facility of the United States, including any individual assigned to be temporarily in charge of such a facility. The chief of mission is the personal representative of the President to the country of accreditation. The chief of mission is responsible for the direction, coordination, and supervision of all US Government executive branch employees in that country (except those under the command of a US area military commander). The security of the diplomatic post is the chief of mission's direct responsibility. Also called COM. (JP 3-08) See FM 3-53.

chief train dispatcher – Supervises train movement, reroutes rail traffic in emergencies, determines train tonnage, orders motive power, determines rail line capacity, and establishes train movement priority. (ATP 4-14)

civil administration – (DOD) An administration established by a foreign government in (1) friendly territory, under an agreement with the government of the area concerned, to exercise certain authority normally the function of the local government; or (2) hostile territory, occupied by United States forces, where a foreign government exercises executive, legislative, and judicial authority until an indigenous civil government can be established. Also called CA. (JP 3-05) See FM 3-57, ATP 3-57.10, ATP 3-57.20, ATP 3-57.60, ATP 3-57.70, ATP 3-57.80.

civil affairs – (DOD) Designated Active and Reserve Component forces and units organized, trained, and equipped specifically to conduct civil affairs operations and to support civil-military operations. Also called CA. (JP 3-57) See ADRP 3-05, FM 3-05, FM 3-18, FM 3-57, ATP 3-07.31, ATP 3-57.10, ATP 3-57.20, ATP 3-57.30, ATP 3-57.60, ATP 3-57.70, ATP 3-57.80.

civil affairs operations – (DOD) Actions planned, executed, and assessed by civil affairs forces that enhance awareness of and manage the interaction with the civil component of the operational environment; identify and mitigate underlying causes of instability within civil society; or involve the application of functional specialty skills normally the responsibility of civil government. Also called CAO. (JP 3-57) See ADRP 3-05, FM 1-04, FM 3-07, FM 3-18, FM 3-57, ATP 1-06.2, ATP 3-05.2, ATP 3-09.24, ATP 3-57.10, ATP 3-57.20, ATP 3-57.60, ATP 3-57.70, ATP 3-57.80.

civil affairs operations project management – The six step process by which civil affairs forces identify, validate, plan, coordinate, facilitate, and monitor both material and nonmaterial civil affairs operations projects to achieve a supported commander's objectives relating to the civil component of the operational environment. (FM 3-57)

civil assistance – Assistance, based on a commander's decision, in which life-sustaining services are provided, order is maintained, and/or goods and services are distributed within the commander's assigned area of operations. (FM 3-57)

civil authorities – (DOD) Those elected and appointed officers and employees who constitute the government of the United States, the governments of the 50 states, the District of Colombia, the Commonwealth of Puerto Rico, United States possessions and territories, and political subdivisions thereof. (JP 3-28) See ADP 3-28, FM 3-53, ATP 2-91.7.

civil authority information support – (DOD) Department of Defense information activities conducted under a designated lead federal agency or other United States civil authority to support dissemination of public or other critical information during domestic emergencies. Also called CAIS. (JP 3-13.2) See FM 3-53.

civil augmentation program – (DOD) Standing, long-term external support contacts designed to augment Service logistic capabilities with contract support in both preplanned and short notice contingencies. Examples include US Army Logistics Civil Augmentation Program, Air Force Contract Augmentation Program, and US Navy Global Contingency Capabilities Contracts. Also called CAP. (JP 4-10) See ATTP 4-10.

civil considerations – The influence of manmade infrastructure, civilian institutions, and attitudes and activities of the civilian leaders, populations, and organizations within an area of operations on the conduct of military operations. (ADRP 5-0)

civilian internee – (DOD) A civilian who is interned during armed conflict, occupation, or other military operation for security reasons, for protection, or because he or she committed an offense against the detaining power. Also called CI. (DODD 2310.01E) See FM 1-04.

civil information – (DOD) Relevant data relating to the civil areas, structures, capabilities, organizations, people, and events of the civil component of the operational environment used to support the situational awareness of the supported commander. (JP 3-57) See FM 3-57, ATP 3-57.70.

civil information management – (DOD) Process whereby data relating to the civil component of the operational environment is gathered, collated, processed, analyzed, produced into information products, and disseminated. Also called CIM. (JP 3-57) See FM 3-57, ATP 3-57.30, ATP 3-57.70.

civil liaison team – Provides limited civil-military interface capability as a spoke for exchange of information between indigenous populations and institutions, intergovernmental organizations, nongovernmental organizations, and other governmental agencies, and has limited capability to link resources to prioritized requirements. The civil liaison team is a stand-alone team for the civil-military operations center. It provides the supported level civil-military operations center with a storefront for civil affairs operations and civil-military operations coordination capability without interfering with the regular staff functions. Also called CLT. (FM 3-57)

civil-military engagement – A formal program that facilitates the U.S. interagency, host nation indigenous authorities, select intergovernmental and nongovernmental partners, and the private sector to build, replace, repair, and sustain civil capabilities and capacities that eliminate, reduce, or mitigate civil vulnerabilities to local regional populations. Civil-military engagement is a globally synchronized and regionally coordinated program of country-specific and regional actions executed through and with indigenous and U.S. interagency partners to eliminate the underlying conditions and core motivations for local and regional population support to violent extremist organizations and the networks. Also called CME. (FM 3-57)

civil-military operations – (DOD) Activities of a commander performed by designated civil affairs or other military forces that establish, maintain, influence, or exploit relations between military forces, indigenous populations, and institutions, by directly supporting the attainment of objectives relating to the reestablishment or maintenance of stability within a region or host nation. Also called CMO. (JP 3-57) See ADRP 3-05, FM 3-57, ATP 3-07.31, ATP 3-57.10, ATP 3-57.20, ATP 3-57.60, ATP 3-57.70, ATP 3-57.80.

civil-military operations center– (DOD) An organization normally comprised of civil affairs, established to plan and facilitate coordination of activities of the Armed Forces of the United States with indigenous populations and institutions, the private sector, intergovernmental organizations, nongovernmental organizations, multinational forces, and other governmental agencies in support of the joint force commander. Also called CMOC. (JP 3-57) See FM 3-57, ATP 3-57.20, ATP 3-57.60, ATP 3-57.70.

civil-military support element – A tasked-organized civil affairs force that conducts civil-military engagement in a specified country or region. A civil-military support element is composed of a persistent-presence element of civil affairs leaders/planners, and a presence-for-purpose element composed of a civil affairs team(s) that may include enablers (for example, health service support, engineer, etc.) who are task organized for a specific time to execute a coordination mission. Also called CMSE. (FM 3-57)

civil reconnaissance – (DOD) A targeted, planned, and coordinated observation and evaluation of specific civil aspects of the environment such as areas, structures, capabilities, organizations, people, or events. Also called CR. (JP 3-57) See FM 3-57, ATP 3-57.60, ATP 3-57.70.

Civil Reserve Air Fleet – (DOD) A program in which the Department of Defense contracts for the services of specific aircraft, owned by a United States entity or citizen, during national emergencies and defense-oriented situations when expanded civil augmentation of military airlift activity is required. Also called CRAF. (JP 3-17) See FM 4-01.

clandestine – (DOD) Any activity or operation sponsored or conducted by governmental departments or agencies with the intent to assure secrecy and concealment. (JP 1-02) See ATP 3-05.20.

clandestine operation – (DOD) An operation sponsored or conducted by governmental departments or agencies in such a way as to assure secrecy or concealment. (JP 3-05) See FM 3-18, ATP 3-05.1, ATP 3-18.4, ATP 3-53.1.

classes of supply – (DOD) The ten categories into which supplies are grouped in order to facilitate supply management and planning. I. Rations and gratuitous issue of health, morale, and welfare items. II. Clothing, individual equipment, tentage, tool sets, and administrative and housekeeping supplies and equipment. III. Petroleum, oils, and lubricants. IV. Construction materials. V. Ammunition. VI. Personal demand items. VII. Major end items, including tanks, helicopters, and radios. VIII. Medical. IX. Repair parts and components for equipment maintenance. X. Nonstandard items to support nonmilitary programs such as agriculture and economic development. (JP 4-09) [***Note***: Army doctrine also includes a miscellaneous category comprising water, captured enemy material, and salvage material.] See ADRP 4-0, ATP 3-35.

clear – 1. A tactical mission task that requires the commander to remove all enemy forces and eliminate organized resistance within an assigned area. (FM 3-90-1) 2. To eliminate transmissions on a tactical radio net in order to allow a higher-precedence transmission to occur. (FM 6-02.53) 3. The total elimination or neutralization of an obstacle that is usually performed by follow-on engineers and is not done under fire. (ATTP 3-90.4) See also **reduce, tactical mission task**.

clearance of fires – The process by which the supported commander ensures that fires or their effects will have no unintended consequences on friendly units or the scheme of maneuver. (FM 3-09)

close air support – (DOD) Air action by fixed- and rotary-wing aircraft against hostile targets that are in close proximity to friendly forces and that require detailed integration of each air mission with the fire and movement of those forces. Also called CAS. (JP 3-0) See FM 3-09, FM 3-99, ATP 3-04.64, ATP 3-06.1, ATP 3-09.24, ATP 3-09.34, ATP 3-55.6, ATP 3-60.2, ATP 3-91.1, ATP 4-01.45.

close area – In contiguous areas of operations, an area assigned to a maneuver force that extends from its subordinates' rear boundaries to its own forward boundary. (ADRP 3-0)

close combat – Warfare carried out on land in a direct-fire fight, supported by direct and indirect fires, and other assets. (ADRP 3-0)

close quarters battle – Sustained combative tactics, techniques, and procedures employed by small, highly trained special operations forces using special purpose weapons, munitions, and demolitions to recover specified personnel, equipment, or material. (ADRP 3-05)

close support – (DOD) That action of the supporting force against targets or objectives which are sufficiently near the supported force as to require detailed integration or coordination of the supporting action. (JP 3-31) See FM 3-09, FM 6-0.

coalition – (DOD) An arrangement between two or more nations for common action. (JP 5-0) See ADRP 3-0, FM 4-95, FM 3-16, FM 3-07.

collaborative planning – Commanders, subordinate commanders, staffs, and other partners sharing information, knowledge, perceptions, ideas, and concepts regardless of physical location throughout the planning process. (ADRP 5-0)

collateral damage – (DOD) Unintentional or incidental injury or damage to persons or objects that would not be lawful military targets in the circumstances ruling at the time. (JP 3-60) See FM 3-09, ATP 3-06.1, ATP 3-60.1, ATP 3-60.2.

collection management – (DOD) In intelligence usage, the process of converting intelligence requirements into collection requirements, establishing priorities, tasking or coordinating with appropriate collection sources or agencies, monitoring results, and retasking, as required. (JP 2-0) See ATP 3-55.3, ATP 3-55.6.

collection point(s) (patient or casualty) - A specific location where casualties are assembled to be transported to a medical treatment facility. It is usually predesignated and may or may not be staffed. (FM 4-02)

collective protection – (DOD) The protection provided to a group of individuals that permits relaxation of individual chemical, biological, radiological, and nuclear protection. Also called COLPRO. (JP 3-11) See ATP 3-05.11.

column formation – The column formation is a combat formation in which elements are placed one behind the other. (FM 3-90-1)

combat and operational stress control – A coordinated program for the prevention of and actions taken by military leadership to prevent, identify, and manage adverse combat and operational stress reactions in units. (FM 4-02)

combat lifesaver – A nonmedical Soldier trained to provide enhanced first aid as a secondary mission. Normally, one member of each squad, team, or crew is trained. (FM 4-02)

combat load – The minimum mission-essential equipment and supplies as determined by the commander responsible for carrying out the mission, required for Soldiers to fight and survive immediate combat operations. (FM 4-40)

combatant command – (DOD) A unified or specified command with a broad continuing mission under a single commander established and so designated by the President, through the Secretary of Defense and with the advice and assistance of the Chairman of the Joint Chiefs of Staff. Also called CCMD. (JP 1) See ATP 3-05.11, ATP 3-34.84.

combatant command (command authority) – (DOD) Nontransferable command authority, which cannot be delegated, of a combatant commander to perform those functions of command over assigned forces involving organizing and employing commands and forces; assigning tasks; designating objectives; and giving authoritative direction over all aspects of military operations, joint training, and logistics necessary to accomplish the missions assigned to the command. Also called COCOM. (JP 1) See ATP 3-13.10, ATP 3-27.5, ATP 3-34.84, ATP 3-52.2, ATP 3-52.3, ATP 4-43.

combatant commander – (DOD) A commander of one of the unified or specified combatant commands established by the President. Also called CCDR. (JP 3-0) See FM 3-53, ATP 3-13.10, ATP 3-52.2.

combatant command historian – The senior joint historian with overall staff responsibility for developing historical policy and plans for the combatant command and executing joint historical operations within the combatant command's area of responsibility. (ATP 1-20)

combat camera – (DOD) The acquisition and utilization of still and motion imagery in support of operational and planning requirements across the range of military operations and during joint exercises. Also called COMCAM. (JP 3-61) See ATP 3-07.31, ATP 3-55.12.

combat formation – A combat formation is an ordered arrangement of forces for a specific purpose and the general configuration of a unit on the ground. (ADRP 3-90)

combat identification – (DOD) The process of attaining an accurate characterization of detected objects in the operational environment sufficient to support an engagement decision. Also called CID. (JP 3-09) See FM 3-99, ATP 3-01.15, ATP 3-52.2, ATP 3-60.1.

combat information – (DOD) Unevaluated data, gathered by or provided directly to the tactical commander which, due to its highly perishable nature or the criticality of the situation, cannot be processed into tactical intelligence in time to satisfy the user's tactical intelligence requirements. (JP 2-01) See ADRP 2-0, ADRP 3-90, FM 2-0.

combat observation and lasing team – A field artillery team controlled at the brigade level that is capable of day and night target acquisition and has both laser range finding and laser-designating capabilities. Also called COLT. (ADRP 3-09)

combat outpost – A reinforced observation post capable of conducting limited combat operations. (FM 3-90-2)

combat power – (DOD) The total means of destruction and/or disruptive force which a military unit/formation can apply against the opponent at a given time. (JP 3-0) See FM 3-07. (Army) The total means of destructive, constructive, and information capabilities that a military unit or formation can apply at a given time. (ADRP 3-0)

combat search and rescue – (DOD) The tactics, techniques, and procedures performed by forces to effect the recovery of isolated personnel during combat. Also called CSAR. (JP 3-50) See ADRP 3-05.

combination yard – Yard that is a combination of receiving, classifying, and departure facilities. (ATP 4-14)

combined arms – The synchronized and simultaneous application of arms to achieve an effect greater than if each arm was used separately or sequentially. (ADRP 3-0)

combined arms maneuver – The application of the element of combat power in unified action to defeat enemy ground forces; to seize, occupy, and defend land areas; and to achieve physical, temporal, and psychological advantages over the enemy to seize and exploit the initiative. (ADP 3-0)

combined arms team – (Army) Two or more arms mutually supporting one another, usually consisting of a mixture of infantry, armor, aviation, field artillery, air defense artillery, and engineers. (ADRP 3-90)

command – (DOD) 1. The authority that a commander in the armed forces lawfully exercises over subordinates by virtue of rank or assignment. See ADRP 3-0, ADP 6-0, ADP 6-22, FM 3-24. 2. An order given by a commander; that is, the will of the commander expressed for the purpose of bringing about a particular action. 3. A unit or units, an organization, or an area under the command of one individual. Also called CMD. (JP-1). See FM 3-07.

command and control – (DOD) The exercise of authority and direction by a properly designated commander over assigned and attached forces in the accomplishment of the mission. Also called C2. (JP 1). See FM 3-07, ATP 3-01.15, ATP 3-06.1, ATP 3-52.2, ATP 3-52.3.

command and control system – (DOD) The facilities, equipment, communications, procedures, and personnel essential to a commander for planning, directing, and controlling operations of assigned and attached forces pursuant to the missions assigned. (JP 6-0) See ATP 3-52.2.

commander's critical information requirement – (DOD) An information requirement identified by the commander as being critical to facilitating timely decision making. Also called CCIR. (JP 3-0) See ADRP 5-0, FM 3-13, FM 3-98, FM 4-40, FM 6-0, ATP 2-01, ATP 2-01.3, ATP 2-19.4, ATP 3-07.31, ATP 3-09.24, ATP 3-55.3.

commander's intent – (DOD) A clear and concise expression of the purpose of the operation and the desired military end state that supports mission command, provides focus to the staff, and helps subordinate and supporting commanders act to achieve the commander's desired results without further orders, even when the operation does not unfold as planned. (JP 3-0) See ADP 5-0, ADP 6-0, ADRP 3-0, ADRP 3-28, ADRP 6-0, FM 3-07, FM 3-09, FM 6-0, FM 3-99, ATP 3-57.60.

commander's visualization – The mental process of developing situational understanding, determining desired end state, and envisioning an operational approach by which the force will achieve that end state. (ADP 5-0)

command group – The commander and selected staff members who assist the commander in controlling operations away from a command post. (FM 6-0)

command post – A unit headquarters where the commander and staff perform their activities. Also called CP. (FM 6-0)

command post cell – A grouping of personnel and equipment organized by warfighting function or by planning horizon to facilitate the exercise of mission command. (FM 6-0)

command relationships – (DOD) The interrelated responsibilities between commanders, as well as the operational authority exercised by commanders in the chain of command; defined further as combatant command (command authority), operational control, tactical control, or support. (JP 1) See ATP 3-52.2.

commitment – Resolve to contribute honorable service to the Nation, and accomplish the mission despite adversity, obstacles, and challenges. (ADRP 1)

committed force – A force in contact with an enemy or deployed on a specific mission or course of action which precludes its employment elsewhere. (ADRP 3-90) See also **attack, decisive engagement**.

common deflection – The deflection, which may vary based on the weapon's sight system, corresponding to the firing unit's azimuth of fire. (ATP 3-09.50)

common grid – Refers to all firing and target-locating elements within a unified command located and oriented, to prescribed accuracies, with respect to a single three-dimensional datum. (FM 3-09)

common operational picture – (DOD) A single identical display of relevant information shared by more than one command that facilitates collaborative planning and assists all echelons to achieve situational awareness. (JP 3-0) See ATP 2.22.7, ATP 3-01.15, ATP 3-52.2. (Army) A single display of relevant information within a commander's area of interest tailored to the user's requirements and based on common data and information shared by more than one command. Also called COP. (ADRP 6-0)

common sensor boundary – A line (depicted by a series of grid coordinates, grid line, phase line or major terrain feature) established by the force counterfire headquarters that divides target acquisition search areas into radar acquisition management areas. (FM 3-09)

common servicing – (DOD) Functions performed by one Service in support of another for which reimbursement is not required. (JP 3-34) See ATP 4-32.16.

common tactical picture – (DOD) An accurate and complete display of relevant tactical data that integrates tactical information from the multi-tactical data link network, ground network, intelligence network, and sensor networks. Also called CTP. (JP 3-01) See ATP 3-01.15.

common-user land transportation – (DOD) Point-to-point land transportation service operated by a single Service for common use by two or more Services. Also called CULT. (JP 4-01.5) See ATP 3-35, ATP 4-0.1.

common-user logistics – (DOD) Materiel or service support shared with or provided by two or more Services, Department of Defense agencies, or multinational partners to another Service, Department of Defense agency, non-Department of Defense agency, and/or multinational partner in an operation. Also called CUL. (JP 4-09) See ATP 3-93.

communications security – (DOD) The protection resulting from all measures designed to deny unauthorized persons information of value that might be derived from the possession and study of telecommunications, or to mislead unauthorized persons in their interpretation of the results of such possession and study. Also called COMSEC. (JP 6-0) See FM 6-02, ATP 6-02.75.

company – A company is a unit consisting of two or more platoons, usually of the same type, with a headquarters and a limited capacity for self-support. (ADRP 3-90)

company team – A combined arms organization formed by attaching one or more nonorganic armor, mechanized infantry, Stryker infantry, or light infantry platoons to a tank, mechanized infantry, Stryker, or infantry company either in exchange for, or in addition to its organic platoons. (ADRP 3-90)

competence – Demonstrated ability to successfully perform duty with discipline and to standard. (ADRP 1)

complex terrain – A geographical area consisting of an urban center larger than a village and/or of two or more types of restrictive terrain or environmental conditions occupying the same space. (ATP 3-34.80)

comprehensive approach – An approach that integrates the cooperative efforts of the departments and agencies of the United States Government, intergovernmental and nongovernmental organizations, multinational partners, and private sector entities to achieve unity of effort toward a shared goal. (FM 3-07)

concealment – Protection from observation or surveillance. (FM 3-96)

concept of operations – (DOD) A verbal or graphic statement that clearly and concisely expresses what the joint force commander intends to accomplish and how it will be done using available resources. (JP 5-0) See ADRP 3-90, FM 3-07, ATP 3-52.2, FM 3-53, FM 3-90-1, FM 6-05. (Army) A statement that directs the manner in which subordinate units cooperate to accomplish the mission and establishes the sequence of actions the force will use to achieve the end state. (ADRP 5-0) See also **commander's intent, operation plan**.

concept plan – (DOD) In the context of joint operation planning level 3 planning detail, an operation plan in an abbreviated format that may require considerable expansion or alteration to convert it into a complete operation plan or operation order. Also called CONPLAN. (JP 5-0) See FM 6-0.

conduct human resources planning and operations – The means by which human resources provider envisions a desired human resources end state in support of the operational commander's mission requirement. End state includes the intent, expected requirement, and outcomes to be achieved in the conduct and sustainment of human resources operations. Planning involves the use of the military decisionmaking process and composite risk management to ensure decisions are being made at the proper level of command. The end result is communicated to subordinates through an operation plan or operation order. (ATP 1-0.2)

confined space – An area large enough and so configured that a member can bodily enter and perform assigned work, but which has limited or restricted means for entry and exit and is not designed for continuous human occupancy. (ATP 3-11.23)

confirmation brief – A briefing subordinate leaders give to the higher commander immediately after the operation order is given. It is their understanding of his intent, their specific tasks, and the relationship between their mission and the other units in the operation. (ADRP 5-0) See also **commander's intent, operation order**.

conflict prevention – (DOD) A peace operation employing complementary diplomatic, civil, and, when necessary, military means, to monitor and identify the causes of conflict, and take timely action to prevent the occurrence, escalation, or resumption of hostilities. (JP 3-07.3) See ATP 3-07.31.

conflict transformation – The process of reducing the means and motivations for violent conflict while developing more viable, peaceful alternatives for the competitive pursuit of political and socio-economic aspirations. (FM 3-07)

consolidation – Organizing and strengthening in newly captured position so that it can be used against the enemy. (FM 3-90-1)

constraint – (Army) A restriction placed on the command by a higher command. A constraint dictates an action or inaction, thus restricting the freedom of action a subordinate commander. (FM 6-0)

contact point – (DOD) 1. In land warfare, a point on the terrain, easily identifiable, where two or more units are required to make contact. See FM 3-90-1. 2. In air operations, the position at which a mission leader makes radio contact with an air control agency. 3. In personnel recovery, a location where isolated personnel can establish contact with recovery forces. Also called CP. (JP 3-50) See FM 3-05.231.

contain – A tactical mission task that requires the commander to stop, hold, or surround enemy forces or to cause them to center their activity on a given front and prevent them from withdrawing any part of their forces for use elsewhere. (FM 3-90-1)

container – (DOD) An article of transport equipment that meets American National Standards Institute/International Organization for Standardization standards that is designed to facilitate and optimize the carriage of goods by one or more modes of transportation without intermediate handling of the contents. (JP 4-01) See FM 4-01.

container control officer – (DOD) A designated official (E6 or above or civilian equivalent) within a command, installation, or activity who is responsible for control, reporting, use, and maintenance of all Department of Defense-owned and controlled intermodal containers and equipment. This officer has custodial responsibility for containers from time received until dispatched. Also called CCO. (JP 4-09) See ATP 4-12.

container management – The process of establishing and maintaining visibility and accountability of all cargo containers moving within the Defense Transportation System. (ADP 4-0)

contaminated remains – (DOD) Remains of personnel which have absorbed or upon which have been deposited radioactive material, or biological or chemical agents. (JP 4-06) See ATP 3-05.11.

contamination – (DOD) 1. The deposit, absorption, or adsorption of radioactive material, or of biological or chemical agents on or by structures, areas, personnel, or objects. 2. Food and/or water made unfit for consumption by humans or animals because of the presence of environmental chemicals, radioactive elements, bacteria or organisms, the byproduct or the growth of bacteria or organisms, the decomposing material (to include the food substance itself), or waste in the food or water. (JP 3-11) See ATP 3-05.11, ATP 4-02.84, ATP 4-44.

contamination control – (DOD) A combination of preparatory and responsive measures designed to limit the vulnerability of forces to chemical, biological, radiological, nuclear, and toxic industrial hazards and to avoid, contain, control exposure to, and, where possible, neutralize them. (JP 3-11) See ATP 3-05.11, ATP 4-02.84.

contiguous area of operations – An area of operations where all of a commander's subordinate forces' areas of operations share one or more common boundary. (FM 3-90-1) See also **area of operations, boundary**.

continental system – A diesel or electric locomotive classification system that uses letters and figures to identify them by their axles. (ATP 4-14)

continuity – The uninterrupted provision of sustainment. (ADP 4-0)

continuity of care - Attempt to maintain the role of care during movement between roles at least equal to the role of care at the originating role. (FM 4-02)

continuous tractive effort – The effort required to keep a train rolling after it has started. Also called CTE. (ATP 4-14)

contracting officer – (DOD) The Service member or Department of Defense civilian with the legal authority to enter into, administer, and/or terminate contracts. (JP 4-10) See ATP 1-06.2, ATTP 4-10.

contracting officer representative – (DOD) A Service member or Department of Defense civilian appointed in writing and trained by a contracting officer, responsible for monitoring contract performance and performing other duties specified by their appointment letter. Also called COR. (JP 4-10) See ATTP 4-10.

contracting support operations – The staff section that oversees contracting operations and leads external coordination efforts. Also called CSPO. (ATP 4-92)

control – (DOD) 1. Authority that may be less than full command exercised by a commander over part of the activities of subordinate or other organizations. (JP 1) 2. In mapping, changing, and photogrammetry, a collective term for a system of marks or objects on the Earth or on a map or photograph, whose positions or elevations (or both) have been or will be determined. (JP 2-03) 3. Physical or psychological pressures exerted with the intent to assure that an agent or group will respond as directed. (JP 3-0) 4. An indicator governing the distribution and use of documents, information, or material. Such indicators are the subject of intelligence community agreement and are specifically defined in appropriate regulations. (JP 2-01) See FM 3-07. (Army) 1. The regulation of forces and warfighting functions to accomplish the mission in accordance with the commander's intent. (ADP 6-0) 2. A tactical mission task that requires the commander to maintain physical influence over a specified area to prevent its use by an enemy or to create conditions necessary for successful friendly operations. (FM 3-90-1) 3. An action taken to eliminate a hazard or reduce its risk. (ATP 5-19)

controlled supply rate – The rate of ammunition consumption that can be supported, considering availability, facilities, and transportation. It is expressed in rounds per unit, individual, or vehicle per day. (ATP 3-09.23)

control measure – A means of regulating forces or warfighting functions. (ADRP 6-0)

conventional forces – (DOD) 1. Those forces capable of conducting operations using nonnuclear weapons; 2. Those forces other than designated special operations forces. Also called CF. (JP 3-05) See ADRP 3-05, FM 3-53, FM 6-05, ATP 3-07.10, ATP 3-52.2, ATP 3-53.2.

convoy – (DOD) 2. A group of vehicles organized for the purpose of control and orderly movement with or without escort protection that moves over the same route at the same time and under one commander. (JP 3-02.1) See FM 4-01, ATP 3-18.14, ATP 4-01.45.

convoy escort – (DOD) 2. An escort to protect a convoy of vehicles from being scattered, destroyed, or captured. (JP 4-01.5) See ATP 4-01.45.

convoy security – A specialized area security task conducted to protect convoys. (ATP 3-91)

coordinated fire line – (DOD) A line beyond which conventional and indirect surface fire support means may fire at any time within the boundaries of the establishing headquarters without additional coordination. The purpose of the coordinated fire line is to expedite the surface-to-surface attack of targets beyond the coordinated fire line without coordination with the ground commander in whose area the targets are located. Also called CFL. (JP 3-09) See FM 3-09, FM 3-90-1, ATP 3-09.34.

coordinating altitude – (DOD) An airspace coordinating measure that uses altitude to separate users and as the transition between different airspace coordinating entities. Also called CA. (JP 3-52) See FM 3-09, ATP 3-04.64.

cordon and search – A technique of conducting a movement to contact that involves isolating a target area and searching suspect locations within that target area to capture or destroy possible enemy forces and contraband. (FM 3-90-1)

cordon security – The security provided between two combat outposts positioned to provide mutual support. (ATP 3-91)

core competency – An essential and enduring capability that a branch or an organization provides to Army operations. (ADP 1-01)

counterair – (DOD) A mission that integrates offensive and defensive operations to attain and maintain a desired degree of air superiority and protection by neutralizing or destroying enemy aircraft and missiles, both before and after launch. (JP 3-01) FM 3-01, FM 3-09, ATP 3-01.15.

counterattack – Attack by part or all of a defending force against an enemy attacking force, for such specific purposes as regaining ground lost, or cutting off or destroying enemy advance units, and with the general objective of denying to the enemy the attainment of the enemy's purpose in attacking. In sustained defensive operations, it is undertaken to restore the battle position and is directed at limited objectives. (ADRP 1-02)

counterfire – (DOD) Fire intended to destroy or neutralize enemy weapons. Includes counterbattery and countermortar fire. (JP 3-09) See ADRP 3-09, FM 3-09, FM 3-90-1, ATP 3-09.12, ATP 3-09.24.

counterinsurgency – (DOD) Comprehensive civilian and military efforts designed to simultaneously defeat and contain insurgency and address its root causes. Also called COIN. (JP 3-24) See ADP 3-05, ADRP 3-05, ADRP 3-07, FM 3-05, FM 3-24, FM 3-53, ATP 2-19.3, ATP 3-05.2, ATP 3-57.30, ATP 4-14.

counterintelligence – (DOD) Information gathered and activities conducted to identify, deceive, exploit, disrupt, or protect against espionage, other intelligence activities, sabotage, or assassinations conducted for or on behalf of foreign powers, organizations or persons or their agents, or international terrorist organizations or activities. Also called CI. (JP 1-02) See FM 2-22.2, FM 3-16, ATP 3-05.20. (Army) Counters or neutralizes intelligence collection efforts through collection, counterintelligence investigations, operations analysis, production, and technical services and support. Counterintelligence includes all actions taken to detect, identify, track, exploit, and neutralize the multidiscipline intelligence activities of friends, competitors, opponents, adversaries, and enemies; is the key intelligence community contributor to protect U.S. interests and equities; assists in identifying essential elements of friendly information, identifying vulnerabilities to threat collection, and actions taken to counter collection and operations against U.S. forces. (FM 2-22.2)

counterintelligence insider threat – (DOD) A person, known or suspect, who uses their authorized access to Department of Defense facilities, systems, equipment, information or infrastructure to damage, disrupt operations, commit espionage on behalf of a foreign intelligence entity or support international terrorist organizations. (JP 1-02) See ADRP 2-0.

countermeasures – (DOD) That form of military science that, by the employment of devices and/or techniques, has as its objective the impairment of the operational effectiveness of enemy activity. (JP 3-13.1) See FM 3-38.

countermobility operations – (DOD) The construction of obstacles and emplacement of minefields to delay, disrupt, and destroy the enemy by reinforcement of the terrain. (JP 3-34) See ATP 3-90.8. (Army/Marine Corps) Those combined arms activities that use or enhance the effects of natural and man-made obstacles to deny enemy freedom of movement and maneuver. (ATP 3-90.8) See also **destroy, disrupt**.

counterpreparation fire – Intensive prearranged fire delivered when the imminence of the enemy attack is discovered. (FM 3-09)

counterproliferation – (DOD) Those actions taken to defeat the threat and/or use of weapons of mass destruction against the United States, our forces, allies, and partners. Also called CP. (JP 3-40) See ADRP 3-05, FM 3-05, FM 3-53, ATP 3-05.2.

counterreconnaissance – A tactical mission task that encompasses all measures taken by a commander to counter enemy reconnaissance and surveillance efforts. Counterreconnaissance is not a distinct mission, but a component of all forms of security operations. (FM 3-90-1) See also **tactical mission task**.

counterterrorism – (DOD) Actions taken directly against terrorist networks and indirectly to influence and render global and regional environments inhospitable to terrorist networks. Also called CT. (JP 3-05) See ADP 3-05, ADRP 3-05, FM 3-05, FM 3-53, ATP 3-75.

country container authority – The appointed staff element that is responsible for enforcement of theater container management policy and procedures established by the combatant commander. (ATP 4-12)

country team– (DOD) The senior, in-country United States coordinating and supervising body, headed by the chief of the United States diplomatic mission, and composed of the senior member of each represented United States department or agency, as desired by the chief of the United States diplomatic mission. (JP 3-07.4) See FM 3-07, FM 3-24, FM 3-53, FM 3-57, ATP 3-05.2, ATP 3-57.10, ATP 3-57.20, ATP 3-57.80.

course of action – (DOD) 2. A scheme developed to accomplish a mission. Also called COA. (JP 5-0) See ATP 5-0.1.

cover – (DOD) In intelligence usage, those measures necessary to give protection to a person, plan, operation, formation, or installation from the enemy intelligence effort and leakage of information. (JP 1-02) See ATP 2-01.3. (Army) 1. Protection from the effects of fires. (FM 3-96) 2. A security task to protect the main body by fighting to gain time while also observing and reporting information and preventing enemy ground observation of and direct fire against the main body. (ADRP 3-90) See also **covering force, security operations.**

covered approach – 1. Any route that offers protection against enemy fire. 2. An approach made under the protection furnished by other forces or by natural cover. (FM 3-21.10) See also **cover**.

covering force – (Army) A self-contained force capable of operating independently of the main body, unlike a screen or guard force to conduct the cover task. (FM 3-90-2)

covering force area – The area forward of the forward edge of the battle area out to the forward positions initially assigned to the covering forces. It is here that the covering forces execute assigned tasks. (FM 3-90-2) See also **covering force, forward edge of the battle area**.

covert crossing – The crossing of an inland water obstacle or other gap that is planned and intended to be executed without detection by an adversary. (ATTP 3-90.4)

covert operation – (DOD) An operation that is so planned and executed as to conceal the identity of or permit plausible denial by the sponsor. (JP 3-05) See FM 3-18, ATP 3-05.1, ATP 3-18.4.

crew – Consists of all personnel operating a particular system. (ADRP 3-90)

crime prevention – A direct crime control method that applies to efforts to reduce criminal opportunity, protect potential human victims, and prevent property loss by anticipating, recognizing, and appraising crime risk and initiating actions to remove or reduce it. (ATP 3-39.10)

criminal intelligence – A category of police intelligence derived from the collection, analysis, and interpretation of all available information concerning known potential criminal threats and vulnerabilities of supported organizations. (FM 3-39)

crisis action planning – (DOD) The Adaptive Planning and Execution system process involving the time-sensitive development of joint operation plans and operation orders for the deployment, employment, and sustainment of assigned and allocated forces and resources in response to an imminent crisis. Also called CAP. (JP 5-0) See ATP 3-27.5.

critical asset list – (DOD) A prioritized list of assets, normally identified by phase of the operation and approved by the joint force commander, that should be defended against air and missile threats. (JP 3-01) See ADRP 3-09, ADRP 3-37.

critical asset security – The protection and security of personnel and physical assets or information that is analyzed and deemed essential to the operation and success of the mission and to resources required for protection. (ADRP 3-37)

critical capability – (DOD) A means that is considered a crucial enabler for a center of gravity to function as such and is essential to the accomplishment of the specified or assumed objective(s). (JP 5-0) See ATP 3-05.20.

critical event – An event that directly influences mission accomplishment. (FM 6-0)

critical requirement – (DOD) An essential condition, resource, and means for a critical capability to be fully operational. (JP 5-0) See ATP 3-05.20.

critical vulnerability – (DOD) An aspect of a critical requirement which is deficient or vulnerable to direct or indirect attack that will create decisive or significant effects. (JP 5-0) See ATP 3-05.20.

cross-leveling – (DOD) At the theater strategic and operational levels, it is the process of diverting en route or in-theater materiel from one military element to meet the higher priority of another within the combatant commander's directive authority for logistics. (JP 4-0) See ATP 3-35.

cueing – The integration of one or more types of reconnaissance or surveillance systems to provide information that directs follow-on collection of more detailed information by another system. (FM 3-90-2)

culminating point – (Army) That point in time and space at which a force no longer possesses the capability to continue its current form of operations. (ADRP 3-0)

Cultural Intelligence Element – An organic element of the Military Information Support Operations Command, providing culturally nuanced analyses and intelligence to subordinate unit commanders and their staffs, as well as to other agencies, focused on political, military, economic, social, information, and infrastructure, and other political-military factors. (FM 3-53)

curve resistance – The resistance offered by a curve to the progress of a train. Also called CR. (ATP 4-14)

cyber electromagnetic activities – Activities leveraged to seize, retain, and exploit an advantage over adversaries and enemies in both cyberspace and the electromagnetic spectrum, while simultaneously denying and degrading adversary and enemy use of the same, and protecting the mission command system. (ADRP 3-0)

cyberspace – (DOD) A global domain within the information environment consisting of the interdependent network of information technology infrastructures and resident data, including the Internet, telecommunications networks, computer systems, and embedded processors and controllers. (JP 1-02) See FM 3-24, FM 3-38, FM 6-02.

cyberspace operations – (DOD) The employment of cyberspace capabilities where the primary purpose is to achieve objectives in or through cyberspace. (JP 3-0) See FM 3-38.

cyberspace superiority – (DOD) The degree of dominance in cyberspace by one force that permits the secure, reliable conduct of operations by that force, and its related land, air, maritime, and space forces at a given time and place without prohibitive interference by an adversary. (JP 3-12) See FM 3-38.

—D—

data – (Army) Unprocessed signals communicated between any nodes in an information system, or sensing from the environment detected by a collector of any kind (human, mechanical, or electronic). (ADRP 6-0)

danger close – (DOD) In close air support, artillery, mortar, and naval gunfire support fires, it is the term included in the method of engagement segment of a call for fire which indicates that friendly forces are within close proximity of the target. The close proximity distance is determined by the weapon and munition fired. (JP 3-09.3) See FM 3-09, ATP 3-06.1.

datum (geodetic) – (DOD) 1. A reference surface consisting of five quantities: the latitude and longitude of an initial point, the azimuth of a line from that point, and the parameters of the reference ellipsoid. 2. The mathematical model of the earth used to calculate the coordinates on any map. Different nations use different datum for printing coordinates on their maps. (JP 2-03) See ATP 3-06.1, ATP 3-50.3.

D-day – (DOD) See times. (JP 3-02) See FM 3-99, FM 6-0.

debarkation – (DOD) The unloading of troops, equipment, or supplies from a ship or aircraft. (JP 3-02.1) See FM 4-01.

debriefing – The systematic questioning of individuals to procure information to answer specific collection requirements by direct and indirect questioning techniques. (FM 2-22.3)

decentralized execution – (DOD) Delegation of execution authority to subordinate commanders. (JP 3-30) See FM 4-01.

decision point – (DOD) A point in space and time when the commander or staff anticipates making a key decision concerning a specific course of action. (JP 5-0) See ADRP 5-0, FM 3-98, FM 6-0.

decision support matrix – A written record of a war-gamed course of action that describes decision points and associated actions at those decision points. Also called DSM. (ADRP 5-0) See also **branch, decision point, decision support template, sequel**.

decision support template – (DOD) A combined intelligence and operations graphic based on the results of wargaming. The decision support template depicts decision points, timelines associated with movement of forces and the flow of the operation, and other key items of information required to execute a specific friendly course of action. Also called DST. (JP 2-01.3) See ADRP 5-0, FM 3-98, ATP 2-01.3.

decisive action – (Army) The continuous, simultaneous combinations of offensive, defensive, and stability or defense support of civil authorities tasks. (ADRP 3-0)

decisive engagement – An engagement in which a unit is considered fully committed and cannot maneuver or extricate itself. In the absence of outside assistance, the action must be fought to a conclusion and either won or lost with the forces at hand. (ADRP 3-90)

decisive operation – The operation that directly accomplishes the mission. (ADRP 3-0) See also **battle, engagement, major operation, shaping operation**.

decisive point – (DOD) A geographic place, specific key event, critical factor, or function that, when acted upon, allows commanders to gain a marked advantage over an adversary or contribute materially to achieving success. (JP 5-0) See ADRP 3-0, ADRP 3-07, ADRP 5-0, ATP 5-0.1.

decisive terrain – Decisive terrain, when, present, is key terrain whose seizure and retention is mandatory for successful mission accomplishment. (FM 3-90-1) See also **key terrain**.

deck – The surface of a railcar on which a load rests. (ATP 4-14)

decontamination – (DOD) The process of making any person, object, or area safe by absorbing, destroying, neutralizing, making harmless, or removing chemical or biological agents, or by removing radioactive material clinging to or around it. (JP 3-11) See ATP 3-05.11, ATP 4-02.84.

deep area – In contiguous areas of operation, an area forward of the close area that a commander uses to shape enemy forces before they are encountered or engaged in the close area. (ADRP 3-0)

defeat – A tactical mission task that occurs when an enemy force has temporarily or permanently lost the physical means or the will to fight. The defeated force's commander is unwilling or unable to pursue his adopted course of action, thereby yielding to the friendly commander's will, and can no longer interfere to a significant degree with the actions of friendly forces. Defeat can result from the use of force or the threat of its use. (FM 3-90-1) See also **decisive point, tactical mission task**.

defeat in detail – Concentrating overwhelming combat power against separate parts of a force rather than defeating the entire force at once. (ADRP 3-90) See also **combat power**.

defeat mechanism – The method through which friendly forces accomplish their mission against enemy opposition. (ADRP 3-0)

defended asset list – (DOD) A listing of those assets from the critical asset list prioritized by the joint force commander to be defended with the resources available. (JP 3-01) See ADRP 3-09, ADRP 3-37.

defense coordinating element – (DOD) A staff and military liaison officers who assist the defense coordinating officer in facilitating coordination and support to activated emergency support functions. Also called DCE. (JP 3-28) See ATP 2-91.7.

defense coordinating officer – (DOD) Department of Defense single point of contact for domestic emergencies who is assigned to a joint field office to process requirements for military support, forward mission assignments through proper channels to the appropriate military organizations, and assign military liaisons, as appropriate, to activated emergency support functions. Also called DCO. (JP 3-28) See ATP 2-91.7.

defense design – A strategy for defense based on a compiled list of defensive tasks required to defend against a specific threat or support specific mission operations. Each defensive task is built using intelligence, features such as friendly force lay down, adversary forces lay down, named area of interest or ballistic missile operations areas, and characteristics such as defended assets, terrain, system location or orientation, and limitations. (FM 3-27)

defense industrial base – (DOD) The Department of Defense, government, and private sector worldwide industrial complex with capabilities to perform research and development, design, produce, and maintain military weapon systems, subsystems, components, or parts to meet military requirements. Also called DIB. (JP 3-27) See FM 4-95, ATP 4-0.1.

defense plan – Multiple defense designs combined together to create a cohesive plan for defending a broad area. (FM 3-27)

defense support of civil authorities – (DOD) Support provided by U.S. Federal military forces, Department of Defense civilians, Department of Defense contract personnel, Department of Defense component assets, and National Guard forces (when the Secretary of Defense, in coordination with the governors of the affected States, elects and requests to use those forces in Title 32, United States Code, status) in response to requests for assistance from civil authorities for domestic emergencies, law enforcement support, and other domestic activities, or from qualifying entities for special events. Also called DSCA. Also known as civil support. (DODD 3025.18) See ADP 3-28, ADRP 3-28, FM 1-04, FM 3-09, FM 3-52, FM 3-57, ATP 2-01, ATP 2-91.7, ATP 3-28.1.

Defense Transportation System – (DOD) That portion of the worldwide transportation infrastructure that supports Department of Defense transportation needs in peace and war. Also called DTS. See also common-user transportation; transportation system. (JP 4-01) See FM 4-01, ATP 4-13, ATP 4-15.

defensive counterair – (DOD) All defensive measures designed to neutralize or destroy enemy forces attempting to penetrate or attack through friendly airspace. Also called DCA. (JP 3-01) See ADRP 3-09, FM 3-01, ATP 3-14.5, ATP 3-27.5, ATP 3-55.6.

defensive cyberspace operation response action – (DOD) Deliberate, authorized defensive measures or activities taken outside of the defended network to protect and defend Department of Defense cyberspace capabilities or other designated systems. Also called DCO-RA. (JP 3-12) See FM 3-38.

defensive cyberspace operations – (DOD) Passive and active cyberspace operations intended to preserve the ability to utilize friendly cyberspace capabilities and protect data, networks, net-centric capabilities, and other designated systems. Also called DCO. (JP 1-02) See FM 3-38, FM 6-02.

defensive fires – Fires that protect friendly forces, population centers, and critical infrastructure. (FM 3-09)

defensive task – A task conducted to defeat an enemy attack, gain time, economize forces, and develop conditions favorable for offensive or stability tasks. (ADRP 3-0)

definitive care – (1) That care which returns an ill or injured Soldier to full function, or the best possible function after a debilitating illness or injury. Definitive care can range from self-aid when a Soldier applies a dressing to a grazing bullet wound that heals without further intervention, to two weeks bed-rest in theater for Dengue fever, to multiple surgeries and full rehabilitation with a prosthesis at a continental United States medical center or Department of Veterans Affairs hospital after a traumatic amputation. (2) That treatment required to return the Service member to health from a state of injury or illness. The Service member's disposition may range from return to duty to medical discharge from the military. It can be provided at any role depending on the extent of the Service member's injury or illness. It embraces those endeavors which complete the recovery of the patient. (FM 4-02)

definitive identification – The employment of multiple state-of-the-art, independent, established protocols and technologies by scientific experts in a nationally recognized laboratory to determine the unambiguous identity of a chemical, biological, radiological, and/or nuclear hazard with the highest level of confidence and degree of certainty necessary to support strategic-level decisions. (ATP 3-11.37)

definitive treatment – The final role of comprehensive care provided to return the patient to the highest degree of mental and physical health possible. It is not associated with a specific role or location in the continuum of care; it may occur in different roles depending upon the nature of the injury or illness. (FM 4-02)

delay – To slow the time of arrival of enemy forces or capabilities or alter the ability of the enemy or adversary to project forces or capabilities. (FM 3-09)

deflection – A horizontal clockwise angle measured from the line of fire or the rearward extension line of fire to the line of sight to a given aiming point with the vertex of the angle at the instrument. (ATP 3-09.50)

delay line – A phase line where the date and time before which the enemy is not allowed to cross the phase line is depicted as part of the graphic control measure. (FM 3-90-1) See also **control measure, phase line.**

delaying operation – (DOD) An operation in which a force under pressure trades space for time by slowing down the enemy's momentum and inflicting maximum damage on the enemy without, in principle, becoming decisively engaged. (JP 3-04) See ADP 3-90, ADRP 3-90, FM 3-90-1, ATP 3-91.

deliberate crossing – The crossing of an inland water obstacle or other gap that requires extensive planning and detailed preparations. (ATTP 3-90.4)

deliberate operation – An operation in which the tactical situation allows the development and coordination of detailed plans, including multiple branches and sequels. (ADRP 3-90)

deliberate planning – (DOD) 1. The Adaptive Planning and Execution system process involving the development of joint operation plans for contingencies identified in joint strategic planning documents. 2. A planning process for the deployment and employment of apportioned forces and resources that occurs in response to a hypothetical situation. (JP 5-0) See ATP 3-05.11, ATP 3-27.5.

demobilization – (DOD) The process of transitioning a conflict or wartime military establishment and defense-based civilian economy to a peacetime configuration while maintaining national security and economic vitality. (JP 4-05) See ADRP 3-07, FM 3-07.

demonstration – (DOD) 2. In military deception, a show of force in an area where a decision is not sought that is made to deceive an adversary. It is similar to a feint but no actual contact with the adversary is intended. (JP 3-13.4) See FM 3-90-1, FM 6-0.

denial operations – Actions to hinder or deny the enemy the use of space, personnel, supplies, or facilities. (FM 3-90-1)

denied area – (DOD) An area under enemy or unfriendly control in which friendly forces cannot expect to operate successfully within existing operational constraints and force capability. (JP 3-05) See ADRP 3-05 and ATP 3-05.1. (Army) An area that is operationally unsuitable for conventional forces due to political, tactical, environmental, or geographical reasons. It is a primary area for special operations forces. (FM 3-05)

Department of Defense information networks – (DOD) The globally interconnected, end-to-end set of information capabilities, and associated processes for collecting, processing, storing, disseminating, and managing information on-demand to warfighters, policy makers, and support personnel, including owned and leased communications and computing systems and services, software (including applications), data, and security. (JP 1-02) See FM 3-38, FM 6-02.

Department of Defense information network operations – (DOD) Operations to design, build, configure, secure, operate, maintain, and sustain Department of Defense networks to create and preserve information assurance on the Department of Defense information networks. (JP 1-02) See FM 3-38, FM 6-02.

departure airfield – (DOD) An airfield on which troops and/or materiel are enplaned for flight. (JP 3-17) See ATP 3-18.11.

departure yard – Yard where classified cars are made up into trains. (ATP 4-14)

deployment – (DOD) The rotation of forces into and out of an operational area. (JP 3-35) See ATP 3-35, ATP 3-91.

derailer – Safety devices designed to limit unauthorized movement of a car or locomotive beyond a specific point. (ATP 4-14)

destroy – A tactical mission task that physically renders an enemy force combat-ineffective until it is reconstituted. Alternatively, to destroy a combat system is to damage it so badly that it cannot perform any function or be restored to a usable condition without being entirely rebuilt. (FM 3-90-1) See also **reconstitution, tactical mission task**.

destruction – 1. In the context of the computed effects of field artillery fires, destruction renders a target out of action permanently or ineffective for a long period of time, producing 30-percent casualties or materiel damage. 2. A type of adjustment for destroying a given target. (FM 3-09)

detachment – A detachment is a tactical element organized on either a temporary or permanent basis for special duties. (ADRP 3-90) See also **assign, attach, command relationships, operational control**.

detachment left in contact – An element left in contact as part of the previously designated (usually rear) security force while the main body conducts its withdrawal. Also called DLIC. (FM 3-90-1) See also **breakout, retrograde**.

detainee – (DOD) A term used to refer to any person captured or otherwise detained by an armed force. (JP 3-63) See FM 1-04, ATP 3-07.31, ATP 3-91.

detection – (DOD) 2. In surveillance, the determination and transmission by a surveillance system that an event has occurred. (JP 3-11) See ATP 3-55.6. 4. In chemical, biological, radiological, and nuclear environments, the act of locating chemical, biological, radiological, and nuclear hazards by use of chemical, biological, radiological, and nuclear detectors or monitoring and/or survey teams. (JP 3-11) See ATP 3-05.11.

detention – A charge made on a carrier conveyance held by or otherwise delayed through the cause of the United States Government. (ATP 4-12)

diamond formation – A diamond formation is a variation of the box combat formation with one maneuver unit leading, maneuver units positioned on each flank, and the remaining maneuver unit to the rear. (ADRP 3-90) See also **box formation, flank.**

direct action – (DOD) Short-duration strikes and other small-scale offensive actions conducted as a special operation in hostile, denied, or politically sensitive environments and which employ specialized military capabilities to seize, destroy, capture, exploit, recover, or damage designated targets. (JP 3-05) See ADP 3-05, ADRP 3-05, FM 3-05, FM 3-53, ATP 3-75.

direct air support center – (DOD) The principal air control agency of the US Marine air command and control system responsible for the direction and control of air operations directly supporting the ground combat element. It processes and coordinates requests for immediate air support and coordinates air missions requiring integration with ground forces and other supporting arms. It normally collocates with the senior fire support coordination center within the ground combat element and is subordinate to the tactical air command center. Also called DASC. (JP 3-09.3) See ATP 3-60.2.

direct approach – The manner in which a commander attacks the enemy's center of gravity or principal strength by applying combat power directly against it. (ADRP 3-90)

direct exchange – A supply method of issuing serviceable materiel in exchange for unserviceable materiel on an item-for-item basis. (FM 4-40)

directed energy – (DOD) An umbrella term covering technologies that relate to the production of a beam of concentrated electromagnetic energy or atomic or subatomic particles. Also called DE. (JP 3-13.1) See FM 3-38.

directed obstacle – An obstacle directed by a higher commander as a specified task to a subordinate unit. (ATP 3-90.8)

direct fire – (DOD) Fire delivered on a target using the target itself as a point of aim for either the weapon or the director. (JP 3-09.3) See ATP 3-06.1.

direct haul – Single transport mission completed by the same vehicle(s). (ATP 4-11)

direction of attack – A specific direction or assigned route a force uses and does not deviate from when attacking. (ADRP 3-90) See also **axis of advance**.

direction of fire – The direction on which a fire unit is laid to the most significant threat in the target area, to the chart direction to the center of the zone of fire, or to the target. (ATP 3-09.50)

directive authority for logistics – (DOD) Combatant command authority to issue directives to subordinate commanders, including peacetime measures, necessary to ensure effective execution of approved operation plans. Essential measures include the optimized use or reallocation of available resources and prevention of elimination of redundant facilities and/or overlapping functions among the Service component commands. Also called DAFL. (JP 1) See ADRP 4-0.

direct liaison authorized – (DOD) That authority granted by a commander (any level) to a subordinate to directly consult or coordinate an action with a command or agency within or outside of the granting command. Also called DIRLAUTH. (JP 1) See FM 6-0, ATP 4-32.16.

direct pressure force –A force employed in a pursuit operation that orients on the enemy main body to prevent enemy disengagement or defensive reconstitution prior to envelopment by the encircling force. It normally conducts a series of attacks to slow the enemy's retirement by forcing him to stand and fight. (FM 3-90-1) See also **disengage, encircling force, envelopment, reconstitution**.

direct support – (DOD) A mission requiring a force to support another specific force and authorizing it to answer directly to the supported force's request for assistance. Also called DS. (JP 3-09.3) See FM 6-0, ATP 2-01, ATP 4-32.16, ATP 4-43. (Army) A support relationship requiring a force to support another specific force and authorizing it to answer directly to the supported force's request for assistance. (ADRP 5-0)

disarmament – (Army) The collection, documentation, control, and disposal of small arms, ammunition, explosives, and light and heavy weapons of former combatants, belligerents, and the local populace. (FM 3-07)

disease and nonbattle injury – (DOD) All illnesses and injuries not resulting from enemy or terrorist action or caused by conflict. Also called DNBI. (JP 4-02) See ATP 4-02.55.

disengage – A tactical mission task where a commander has his unit break contact with the enemy to allow the conduct of another mission or to avoid decisive engagement. (FM 3-90-1) See also **decisive engagement, tactical mission task**.

disengagement line – A phase line located on identifiable terrain that, when crossed by the enemy, signals to defending elements that it is time to displace to their next position. (ADRP 3-90) See also **phase line**.

dislocated civilian – (DOD) A broad term primarily used by the Department of Defense that includes displaced person, an evacuee, an internally displaced person, a migrant, a refugee, or a stateless person. Also called DC. (JP 3-29) See FM 3-07, FM 3-57, ATP 3-57.10, ATP 3-57.60.

dismounted march – Movement of troops and equipment mainly by foot, with limited support by vehicles. Also called foot march. (FM 3-90-2)

displaced person – (DOD) A broad term used to refer to internally and externally displaced persons collectively. (JP 3-29) See FM 3-57, ATP 3-07.31.

display – (DOD) In military deception, a static portrayal of an activity, force, or equipment intended to deceive the adversary's visual observation. (JP 3-13.4) See FM 6-0.

disrupt – 1. A tactical mission task in which a commander integrates direct and indirect fires, terrain, and obstacles to upset an enemy's formation or tempo, interrupt his timetable, or cause enemy forces to commit prematurely or attack in piecemeal fashion. 2. An obstacle effect that focuses fire planning and obstacle effort to cause the enemy to break up his formation and tempo, interrupt his timetable, commit breaching assets prematurely, and attack in a piecemeal effort. (FM 3-90-1)

distribution – (DOD) 5. The operational process of synchronizing all elements of the logistic system to deliver the "right things" to the "right place" at the "right time" to support the geographic combatant commander. (JP 4-0) See FM 4-01, FM 4-95, ATP 4-48.

distribution management – The function of synchronizing and coordinating a complex of networks (physical, communications, information, and financial) and the sustainment functions (logistics, personnel services, and health service support) to achieve responsive support to operational requirements. (ATP 4-0.1)

distribution manager – (DOD) The executive agent for managing distribution with the combatant commander's area of responsibility. (JP 4-09) See ATP 4-0.1.

distribution system – (DOD) That complex of facilities, installations, methods, and procedures designed to receive, store, maintain, distribute, and control the flow of military materiel between the point of receipt into the military system and the point of issue to using activities and units. (JP 4-09) See FM 4-01, FM 4-40, FM 4-95.

diversion – (DOD) 1. The act of drawing the attention and forces of an enemy from the point of the principal operation; an attack, alarm, or feint that diverts attention. (JP 3-03) See FM 3-09.

division – An Army echelon of command above brigade and below corps. It is a tactical headquarters which employs a combination of brigade combat teams, multifunctional brigades, and functional brigades in land operations. (ADRP 3-90)

doctrine – (DOD) Fundamental principles by which the military forces or elements thereof guide their actions in support of national objectives. It is authoritative but requires judgment in application, (JP 1-02) See ADP 1-01, ADP 1-02.

document and media exploitation – The processing, translation, analysis, and dissemination of collected hardcopy documents and electronic media that are under the U.S. Government's physical control and are not publicly available. (ATP 2-91.8)

double envelopment – This results from simultaneous maneuvering around both flanks of a designated enemy force. (FM 3-90-1)

drawbar pull – The actual pulling ability of a locomotive after deducting from tractive effort, the energy required to move the locomotive itself. Also called DBP. (ATP 4-14)

drop zone – (DOD) A specific area upon which airborne troops, equipment, or supplies are airdropped. Also called DZ. (JP 3-17) See ATP 3-06.1, ATP 3-18.11.

dynamic target – Any target that is identified too late or not selected for action during the deliberate targeting process. (ATP 3-60.1)

dynamic targeting – (DOD) Targeting that prosecutes targets identified too late, or not selected for action in time to be included in deliberate targeting. (JP 3-60) See ATP 3-60, ATP 3-60.1, ATP 3-91.

—E—

early-entry command post – A lead element of a headquarters designed to control operations until the remaining portions of the headquarters are deployed and operational. (FM 6-0)

earliest arrival date – (DOD) A day, relative to C-day, that is specified as the earliest date when a unit, a resupply shipment, or replacement personnel can be accepted at a port of debarkation during a deployment. Also called EAD. See also latest arrival date. (JP 5-0) See FM 4-01.

echelon – Separate level of command. (ADRP 1-02)

echelon formation – A unit formation with subordinate elements arranged on an angle to the left or to the right of the direction of attack (echelon left, echelon right). This formation provides for firepower forward and to the flank of the direction of the echelon. It facilitates control in open areas. It provides minimal security to the opposite flank of the direction of the echeloning. (FM 3-90-1) See also **box formation**.

echelon support – The method of supporting an organization arrayed within an area of an operation. (ATP 4-90)

economy – Providing sustainment resources in an efficient manner to enable a commander to employ all assets to achieve the greatest effect possible. (ADP 4-0)

effect – (DOD) 1. The physical or behavioral state of a system that results from an action, a set of actions, or another effect. 2. The result, outcome, or consequence of an action. 3. A change to a condition, behavior, or degree of freedom. (JP 3-0) See FM 3-53, FM 3-57, ATP 3-09.24, ATP 3-53.2, ATP 3-57.60, ATP 3-57.70, ATP 3-57.80, ATP 3-60.

electromagnetic compatibility – (DOD) The ability of systems, equipment, and devices that use the electromagnetic spectrum to operate in their intended environments without causing or suffering unacceptable or unintentional degradation because of electromagnetic radiation or response. Also called EMC. (JP 3-13.1) See FM 3-38.

electromagnetic hardening – (DOD) Action taken to protect personnel, facilities, and/or equipment by blanking, filtering, attenuating, grounding, bonding, and/or shielding against undesirable effects of electromagnetic energy. See also electronic warfare. (JP 3-13.1) See FM 3-38.

electromagnetic interference – (DOD) Any electromagnetic disturbance, induced intentionally or unintentionally, that interrupts, obstructs, or otherwise degrades or limits the effective performance of electronics and electrical equipment. Also called EMI. (JP 3-13.1) See FM 3-38.

electromagnetic intrusion – (DOD) The intentional insertion of electromagnetic energy into transmission paths in any manner, with the objective of deceiving operators or of causing confusion. (JP 3-13.1) See FM 3-38.

electromagnetic jamming – (DOD) The deliberate radiation, reradiation, or reflection of electromagnetic energy for the purpose of preventing or reducing an enemy's effective use of the electromagnetic spectrum, and with the intent of degrading or neutralizing the enemy's combat capability. (JP 3-13.1) See FM 3-38.

electromagnetic operational environment – (DOD) The background electromagnetic environment and the friendly, neutral, and adversarial electromagnetic order of battle within the electromagnetic area of influence associated with a given operational area. Also called EMOE. (JP 6-01) See FM 3-99.

electromagnetic pulse – (DOD) The electromagnetic radiation from a strong electronic pulse, most commonly caused by a nuclear explosion that may couple with electrical or electronic systems to produce damaging current and voltage surges. Also called EMP. (JP 3-13.1) See FM 3-38, ATP 3-05.11.

electromagnetic radiation – (DOD) Radiation made up of oscillating electric and magnetic fields and propagated with the speed of light. (JP 6-01) See ATP 4-32.

electromagnetic spectrum – (DOD) The range of frequencies of electromagnetic radiation from zero to infinity. It is divided into 26 alphabetically designated bands. (JP 3-13.1) See FM 3-38, ATP 3-13.10.

electromagnetic spectrum management – (DOD) Planning, coordinating, and managing use of the electromagnetic spectrum through operational, engineering, and administrative procedures. (JP 6-01) See FM 3-38, FM 3-99.

electronic attack – (DOD) Division of electronic warfare involving the use of electromagnetic energy, directed energy, or antiradiation weapons to attack personnel, facilities, or equipment with the intent of degrading, neutralizing, or destroying enemy combat capability and is considered a form of fires. Also called EA. (JP 3-13.1) See FM 3-09, FM 3-38, ATP 3-13.10.

electronic intelligence – (DOD) Technical and geolocation intelligence derived from foreign noncommunications electromagnetic radiations emanating from other than nuclear detonations or radioactive sources. Also called ELINT. (JP 3-13.1) See FM 3-38, ATP 3-05.20, ATP 3-13.10.

electronic masking – (DOD) The controlled radiation of electromagnetic energy on friendly frequencies in a manner to protect the emissions of friendly communications and electronic systems against enemy electronic warfare support measures/signals intelligence without significantly degrading the operation of friendly systems. (JP 3-13.1) See FM 3-38.

electronic probing – (DOD) Intentional radiation designed to be introduced into the devices or systems of potential enemies for the purpose of learning the functions and operational capabilities of the devices or systems. (JP 3-13.1) See FM 3-38.

electronic protection – (DOD) Division of electronic warfare involving actions taken to protect personnel, facilities, and equipment from any effects of friendly or enemy use of the electromagnetic spectrum that degrade, neutralize, or destroy friendly combat capability. Also called EP. (JP 3-13.1) See FM 3-38, ATP 3-13.10.

electronic reconnaissance – (DOD) The detection, location, identification, and evaluation of foreign electromagnetic radiations. (JP 3-13.1) See FM 3-38.

electronics security – (DOD) The protection resulting from all measures designed to deny unauthorized persons information of value that might be derived from their interception and study of noncommunications electromagnetic radiations, e.g., radar. (JP 3-13.1) See FM 3-38.

electronic warfare – (DOD) Military action involving the use of electromagnetic and directed energy to control the electromagnetic spectrum or to attack the enemy. Also called EW. (JP 3-13.1) See FM 3-38, ATP 3-13.10, ATP 4-32.

electronic warfare reprogramming – (DOD) The deliberate alteration or modification of electronic warfare or target sensing systems, or the tactics and procedures that employ them, in response to validated changes in equipment, tactics, or the electromagnetic environment. (JP 3-13.1) See FM 3-38, ATP 3-13.10.

electronic warfare support – (DOD) Division of electronic warfare involving actions tasked by, or under direct control of, an operational commander to search for, intercept, identify, and locate or localize sources of intentional and unintentional radiated electromagnetic energy for the purpose of immediate threat recognition, targeting, planning and conduct of future operations. Also called ES. (JP 3-13.1) See FM 3-38, ATP 3-13.10.

electro-optical-infrared countermeasure – (DOD) A device or technique employing electro-optical-infrared materials or technology that is intended to impair the effectiveness of enemy activity, particularly with respect to precision guided weapons and sensor systems. Also called EO-IR CM. (JP 3-13.1) See FM 3-38.

embarkation – (DOD) The process of putting personnel and/or vehicles and their associated stores and equipment into ships and/or aircraft. (JP 3-02.1) See FM 4-01.

emergency management – Emergency management, as a subset of incident management and concerns the coordination and integration of activities that are necessary to build, sustain, and improve the capability to prepare for, protect against, respond, recover from, or mitigate threatened or actual natural disaster, acts of terrorism, or other manmade disasters. (FM 3-11)

emergency medical treatment – The immediate application of medical procedures to the wounded, injured, or sick by specially trained medical personnel. (FM 4-02)

emergency operations center – (DOD) A temporary or permanent facility where the coordination of information and resources to support domestic incident management activities normally takes place. Also called EOC. (JP 3-41) See ADRP 3-28.

emergency preparedness liaison officer – (DOD) A senior reserve officer who represents their Service at the appropriate joint field office conducting planning and coordination responsibilities in support of civil authorities. Also called EPLO. (JP 3-28) See ATP 2-91.7.

emergency support functions – (DOD) A grouping of government and certain private-sector capabilities into an organizational structure to provide the support, resources, program implementation, and services that are most likely to be needed to save lives, protect property and the environment, restore essential services and critical infrastructure, and help victims and communities return to normal, when feasible, following domestic incidents. Also called ESF. (JP 3-28) See ATP 2-91.7, ATP 3-28.1.

emerging target – Detection that meets sufficient criteria to be developed as a potential target using dynamic targeting. The criticality and time sensitivity of an emerging target, and its probability of being a potential target, is initially undetermined. (ATP 3-60.1)

emission control – (DOD) The selective and controlled use of electromagnetic, acoustic, or other emitters to optimize command and control capabilities while minimizing, for operations security: a. detection by enemy sensors; b. mutual interference among friendly systems; and/or c. enemy interference with the ability to execute a military deception plan. Also called EMCON. (JP 3-13.1) See FM 3-38.

encirclement operations – Operations where one force loses its freedom of maneuver because an opposing force is able to isolate it by controlling all ground lines of communications and reinforcement. (ADRP 3-90)

encircling force – In pursuit operations, the force which maneuvers to the rear or flank of the enemy to block his escape so that the enemy can be destroyed between the direct pressure force and encircling force. This force advances or flies along routes parallel to the enemy's line of retreat. If the encircling force cannot outdistance the enemy to cut the enemy off, the encircling force may also attack the flank of a retreating enemy. (FM 3-90-1) See also **block, destroy, direct pressure force, envelopment**.

end delivery tonnage – The through tonnage, in short tons, of payload that is delivered at the end of the railway line (railhead) each day. Also called EDT. (ATP 4-14)

end evening civil twilight – (DOD) The point in time when the sun has dropped 6 degrees beneath the western horizon, and is the instant at which there is no longer sufficient light to see objects with the unaided eye. Also called EECT. (JP 2-01.3) See ATP 2-01.3.

end of evening nautical twilight – (DOD) The point in time when the sun has dropped 12 degrees below the western horizon, and is the instant of last available daylight for the visual control of limited military operations. Also called EENT. (JP 2-01.3) See ATP 2-01.3.

end state – (DOD) The set of required conditions that defines achievement of the commander's objectives. (JP 3-0) See FM 3-07, FM 3-24.

enemy – A party identified as hostile against which the use of force is authorized. (ADRP 3-0)

enemy combatant – (DOD) In general, a person engaged in hostilities against the United States or its coalition partners during an armed conflict. Also called EC. (DODD 2310.01E) See FM 1-04.

engagement – (DOD) 1. In air defense, an attack with guns or air-to-air missiles by an interceptor aircraft, or the launch of an air defense missile by air defense artillery and the missile's subsequent travel to intercept. (JP 3-01) See FM 3-07. 2. A tactical conflict, usually between opposing, lower echelon maneuver forces. (JP 3-0) See ADP 3-90, ADRP 3-90, FM 3-07, ATP 3-20.15.

engagement area – An area where the commander intends to contain and destroy an enemy force with the massed effects of all available weapons and supporting systems. Also called EA. (FM 3-90-1) See also **contain, destroy**.

engagement criteria – Protocols that specify those circumstances for initiating engagement with an enemy force. (FM 3-90-1) See also **decision point, engagement, engagement area**.

engagement priority – Specifies the order in which the unit engages enemy systems or functions. (FM 3-90-1)

engineer regulating point – Checkpoint to ensure that vehicles do not exceed the capacity of the crossing means and to give drivers final instructions on site-specific procedures and information, such as speed and vehicle interval. Also called ERP. (ATTP 3-90.4)

engineer work line – A coordinated boundary or phase line used to compartmentalize an area of operations to indicate where specific engineer units have primary responsibility for the engineer effort. (FM 3-34)

en route care – The care required to maintain the phased treatment initiated prior to evacuation and the sustainment of the patient's medical condition during evacuation. (FM 4-02)

envelopment – A form of maneuver in which an attacking force seeks to avoid the principal enemy defenses by seizing objectives behind those defenses that allow the targeted enemy force to be destroyed in their current positions. (FM 3-90-1)

environmental assessment – A study to determine if significant environmental impacts are expected from a proposed action. (ATP 3-34.5)

environmental compliance – The unconditional obeying of international, foreign nation, federal, state, and local environmental rules, regulations, and guidelines that affect current operations. (ATP 3-34.5)

environmental conditions report – A concise summary of events or situations that created a negative or positive change in environmental conditions at a base camp site. (ATP 3-34.5)

environmental protection level – The varying level of environmental protection that can reasonably be afforded at any particular time during military operations, given the absolute requirement that such a diversion of resources away from the mission at hand does not adversely affect that mission, any friendly personnel, or indigenous or refugee populations. (ATP 3-34.5)

environmental reconnaissance – The systematic observation and recording of site or area data collected by visual or physical means, dealing specifically with environmental conditions as they exist, and identifying areas that are environmentally sensitive or of relative environmental concern, for information and decisionmaking purposes. (ATP 3-34.5)

environmental restoration – The systematic removal of pollution or contaminants from the environment, especially from the soil or groundwater, by physical, chemical, or biological means; also known as remediation or environmental cleanup. (ATP 3-34.5)

environmental services – Environmental services are the various combinations of scientific, technical, and advisory activities (including modification processes and the influence of man-made and natural factors) required to acquire, produce, and supply information on the past, present, and future states of space, atmospheric, oceanographic, and terrestrial surroundings for use in military planning and decisionmaking processes or to modify those surroundings to enhance military operations. (ATP 3-34.5)

environmental stewardship – The integration and application of environmental values into the military mission to sustain readiness, improve the quality of life, strengthen civil relations, and preserve valuable natural resources. (ATP 3-34.5)

esoteric communications – Public statements whose surface meaning (manifest content) does not reveal the real purpose, meaning, or significance (latent content) of the author. (ATP 2-22.9)

esprit de corps – A traditional military expression that denotes the Army's common spirit, a collective ethos of camaraderie and cohesion within the team. (ADRP 1)

essential care – Medical care and treatment within the theater of operations and which is mission, enemy, terrain and weather, troops and support available, time available, and civil considerations-dependent. It includes first responder care, initial resuscitation and stabilization as well as treatment and hospitalization. Forward care may include stabilizing surgery to ensure the patient can tolerate further evacuation as well as en route care during evacuation. The objective is to either return the patient to duty within the theater evacuation policy, or to begin initial treatment required for optimization of outcome. (FM 4-02)

essential element of friendly information – (Army) A critical aspect of a friendly operation that, if known by the enemy, would subsequently compromise, lead to failure, or limit success of the operation and therefore should be protected from enemy detection. Also called EEFI. (ADRP 5-0) See also **commander's critical information requirement**.

essential personnel services – Essential personnel services include customer service, awards and decorations, evaluation reports, promotions and reductions, transfers and discharges, officer procurement, leaves and passes, military pay, personnel action request, line of duty investigations, AR 15-6 investigations, suspension of favorable actions and bars to reenlistment, citizenship and naturalization, congressional inquiries, and common access card and identification tags. (ATP 1-0.2)

essential task – (Army) A specified or implied task that must be executed to accomplish the mission. (FM 6-0) See also **implied task, specified task.**

evacuation – (DOD) 2. The clearance of personnel, animals, or materiel from a given locality. 4. The ordered or authorized departure of noncombatants from a specific area by Department of State, Department of Defense, or appropriate military commander. This refers to the movement from one area to another in the same or different countries. The evacuation is caused by unusual or emergency circumstances and applies equally to command or non-command sponsored family members. (JP 3-68) See ATP 3-05.68.

evacuee – (DOD) A civilian removed from a place of residence by military direction for reasons of personal security or the requirements of the military situation. See also displaced person; refugee. (JP 3-57) See ATP 3-05.68.

evaluating – Using criteria to judge progress toward desired conditions and determining why the current degree of progress exists. (ADRP 5-0)

evasion plan of action – (DOD) A course of action, developed prior to executing a combat mission, that is intended to improve a potential isolated person's chances of successful evasion and recovery by providing the recovery forces with an additional source of information that can increase the predictability of the evader's action and movement. Also called EPA. (JP 3-50) See ATP 3-50.3.

event matrix – (DOD) A cross-referenced description of the indicators and activity expected to occur in each named area of interest. See also activity; area of interest; indicator. (JP 2-01.3) See ATP 2-01.3.

event template – (DOD) A guide for collection planning that depicts the named areas of interest where activity, or its lack of activity, will indicate which course of action the adversary has adopted. See also activity; area of interest; collection planning; course of action. (JP 2-01.3) See ATP 2-01.3.

execution – Putting a plan into action by applying combat power to accomplish the mission. (ADP 5-0) See also **adjustment decision, combat power, execution decision, situational understanding**.

execution matrix – A visual and sequential representation of the critical tasks and responsible organizations by time. (ADRP 5-0)

executive agent – (DOD) A term used to indicate a delegation of authority by the Secretary of Defense or Deputy Secretary of Defense to a subordinate to act on behalf of the Secretary of Defense. Also called EA. (JP 1) See FM 4-95, ATP 4-44.

exfiltrate – A tactical mission task where a commander removes Soldiers or units from areas under enemy control by stealth, deception, surprise, or clandestine means. (FM 3-90-1)

exfiltration – (DOD) The removal of personnel or units from areas under enemy control by stealth, deception, surprise, or clandestine means. See also special operations; unconventional warfare. (JP 3-50) See ATP 3-18.4.

explicit knowledge - Codified or formally documented knowledge organized and transferred to others through digital or non-digital means. (ATP 6-01.1)

exploitation – (DOD) 1. Taking full advantage of success in military operations, following up initial gains, and making permanent the temporary effects already created. 2. Taking full advantage of any information that has come to hand for tactical, operational, or strategic purposes. 3. An offensive operation that usually follows a successful attack and is designed to disorganize the enemy in depth. (JP 2-01.3) See ATP 3-90.15, ATP 4-32. (Army) An offensive task that usually follows a successful attack and is designed to disorganize the enemy in depth. (ADRP 3-90) See also **attack, offensive operations**.

explosive hazards – A condition where danger exists because explosives are present that may react in a mishap with potential unacceptable effects to people, property, operational capability, or the environment. (FM 4-30)

explosive ordnance – (DOD) All munitions containing explosives, nuclear fission or fusion materials, and biological and chemical agents. (JP 3-34) See ATP 4-32, ATP 4-32.2.

explosive ordnance disposal – (DOD) The detection, identification, on-site evaluation, rendering safe, recovery, and final disposal of unexploded explosive ordnance. Also called EOD. (JP 3-34) See ATP 4-32.2, ATP 4-32.16. (Army) The detection, identification, on-site evaluation, rendering safe, exploitation, recovery, and final disposal of explosive ordnance. (FM 4-30)

explosive ordnance disposal unit – (DOD) Personnel with special training and equipment who render explosive ordnance safe, make intelligence reports on such ordnance, and supervise the safe removal thereof. (JP 3-34) See ATP 4-32.

explosive ordnance disposal incident – (NATO) The suspected or detected presence of unexploded explosive ordnance, or damaged explosive ordnance, which constitutes a hazard to operations, installations, personnel or material. Not included in this definition are the accidental arming or other conditions that develop during the manufacture of high explosive material, technical service assembly operations or the laying of mines and demolition charges. (STANAG 3680) See ATP 4-32.

explosive ordnance disposal procedures – (NATO) Those particular courses or modes of action taken by explosive ordnance disposal personnel for access to, diagnosis, rendering safe, recovery and final disposal of explosive ordnance or any hazardous material associated with an explosive ordnance disposal incident. a. Access procedures – Those actions taken to locate exactly and to gain access to unexploded explosive ordnance. b. Diagnostic procedures – Those actions taken to identify and evaluate unexploded explosive ordnance. c. Render safe procedures – The portion of the explosive ordnance disposal procedures involving the application of special explosive ordnance disposal methods and tools to provide for the interruption of functions or separation of essential components of unexploded explosive ordnance to prevent an unacceptable detonation. d. Recovery procedures – Those actions taken to recover unexploded explosive ordnance. e. Final disposal procedures – The final disposal of explosive ordnance which may include demolition or burning in place, removal to a disposal area or other appropriate means. (STANAG 3680) See ATP 4-32.

exterior lines – Lines on which a force operates when its operations converge on the enemy. (ADRP 3-0)

external trust – The confidence and faith that the American people have in the Army to serve the Nation ethically, effectively, and efficiently. (ADRP 1)

—F—

facility – (DOD) A real property entity consisting of one or more of the following: a building, a structure, a utility system, pavement, and underlying land. (JP 3-34) See ATP 3-22.40, ATP 4-13.

farside objective – A defined location on the far side of an obstacle that an assaulting force seizes to eliminate enemy direct fires and observed indirect fires onto a reduction area or a crossing site to prevent the enemy from interfering with the reduction of the obstacle and allow follow-on forces to move securely through the created lanes. It can be oriented on the terrain or on an enemy force. (ATTP 3-90.4)

federal service – (DOD) A term applied to National Guard members and units when called to active duty to serve the United States Government under Article I, Section 8 and Article II, Section 2 of the Constitution and the Title 10, United States Code, Sections 12401 to 12408. (JP 4-05) See ADRP 3-28.

feint – (DOD) In military deception, an offensive action involving contact with the adversary conducted for the purpose of deceiving the adversary as to the location and/or time of the actual main offensive action. (JP 3 13.4) See FM 3-90-1, FM 6-0.

F-hour – (DOD) Effective time of announcement by the Secretary of Defense to the Military Departments of a decision to mobilize Reserve units. (JP 1-02) See FM 6-0.

field artillery – (DOD) Equipment, supplies, ammunition, and personnel involved in the use of cannon, rocket, or surface-to-surface missile launchers. Field artillery cannons are classified according to caliber as follows. Light — 120mm and less. Medium — 121-160mm. Heavy — 161-210mm. Very heavy — greater than 210mm. Also called FA. (JP 3-09) See FM 3-09. (Army) The equipment, supplies, ammunition, and personnel involved in the use of indirect fire cannon, rocket, or surface-to-surface missile launchers. (ADRP 3-09)

field confirmatory identification – The employment of technologies with increased specificity and sensitivity by technical forces in a field environment to identify chemical, biological, radiological, and/or nuclear hazard with a moderate level of confidence and degree of certainty necessary to support follow-on tactical decisions. (ATP 3-11.37)

field force engineering – The application of the Engineer Regiment capabilities from the three engineer disciplines (primarily general engineering) to support operations through reachback and forward presence. (FM 3-34)

field historian – An Army historian, military or civilian, that serves outside of the Center of Military History documenting, recording, and reporting the official history of the Army at the command and unit levels. (ATP 1-20)

field maintenance – On system maintenance, repair and return to the user including maintenance actions performed by operators. (FM 4-30)

field of fire – The area that a weapon or group of weapons may cover effectively from a given position. (FM 3-90-1)

field services – Includes aerial delivery, clothing and light-textile repair, food service, shower and laundry, mortuary affairs, and water purification. These services enhance unit effectiveness and mission success by providing for Soldier basic needs. (ADRP 4-0)

fighter engagement zone – (DOD) In air defense, that airspace of defined dimensions within which the responsibility for engagement of air threats normally rests with fighter aircraft. Also called FEZ. (JP 3-01) See ATP 3-01.15.

final coordination line – A phase line close to the enemy position used to coordinate the lifting or shifting of supporting fires with the final deployment of maneuver elements. Also called FCL. (ADRP 3-90) See also **assault, assault echelon, phase line**.

final protective fire – (DOD, NATO) An immediately available prearranged barrier of fire designed to impede enemy movement across defensive lines or areas. (JP 1-02) See ADRP 3-90, FM 3-09.

final protective line – A selected line of fire where an enemy assault is to be checked by interlocking fire from all available weapons and obstacles. Also called FPL. (ADRP 1-02) See also **field of fire, final protective fire**.

finance operations – The execution of the joint financial management mission to provide financial advice and guidance, support the procurement process, provide pay support, and provide banking and disbursing support. (FM 1-06)

financial management – The sustainment of U.S. Army, joint, interagency, interdepartmental, and multinational operations through the execution of two mutually supporting core functions, resource management and finance operations. These two functions are comprised of the following core competencies: fund the force, banking and disbursing support, pay support, accounting support and cost management, financial management planning and operations, and management internal controls. (FM 1-06)

fire and movement – The concept of applying fires from all sources to suppress, neutralize, or destroy the enemy, and the tactical movement of combat forces in relation to the enemy (as components of maneuver, applicable at all echelons). At the squad level, it entails a team placing suppressive fire on the enemy as another team moves against or around the enemy. (FM 3-96)

fire direction center – (DOD) That element of a command post, consisting of gunnery and communications personnel and equipment, by means of which the commander exercises fire direction and/or fire control. The fire direction center receives target intelligence and requests for fire, and translates them into appropriate fire direction. The fire direction center provides timely and effective tactical and technical fire control in support of current operations. Also called FDC. (JP 3-09.3) See FM 3-09.

fire plan – A tactical plan for using the weapons of a unit or formation so that their fire will be coordinated. (FM-3-09)

fire strike – The massed, synchronized, and nearly simultaneous delivery of primarily terminally guided indirect fire and area munitions. (FM 3-90-2)

fire superiority – That degree of dominance in the fires of one force over another that permits that force to conduct maneuver at a given time and place without prohibitive interference by the enemy. (FM 3-90-1) See also **maneuver**.

fire support – (DOD) Fires that directly support land, maritime, amphibious, and special operations forces to engage enemy forces, combat formations, and facilities in pursuit of tactical and operational objectives. (JP 3-09) See ADP 3-09, ADRP 3-09, ATP 3-04.64, ATP 3-06.1, ATP 3-09.24.

fire support area – (DOD) An appropriate maneuver area assigned to fire support ships by the naval force commander from which they can deliver gunfire support to an amphibious operation. Also called FSA. (JP 3-09) See FM 3-09.

fire support coordination – (DOD) The planning and executing of fire so that targets are adequately covered by a suitable weapon or group of weapons. (JP 3-09) See ADP 3-09, ADRP 3-09, FM 3-09, ATP 3-09.24.

fire support coordination center – (DOD) A single location in which are centralized communications facilities and personnel incident to the coordination of all forms of fire support. Also called FSCC. (JP 3-09) See ATP 3-60.2.

fire support coordination line – (DOD) A fire support coordination measure established by the land or amphibious force commander to support common objectives within an area of operation; beyond which all fires must be coordinated with affected commanders prior to engagement, and short of the line, all fires must be coordinated with the establishing commander prior to engagement. Also called FSCL. (JP 3-09) See FM 3-09, FM 3-90-1, ATP 3-09.34, ATP 3-60.2.

fire support coordination measure – (DOD) A measure employed by commanders to facilitate the rapid engagement of targets and simultaneously provide safeguards for friendly forces. Also called FSCM. (JP 3-0) See FM 3-09, FM 3-99, ATP 3-09.24, ATP 3-09.34, ATP 3-52.1, ATP 3-60.2.

fire support coordinator – The brigade combat team's organic field artillery battalion commander; if a fires brigade is designated as the division force field artillery headquarters, the fires brigade commander is the division's fire support coordinator and is assisted by the chief of fires who then serves as the deputy fire support coordinator during the period the force field artillery headquarters is in effect. (ADRP 3-09) See also **fire support**.

fire support officer – (Army) The field artillery officer from the operational to tactical level responsible for advising the supported commander or assisting the senior fires officer of the organization on fires functions and fire support. (ADRP 3-09)

fire support plan – A plan that that addresses each means of fire support available and describes how Army indirect fires, joint fires, and target acquisition are integrated with maneuver to facilitate operational success. (FM 3-09)

fire support planning – The continuing process of analyzing, allocating, and scheduling fires to describe how fires are used to facilitate the actions of the maneuver force. (FM 3-09)

fire support station – (DOD) An exact location at sea within a fire support area from which a fire support ship delivers fire. Also called FSS. (JP 3-02) See FM 3-09.

fire support team – A field artillery team organic to each maneuver battalion and selected units to plan and coordinate all available company supporting fires, including mortars, field artillery, naval surface fire support, and close air support integration. (ADRP 3-09)

fire team – A small military unit. (ADRP 3-90)

fires – (DOD) The use of weapons systems to create a specific lethal or nonlethal effect on a target. (JP 3-0) See ADP 3-09, FM 3-09, FM 6-05, ATP 3-04.64, ATP 3-09.24

fires warfighting function – The related tasks and systems that provide collective and coordinated use of Army indirect fires, air and missile defense, and joint fires through the targeting process. (ADRP 3-0)

first aid (self-aid/buddy aid) – Urgent and immediate lifesaving and other measures which can be performed for casualties (or performed by the victim himself) by nonmedical personnel when medical personnel are not immediately available. (FM 4-02)

fix – A tactical mission task where a commander prevents the enemy from moving any part of his force from a specific location for a specific period. Fix is also an obstacle effect that focuses fire planning and obstacle effort to slow an attacker's movement within a specified area, normally an engagement area. (FM 3-90-1) See also **block, contain, disrupt, support by fire, tactical mission task, turn.**

flank – The right or left limit of a unit. (ADRP 3-90)

flank attack – A form of offensive maneuver directed at the flank of an enemy. (FM 3-90-1)

flanking position – A geographical location on the flank of the force from which effective fires can be placed on that flank. (ADRP 3-90)

follow and assume – A tactical mission task in which a second committed force follows a force conducting an offensive task and is prepared to continue the mission if the lead force is fixed, attrited, or unable to continue. (FM 3-90-1) See also **attack, fix, follow and support, offensive operations, tactical mission task.**

follow and support – A tactical mission task in which a committed force follows and supports a lead force conducting an offensive task. (FM 3-90-1) See also **direct pressure force, encircling force, exploitation, follow and assume, offensive operations, tactical mission task.**

follow-on echelon – Those additional forces moved into the objective area after the assault echelon. (FM 3-99) See also **air assault operation, assault echelon.**

force closure – (DOD) The point in time when a supported joint force commander determines that sufficient personnel and equipment resources are in the assigned operational area to carry out assigned tasks. (JP 3-35) See ATP 3-35.

force field artillery headquarters – If designated by the supported commander, is normally the senior field artillery headquarters organic, assigned, attached, or placed under the operational control of that command. The supported commander specifies the commensurate responsibilities of the force field artillery headquarters and the duration of those responsibilities. (ADRP 3-09)

force health protection – (DOD) 1. Measures to promote, improve, or conserve the behavioral and physical well-being of Service members to enable a healthy and fit force, prevent injury and illness, and protect the force from health hazards. Also called FHP. (JP 4-02) See FM 4-02, ATP 4-02.55, ATP 4-02.84. (Army) 2. Encompasses measures to promote, improve, conserve or restore the mental or physical well-being of Soldiers. These measures enable a healthy and fit force, prevent injury and illness, and protect the force from health hazards. These measures also include the prevention aspects of a number of Army Medical Department functions (preventive medicine, including medical surveillance and occupational and environmental health surveillance; veterinary services, including the food inspection and animal care missions, and the prevention of zoonotic disease transmissible to man; combat and operational stress control; dental services (preventive dentistry); and laboratory services [area medical laboratory support]. (FM 4-02)

force projection – (DOD) The ability to project the military instrument of national power from the United States or another theater, in response to requirements for military operations. (JP 3-0) See ADP 4-0, FM 4-95.

force protection – (DOD) Preventive measures taken to mitigate hostile actions against Department of Defense personnel (to include family members), resources, facilities, and critical information. Also called FP. (JP 3-0) See ADRP 3-37, FM 4-01, ATP 3-07.31.

force protection condition – (DOD) A Chairman of the Joint Chiefs of Staff-approved standard for identification of and recommended responses to terrorist threats against United States personnel and facilities. Also called FPCON. (JP 3-07.2) See ATP 3-07.31.

force tailoring – The process of determining the right mix of forces and the sequence of their deployment in support of a joint force commander. (ADRP 3-0)

forcible entry – (DOD) Seizing and holding of a military lodgment in the face of armed opposition. (JP 3-18) See FM 3-99, ATP 3-91.

ford – A shallow part of a body of water or wet gap that can be crossed without bridging, boats, ferries, or rafts. It is a location in a water barrier where the physical characteristics of current, bottom, and approaches permit the passage of personnel, vehicles, and other equipment where the wheels or tracks remain in contact with the bottom at all times. (ATTP 3-90.4) See also **gap**.

foreign disaster relief – (DOD) Prompt aid that can be used to alleviate the suffering of foreign disaster victims. Normally, it includes humanitarian services and transportation; provision of food, clothing, medicine, beds, and bedding; temporary shelter and housing; the furnishing of medical materiel and medical and technical personnel; and making repairs to essential services. (JP 3-29) See ATP 3-57.20.

foreign humanitarian assistance – (DOD) Department of Defense activities conducted outside the United States and its territories to directly relieve or reduce human suffering, disease, hunger, or privation. Also called FHA. (JP 3-29) See FM 3-57, ATP 1-06.2, ATP 3-05.2, ATP 3-07.5, ATP 3-07.31, ATP 3-57.10, ATP 3-57.20, ATP 3-57.30, ATP 3-57.60, ATP 3-57.70.

foreign instrumentation signals intelligence – (DOD) A subcategory of signals intelligence, consisting of technical information and intelligence derived from the intercept of foreign electromagnetic emissions associated with the testing and operational deployment of non-US aerospace, surface, and subsurface systems. Foreign instrumentation signals include but are not limited to telemetry, beaconry, electronic interrogators, and video data links. Also called FISINT. (JP 2-01) See ATP 3-05.20.

foreign internal defense – (DOD) Participation by civilian and military agencies of a government in any of the action programs taken by another government or other designated organization to free and protect its society from subversion, lawlessness, insurgency, terrorism, and other threats to its security. Also called FID. (JP 3-22) See ADP 3-05, ADRP 3-05, ADRP 3-07, FM 3-05, FM 3-07, FM 3-24, FM 3-53, FM 3-57, ATP 3-05.2, ATP 3-07.10, ATP 3-57.20, ATP 3-57.30, ATP 3-57.80, ATP 3-93.

foreign military sales – (DOD) That portion of United States security assistance authorized by the Foreign Assistance Act of 1961, as amended, and the Arms Export Control Act of 1976, as amended. This assistance differs from the Military Assistance Program and the International Military Education and Training Program in that the recipient provides reimbursement for defense articles and services transferred. Also called FMS. (JP 4-08) See ATP 3-57.30.

foreign security forces – Forces, including, but not limited to military, paramilitary, police, and intelligence forces; border police, coast guard, and customs officials; and prison guards and correctional personnel, that provide security for a host nation and its relevant population or support a regional security organization's mission. (FM 3-22)

forms of maneuver – Distinct tactical combinations of fire and movement with a unique set of doctrinal characteristics that differ primarily in the relationship between the maneuvering force and the enemy. (ADRP 3-90)

forward air controller (airborne) – (DOD) A specifically trained and qualified aviation officer who exercises control from the air of aircraft engaged in close air support of ground troops. The forward air controller (airborne) is normally an airborne extension of the tactical air control party. A qualified and current forward air controller (airborne) will be recognized across the Department of Defense as capable and authorized to perform terminal attack control. Also called FAC(A). (JP 3-09.3) See FM 3-09, ATP 4-01.45.

forward arming and refueling point – (DOD) A temporary facility, organized, equipped, and deployed to provide fuel and ammunition necessary for the employment of aviation maneuver units in combat. Also called **FARP**. (JP 3-09.3) See ATP 3-17.2.

forward boundary – A boundary of an echelon that is primarily designated to divide responsibilities between it and its next higher echelon (FM 3-90-1)

forward edge of the battle area – (DOD) The foremost limits of a series of areas in which ground combat units are deployed, excluding the areas in which the covering or screening forces are operating, designated to coordinate fire support, the positioning of forces, or the maneuver of units. Also called FEBA. (JP 3-09.3) See ADRP 3-90, FM 3-90-1.

forward line of own troops – (DOD) A line that indicates the most forward positions of friendly forces in any kind of military operation at a specific time. Also called FLOT. (JP 3-03) See FM 3-90-1.

forward logistics element – Comprised of task-organized multifunctional logistics assets designed to support fast-moving offensive operations in the early phases of decisive action. Also called FLE. (ATP 4-90)

forward-looking infrared – (DOD) An airborne, electro-optical thermal imaging device that detects far-infrared energy, converts the energy into an electronic signal, and provides a visible image for day or night viewing. Also called FLIR. (JP 3-09.3) See ATP 3-06.1.

forward observer – (DOD) An observer operating with front line troops and trained to adjust ground or naval gunfire and pass back battlefield information. In the absence of a forward air controller, the observer may control close air support strikes. Also called FO. (JP 3-09) See FM 3-09.

forward operating site – (DOD) A scaleable location outside the United States and US territories intended for rotational use by operating forces. Such expandable "warm facilities" may be maintained with a limited US military support presence and possibly pre-positioned equipment. Forward operating sites support rotational rather than permanently stationed forces and are a focus for bilateral and regional training. Also called **FOS**. (JP 1-02) See ATP 3-17.2.

forward passage of lines – Occurs when a unit passes through another unit's positions while moving toward the enemy. (ADRP 3-90) See also **passage of lines, rearward passage of lines**.

forward resuscitative surgery – Urgent initial surgery required to render a patient transportable for further evacuation to a medical treatment facility staffed and equipped to provide for the patient's care. (FM 4-02)

463L system – (DOD) A material handling system that consists of military and civilian aircraft cargo restraint rail systems, aircraft pallets, nets, tie down, coupling devices, facilities, handling equipment, procedures, and other components designed to efficiently accomplish the air logistics and aerial delivery mission. (JP 4-09) See FM 4-01.

fratricide – The unintentional killing or wounding of friendly or neutral personnel by friendly firepower. (ADRP 3-37)

fragmentary order – (DOD) An abbreviated form of an operation order issued as needed after an operation order to change or modify that order or to execute a branch or sequel to that order. Also called FRAGORD. (JP 5-0) See FM 6-0.

free-fire area – (DOD) A specific area into which any weapon system may fire without additional coordination with the establishing headquarters. Also called FFA. (JP 3-09) See ATP 3-09.34, FM 3-09, FM 3-90-1.

frequency deconfliction – (DOD) A systematic management procedure to coordinate the use of the electromagnetic spectrum for operations, communications, and intelligence functions. Frequency deconfliction is one element of electromagnetic spectrum management. (JP 3-13.1) See FM 3-38.

friendly – (DOD) A contact positively identified as friendly. (JP 3-01) See ADRP 3-37.

friendly force information requirement – (DOD) Information the commander and staff need to understand the status of friendly and supporting capabilities. Also called FFIR. (JP 3-0) See ADRP 5-0, FM-3-98, FM 6-0, ATP 2-19.4.

frontal attack – A form of maneuver in which the attacking force seeks to destroy a weaker enemy force or fix a larger enemy force in place over a broad front. (FM 3-90-1)

function – (Army) A practical grouping of tasks and systems (people, organizations, information, and processes) united by a common purpose. (ADP 1-01)

fusion – Consolidating, combining, and correlating information together. (ADRP 2-0)

—G—

gap – (Army) 1. An area free of armed mines or obstacles whose width and direction allow a friendly force to pass through while dispersed in a tactical formation. (ADRP 1-02) 2. Any break or breach in the continuity of tactical dispositions or formations beyond effective small arms coverage. Gaps (soft spots, weaknesses) may in fact be physical gaps in the enemy's disposition, but they also may be any weakness in time, space, or capability; a moment in time when the enemy is overexposed and vulnerable, a seam in an air defense umbrella, an infantry unit caught unprepared in open terrain, or a boundary between two units. 3. A ravine, mountain pass, river, or other terrain feature that presents an obstacle that may be bridged. (ATTP 3-90.4) See also **lane**.

gap crossing – The projection of combat power across a linear obstacle (wet or dry gap). (ATTP 3-90.4)

gap-crossing operation – A mobility operation consisting of river crossing, brigade-level crossing, and special gap-crossing operations conducted to project combat power across a linear obstacle (wet or dry gap). (ATTP 3-90.4)

general support – (DOD) 1. That support which is given to the supported force as a whole and not to any particular subdivision thereof. Also called GS. (JP 3-09.3) See ADRP 5-0, FM 3-09, FM 6-0, ATP 2-01, ATP 4-02.2.

general support–reinforcing – (Army) A support relationship assigned to a unit to support the force as a whole and to reinforce another similar-type unit. (ADRP 5-0)

generated obscuration – Obscuration produced by generator systems, smoke pots, and hand grenades. (ATP 3-11.50)

geospatial data and information – The geographic-referenced and tactical objects and events that support the unit mission, task, and purpose. (ATP 3-34.80)

geospatial engineering – (DOD) Those engineering capabilities and activities that contribute to a clear understanding of the physical environment by providing geospatial information and services to commanders and staffs. (JP 3-34) See ATP 2-22.7

geospatial information – (DOD) Information that identifies the geographic location and characteristics of natural or constructed features and boundaries on the Earth, including: statistical data and information derived from, among other things, remote sensing, mapping, and surveying technologies; and mapping, charting, geodetic data and related products. (JP 2-03) See ATP 2-22.7.

geospatial information and services – (DOD) The collection, information extraction, storage, dissemination, and exploitation of geodetic, geomagnetic, imagery, gravimetric, aeronautical, topographic, hydrographic, littoral, cultural, and toponymic data accurately referenced to a precise location on the Earth's surface. Also called GI&S. (JP 2-03) See ATP 2-22.7.

geospatial intelligence – (DOD) The exploitation and analysis of imagery and geospatial information to describe, assess, and visually depict physical features and geographically referenced activities on the Earth. Geospatial intelligence consists of imagery, imagery intelligence, and geospatial information. Also called GEOINT. (JP 2-03) See ADRP 2-0, FM 2-0, FM 3-16, ATP 2-22.7, ATP 3-05.20, ATP 3-60.1.

global ballistic missile defense – (DOD) Defense against ballistic missile threats that cross one or more geographical combatant command boundaries and requires synchronization among the affected combatant commands. Also called GBMD. (JP 3-01) See ATP 3-14.5, ATP 3-27.5.

global engagement manager – Provides automated tools and decision aids that enable commanders to exercise mission command of ballistic missile defense forces deployed within the combatant command area of responsibility. (ATP 3-27.5)

global force management – (DOD) 1. A process that provides near-term sourcing solutions while providing the integrating mechanism between force apportionment, allocation, and assignment. Also called GFM. (JP 3-35) See ATP 3-35.

governance – (DOD) The state's ability to serve the citizens through the rules, processes, and behavior by which interests are articulated, resources are managed, and power is exercised in a society, including the representative participatory decision-making processes typically guaranteed under inclusive, constitutional authority. (JP 3-24) See FM 3-07.

government-in-exile – A government that has been displaced from its country, but remains recognized as the legitimate sovereign authority. (ATP 3-05.1)

government-owned containers – Containers purchased by the U.S. Government identified by ISO numbers starting with USAU or USAX. (ATP 4-12)

grade resistance – The resistance offered by a grade to the progress of a train. Also called GR. (ATP 4-14)

graphic control measure – A symbol used on maps and displays to regulate forces and warfighting functions. (ADRP 6-0)

ground-based midcourse defense – (DOD) A surface-to-air ballistic missile defense system for exo-atmospheric midcourse phase interception of long-range ballistic missiles using the ground-based interceptors. Also called GMD. (JP 3-01) See ATP 3-27.5.

group – (DOD) 3. A long-standing functional organization that is formed to support a broad function within a joint force commander's headquarters. Also called GP. (JP 3-33) See ATP 4-32.16.

gross trailing load – The maximum tonnage that a locomotive can move under given conditions. Also called GTL. (ATP 4-14)**guard** – A security task to protect the main force by fighting to gain time while also observing and reporting information and preventing enemy ground observation of and direct fire against the main body. Units conducting a guard mission cannot operate independently because they rely upon fires and functional and multifunctional support assets of the main body. (ADRP 3-90)

guard rail – A rail or series of rails that lay parallel to the running rails of a track that help prevent derailments by holding wheels in alignment and keeping derailed wheels on the ties. (ATP 4-14)

guerrilla – An irregular, predominantly indigenous member of a guerrilla force organized similar to military concepts and structure in order to conduct military and paramilitary operations in enemy-held, hostile, or denied territory. Although a guerrilla and guerrilla forces can exist independent of an insurgency, guerrillas normally operate in covert and overt resistance operations of an insurgency. (ATP 3-05.1)

guerrilla base – A temporary site where guerrilla installations, headquarters, and some guerrilla units are located. A guerrilla base is considered to be transitory and must be capable of rapid displacement by personnel within the base. (ATP 3-05.1)

guerrilla force – (DOD) A group of irregular, predominantly indigenous personnel organized along military lines to conduct military and paramilitary operations in enemy-held, hostile, or denied territory. (JP 3-05) See ADRP 3-05, ATP 3-05.1.

gun-target line – (DOD, NATO) An imaginary straight line from gun to target. Also called GTL. (JP 1-02) See ATP 3-09.30.

—H—

hasty crossing – The crossing of an inland water obstacle or other gap using the crossing means at hand or those readily available, and made without pausing for elaborate preparations. (ATTP 3-90.4)

hasty operation – An operation in which a commander directs immediately available forces, using fragmentary orders, to perform activities with minimal preparation, trading planning and preparation time for speed of execution. (ADRP 3-90) See also **fragmentary order**.

hazard – (DOD) A condition with the potential to cause injury, illness, or death of personnel; damage to or loss of equipment or property; or mission degradation. (JP 3-33) See ADRP 3-37.

hazardous waste – A solid waste that is listed as such in federal law or exhibits any of the hazardous characteristics of ignitability, corrosiveness, reactivity, or toxicity. Also called HW. (ATP 3-34.5)

hazardous waste accumulation site – A specially designated site for the temporary collection of hazardous wastes where no container may remain on site without permit for more than a specified duration, of which is correlative to the amount of refuse stored. (ATP 3-34.5)

health service support – (DOD) All services performed, provided, or arranged to promote, improve, conserve, or restore the mental or physical wellbeing of personnel, which include, but are not limited to, the management of health services resources, such as manpower, monies, and facilities; preventive and curative health measures; evacuation of wounded, injured, or sick; selection of the medically fit and disposition of the medically unfit; blood management; medical supply, equipment, and maintenance thereof; combat and operational stress control; medical, dental, veterinary, laboratory, optometric, nutrition therapy, and medical intelligence services. Also called HSS. (JP 4-02) See FM 4-02, ATP 4-02.55, ATP 4-02.84. (Army) Health service support encompasses all support and services performed, provided, and arranged by the Army Medical Department to promote, improve, conserve, or restore the mental and physical well-being of personnel in the Army. Additionally, as directed, provide support in other Services, agencies, and organizations. This includes casualty care (encompassing a number of Army Medical Department functions—organic and area medical support, hospitalization, the treatment aspects of dental care and behavioral/neuropsychiatric treatment, clinical laboratory services, and treatment of chemical, biological, radiological, and nuclear patients), medical evacuation, and medical logistics. (FM 4-02)

H-hour – (DOD) The specific hour on D-day at which a particular operation commences. (JP 5-0) See FM 3-99, FM 6-0.

high-payoff target – (DOD) A target whose loss to the enemy will significantly contribute to the success of the friendly course of action. Also called HPT. (JP 3-60) See FM 3-09, ATP 2-19.3, ATP 2-19.4, ATP 3-09.24, ATP 3-60, ATP 3-60.1.

high-payoff target list – A prioritized list of high-payoff targets by phase of the operation. (FM 3-09)

high-risk personnel – (DOD) Personnel who, by their grade, assignment, symbolic value, or relative isolation, are likely to be attractive or accessible terrorist targets. Also called HRP. (JP 3-07.2) See ADRP 3-37.

high-value airborne asset protection – (DOD) A defensive counterair mission using fighter escorts that defend airborne national assets which are so important that the loss of even one could seriously impact United States warfighting capabilities or provide the enemy with significant propaganda value. Also called HVAA protection. (JP 3-01) See ATP 3-55.6.

high-value individual – A person of interest who is identified, surveilled, tracked, influenced, or engaged. Also called HVI. (ATP 3-60)

high-value target – (DOD) A target the enemy commander requires for the successful completion of the mission. Also called HVT. (JP 3-60) See FM 3-09, ATP 2-01.3, ATP 2-19.3, ATP 2-19.4, ATP 3-09.24, ATP 3-60, ATP 3-60.1.

historical documents – Documents, materials, and data collected by the field historian to supplement the official record. (ATP 1-20)

historical monograph – An in-depth, systematically researched and presented historical work that focuses on a single subject or event. (ATP 1-20)

homeland – (DOD) The physical region that includes the continental United States, Alaska, Hawaii, United States possessions and territories, surrounding waters and air space. (JP 3-28) See ADP 3-28, ATP 2-91.7.

homeland defense – (DOD) The protection of United States sovereignty, territory, domestic population, and critical defense infrastructure against external threats and aggression or other threats as directed by the President. Also called HD. (JP 3-27) See ADRP 3-0, ADRP 3-28, ATP 4-32.

homeland security – (DOD) A concerted national effort to prevent terrorist attacks within the United States; reduce America's vulnerability to terrorism, major disasters, and other emergencies; and minimize the damage and recover from attacks, major disasters, and other emergencies that occur. Also called HS. (JP 3-28) See ADRP 3-28.

honorable service – Support and defense of the Constitution, the American people, and the national interest in a manner consistent with the Army Ethic. (ADRP 1)

hospital – A medical treatment facility capable of providing inpatient care. It is appropriately staffed and equipped to provide diagnostic and therapeutic services, as well as the necessary supporting services required to perform its assigned mission and functions. A hospital may, in addition, discharge the functions of a clinic. (FM 4-02)

hostile act – (DOD) An attack or other use of force against the United States, United States forces, or other designated persons or property to preclude or impede the mission and/or duties of United States forces, including the recovery of United States personnel or vital United States Government property. (JP 3-28) See ATP 3-22.40.

hostile casualty – (DOD) A person who is the victim of a terrorist activity or who becomes a casualty "in action." "In action" characterizes the casualty as having been the direct result of hostile action, sustained in combat or relating thereto, or sustained going to or returning from a combat mission provided that the occurrence was directly related to hostile action. Included are persons killed or wounded mistakenly or accidentally by friendly fire directed at a hostile force or what is thought to be a hostile force. However, not to be considered as sustained in action and not to be interpreted as hostile casualties are injuries or death due to the elements, self-inflicted wounds, combat fatigue, and except in unusual cases, wounds or death inflicted by a friendly force while the individual is in an absent-without-leave, deserter, or dropped-from-rolls status or is voluntarily absent from a place of duty. (JP 1-02) See ATP 4-02.55.

hostile criteria – Description of conditions under which an aircraft or a vehicle may be identified as hostile for engagement purposes. (FM 3-01.7) See also **rules of engagement**.

hostile intent – (DOD) The threat of imminent use of force against the United States, United States forces, or other designated persons or property. (JP 3-01) See ATP 3-22.40, ATP 4-02.55.

host nation – (DOD) A nation which receives the forces and/or supplies of allied nation and/or NATO organizations to be located on, operate in, or to transit through its territory. Also called HN. (JP 3-57) See FM 3-38, FM 3-53, FM 3-57, FM 4-01, ATP 3-05.2, ATP 3-07.31, ATP 3-57.10, ATP 3-57.20, ATP 3-57.30, ATP 3-57.60, ATP 3-57.70, ATP 3-57.80, ATP 4-15, ATP 4-43.

host-nation support – (DOD) Civil and/or military assistance rendered by a nation to foreign forces within its territory during peacetime, crisis or emergencies, or war based agreements mutually concluded between nations. Also called HNS. (JP 4-0) See FM 3-57, FM 4-01, ATP 3-05.2, ATP 4-15.

hub – (DOD) An organization that sorts and distributes inbound cargo from wholesale supply sources (airlifted, sealifted, and ground transportable) and/or from within the theater. See also hub and spoke distribution (JP 4-09) See FM 4-01. (Army) An organization that sorts and distributes inbound cargo from multiple supply sources. (ATP 4-11)

hub and spoke distribution – (DOD) A physical distribution system, in which a major port serves as a central point from which cargo is moved to and from several radiating points to increase transportation efficiencies and in-transit visibility. (JP 4-09) See FM 4-01.

human intelligence – (DOD) A category of intelligence derived from information collected and provided by human sources. (JP 2-0) See FM 3-16, FM 3-24, ATP 3-05.20, ATP 3-07.31. (Army) The collection by a trained human intelligence collector of foreign information from people and multimedia to identify elements, intentions, composition, strength, dispositions, tactics, equipment, and capabilities. Also called HUMINT. (FM 2-22.3) See also **intelligence**.

humanitarian and civic assistance – (DOD) Assistance to the local populace is specifically authorized by Title 10, United States Code, Section 401, and funded under separate authorities, provided by predominantly United States forces in conjunction with military operations. Also called HCA. (JP 3-29) See FM 3-57, ATP 3-05.2, ATP 3-07.31, ATP 3-57.20, ATP 3-57.30, ATP 3-57.80.

humanitarian assistance coordination center – (DOD) A temporary center established by a geographic combatant commander to assist with interagency coordination and planning. A humanitarian assistance coordination center operates during the early planning and coordination stages of foreign humanitarian assistance operations by providing the link between the geographic combatant commander and other United States Government agencies, nongovernmental organizations, and international and regional organizations at the strategic level. Also called HACC. (JP 3-29) See FM 3-57.

humanitarian mine action – (DOD) Activities that strive to reduce the social, economic, and environmental impact of land mines, unexploded ordnance and small arms ammunition - also characterized as explosive remnants of war. (JP 3-15) See ATP 4-32.

human resources services – Includes the key functions of essential personnel services, casualty operations, and postal operations. Human resources services function directly impact a Soldier's status assignment, qualifications, financial status, career progression, and quality of life which allow the Army leadership to effectively manage the force. (ATP 1-0.2)

human resources support – The functions and tasks executed within the Army Personnel Life Cycle Model (acquire, develop, distribute, structure, deploy, compensate, transition, and sustain) provides human resources services and support to Soldiers, their families, Department of Defense civilians, and other individuals authorized to accompany the force. Key functions include Man the Force, provide human resources services, coordinate personnel support, and conduct human resources planning and operations. (ATP 1-0.2)

hybrid threat – The diverse and dynamic combination of regular forces, irregular forces, terrorist forces, and/or criminal elements unified to achieve mutually benefitting effects. (ADRP 3-0)

hypo-chlorination – The application of a hypo-chlorinator to feed calcium or sodium hypochlorite. (ATP 4-44)

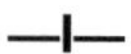

identification – (DOD) 1. The process of determining the friendly or hostile character of an unknown detected contact. See ATP 3-22.40. 2. In arms control, the process of determining which nation is responsible for the detected violations of any arms control measure. 3. In ground combat operations, discrimination between recognizable objects as being friendly or enemy, or the name that belongs to the object as a member of a class. Also called ID. (JP 3-01) See ATP 3-01.15, ATP 3-60.2.

imagery – (DOD) A likeness or presentation of any natural or man-made feature or related object or activity, and the positional data acquired at the same time the likeness or presentation was acquired, including: product produced by space-based national intelligence reconnaissance systems; and likeness and presentations produced by satellites, airborne platforms, unmanned aerial vehicles, or other similar means (except that such term does not include handheld or clandestine photography taken by or on behalf of human intelligence collection organizations). (JP 2-03) See ATP 2-22.7, ATP 3-55.6, ATP 3-55.12.

imagery exploitation – (DOD) The cycle of processing, using, interpreting, mensuration and/or manipulating imagery, and any assembly or consolidation of the results of dissemination. (JP 2-03) See ATP 3-55.6.

imagery intelligence – (DOD) The technical, geographic, and intelligence information derived through the interpretation or analysis of imagery and collateral materials. Also called IMINT. (JP 2-03) See ATP 2-22.7, ATP 3-05.20.

immediate decontamination – (DOD) Decontamination carried out by individuals immediately upon becoming contaminated to save lives, minimize casualties, and limit the spread of contamination. (JP 3-11) See ATP 3-05.11.

immediate response authority – (DOD) A Federal military commander's, Department of Defense component head's, and/or responsible Department of Defense civilian official's authority temporarily to employ resources under their control, subject to any supplemental direction provided by higher headquarters, and provide those resources to save lives, prevent human suffering, or mitigate great property damage in response to a request for assistance from civil authority, under imminently serious conditions when time does not permit approval from a higher authority within the United States. Immediate response authority does not permit actions that would subject civilians to the use of military power that is regulatory, prescriptive, proscriptive, or compulsory. (DODD 3025.18) See ADP 3-28, ADRP 3-28, ATP 2-91.7, ATP 3-28.1.

improvisation – The ability to adapt sustainment operations to unexpected situations or circumstances affecting a mission. (ADP 4-0)

improvised explosive device – (DOD) A weapon that is fabricated or emplaced in an unconventional manner incorporating destructive, lethal, noxious, pyrotechnic, or incendiary chemicals designed to kill, destroy, incapacitate, harass, deny mobility, or distract. Also called IED. (JP 3-15.1) See ATP 4-01.45, ATP 4-32, ATP 4-32.2, ATP 4-32.16.

implied task – (Army) A task that must be performed to accomplish a specified task or mission but is not stated in the higher headquarters' order. (FM 6-0) See also **essential task, specified task**.

incident – (DOD) An occurrence, caused by either human action or natural phenomena, that requires action to prevent or minimize loss of life or damage, loss of, or other risks to property, information, and/or natural resources. (JP 3-28) See ADP 3-28, ADRP 3-28, ATP 2-91.7, ATP 3-28.1.

incident awareness and assessment – (DOD) The Secretary of Defense approved use of Department of Defense intelligence, surveillance, reconnaissance, and other intelligence capabilities for domestic non-intelligence support for defense support of civil authorities. Also called IAA. (JP 3-28) See ATP 2-91.7.

incident command system– (DOD) A standard on-scene emergency management construct designed to aid in the management of resources during incidents. Also called ICS. (JP 3-28) See ATP 2-91.7, ATP 3-28.1.

incident management – (DOD) A national comprehensive approach to preventing, preparing for, responding to, and recovering from terrorist attacks, major disasters, and other emergencies. (JP 3-28) See ADP 3-28, ATP 3-28.1.

indications – (DOD) In intelligence usage, information in various degrees of evaluation, all of which bear on the intention of a potential enemy to adopt or reject a course of action. (JP 2-0) See ATP 2-33.4.

indicator – (DOD) In intelligence usage, an item of information which reflects the intention or capability of an adversary to adopt or reject a course of action. (JP 2-0) See FM 3-98, ATP 2-01, ATP 2-19.4. (Army) In the context of assessment, an item of information that provides insight into a measure of effectiveness or measure of performance. (ADRP 5-0) See also **intelligence**.

indigenous populations and institutions – (DOD) The societal framework of an operational environment including citizens, legal and illegal immigrants, dislocated civilians, and governmental, tribal, ethnic, religious, commercial, and private organizations and entities. Also called IPI. (JP 3-57) See FM 3-57, ATP 3-57.30, ATP 3-57.70.

indirect approach – The manner in which a commander attacks the enemy's center of gravity by applying combat power against a series of decisive points while avoiding enemy strength. (ADRP 3-90)

individual initiative – The willingness to act in the absence of orders, when existing orders no longer fit the situation, or when unforeseen opportunities or threats arise. (ADRP 3-0)

individual protective equipment – (DOD) In chemical, biological, radiological, or nuclear operations, the personal clothing and equipment required to protect an individual from chemical, biological, and radiological hazards. Also called IPE. (JP 3-11) See ATP 3-05.11, ATP 4-02.84.

infiltration – (Army) A form of maneuver in which an attacking force conducts undetected movement through or into an area occupied by enemy forces to occupy a position of advantage in the enemy rear while exposing only small elements to enemy defensive fires. (FM 3-90-1)

infiltration lane – A control measure that coordinates forward and lateral movement of infiltrating units and fixes fire planning responsibilities. (FM 3-90-1) See also **infiltration**.

influent – Water flowing into a reservoir, basin, or treatment operation. (ATP 4-44)

information collection – An activity that synchronizes and integrates the planning and employment of sensors and assets as well as the processing, exploitation, and dissemination systems in direct support of current and future operations. (FM 3-55)

information environment – (DOD) The aggregate of individuals, organizations, and systems that collect, process, disseminate, or act on information. (JP 3-13) See FM 3-13, FM 3-16, FM 3-24, FM 3-38, FM 3-99, ATP 2-01.3, ATP 3-05.20, ATP 3-53.2.

information for effect – Factual information used in publication or broadcast to negatively affect perceptions and/or damage credibility and capability of the targeted group. (FM 3-53)

information fratricide – The result of employing information-related capabilities in a way that causes effects in the information environment that impede the conduct of friendly operations or adversely affect friendly forces. (FM 3-13)

information management – (Army) The science of using procedures and information systems to collect, process, store, display, disseminate, and protect data, information, and knowledge products. (ADRP 6-0) See also **situational understanding**.

information operations – (DOD) The integrated employment, during military operations, of information-related capabilities in concert with other lines of operation to influence, disrupt, corrupt, or usurp the decision-making of adversaries and potential adversaries while protecting our own. Also called IO. (JP 3-13) See FM 3-13, FM 3-16, FM 3-24, FM 3-53, FM 3-99, FM 6-05, ATP 3-13.10, ATP 3-53.2, ATP 3-55.12.

information protection – Active or passive measures used to safeguard and defend friendly information and information systems. (ADRP 6-0)

information-related capabilities – Capabilities, techniques, or activities employing information to affect any of the three dimensions within the information environment to generate ends. (FM 3-13)

information-related capability – (DOD) A tool, technique, or activity employed within a dimension of the information environment that can be used to create effects and operationally desirable conditions. Also called IRC. (JP 3-13) See FM 3-07, FM 3-16, FM 3-24.

information requirement – (DOD) In intelligence usage, those items of information regarding the adversary and other relevant aspects of the operational environment that need to be collected and processed in order to meet the intelligence requirements of a commander. (JP 2-0) See ADRP 2-0, ATP 2-19.4ATP 3-55.6. (Army) Any information elements the commander and staff require to successfully conduct operations. (ADRP 6-0)

information superiority – (DOD) The operational advantage derived from the ability to collect, process, and disseminate an uninterrupted flow of information while exploiting or denying an adversary's ability to do the same. (JP 3-13) See FM 3-57, FM 3-99, ATP 3-18.4.

information system – (Army) Equipment that collects, processes, stores, displays, and disseminates information. This includes computers—hardware and software—and communications, as well as policies and procedures for their use. (ADP 6-0)

initial response force – (Army) A unit designated by the commander to respond to threat attacks or emergency situations. (FM 3-39)

inland petroleum distribution system – (DOD) A multi-product system consisting of both commercially available and military standard petroleum equipment that can be assembled by military personnel and, when assembled into an integrated petroleum distribution system, provides the military with the capability required to support an operational force with bulk fuels. The inland petroleum distribution system is comprised of three primary subsystems: tactical petroleum terminal, pipeline segments, and pump stations. Also called IPDS. (JP 4-03) See FM 4-01, ATP 4-43.

inpatient – A person admitted to and treated within a Role 3 and 4 hospital and who cannot be returned to duty within the same calendar day. (FM 4-02)

institutional training domain – The Army's institutional training and education system, which primarily includes training base centers and schools that provide initial training and subsequent professional military education for Soldiers, military leaders, and Army civilians. (ADP 7-0)

instruments of national power – (DOD) All of the means available to the government in its pursuit of national objectives. They are expressed as diplomatic, economic, informational and military. (JP 1) See ATP 3-57.60.

insurgency – (DOD) The organized use of subversion and violence to seize, nullify, or challenge political control of a region. Insurgency can also refer to the group itself. (JP 3-24) See ADRP 3-05, FM 3-07, FM 3-24, FM 3-57, ATP 3-05.1, ATP 3-05.2, ATP 3-57.70, ATP 3-57.80, ATP 4-14.

integrated air and missile defense – (DOD) The integration of capabilities and overlapping operations to defend the homeland and United States national interests, protect the joint force, and enable freedom of action by negating an adversary's ability to create adverse effects from their air and missile capabilities. Also called IAMD. (JP 3-01) See ATP 3-27.5.

integration – (DOD) 1. In force protection, the synchronized transfer of units into an operational commander's force prior to mission execution. See FM 6-05. 2. The arrangement of military forces and their actions to create a force that operates by engaging as a whole. See FM 3-07, FM 3-38, FM 6-05. 3. In photography, a process by which the average radar picture seen on several scans of the time base may be obtained on a print, or the process by which several photographic images are combined into a single image. (JP 1) See FM 6-05. (Army) Combining all of the elements of sustainment (task, functions, systems, processes, organizations) to operations assuring unity of command and effort. (ADP 4-0)

intelligence – (DOD)1. The product resulting from the collection, processing, integration, evaluation, analysis, and interpretation of available information concerning foreign nations, hostile or potentially hostile forces or elements, or areas of actual or potential operations. 2. The activities that result in the product. 3. The organizations engaged in such activities. (JP 2-0) See ADRP 2-0, FM 3-07, FM 3-24, ATP 2-01.3, ATP 2-91.7, ATP 3-04.64, ATP 3-55.6.

intelligence analysis – The process by which collected information is evaluated and integrated with existing information to facilitate intelligence production. (ADRP 2-0)

intelligence community – (DOD) All departments or agencies of a government that are concerned with intelligence activity, either in an oversight, managerial, support, or participatory role. (JP 1-02) See ADRP 2-0.

intelligence estimate – (DOD) The appraisal, expressed in writing or orally, of available intelligence relating to a specific situation or condition with a view to determining the courses of action open to the enemy or adversary and the order of probability of their adoption. (JP 2-0) See ATP 2-01.3.

intelligence operations – (DOD) The variety of intelligence and counterintelligence tasks that are carried out by various intelligence organizations and activities within the intelligence process. (JP 2-01) See ATP 2-01, ATP 3-05.1. (Army) The tasks undertaken by military intelligence units and Soldiers to obtain information to satisfy validated requirements. (ADRP 2-0)

intelligence preparation of the battlefield/battlespace – (Army, Marine Corps) The systematic process of analyzing the mission variables of enemy, terrain, weather, and civil considerations in an area of interest to determine their effect on operations. *(Marine Corps) The systematic, continuous process of analyzing the threat and environment in a specific geographic area.* Also called IPB. (ATP 2-01.3/MCRP 2-3A)

intelligence preparation of the battlespace – (DOD) The analytical methodologies employed by the Services or joint force component commands to reduce uncertainties concerning the enemy, environment, time, and terrain. Also called IPB. (JP 2-01.3) See ATP 3-05.1, ATP 3-05.20.

intelligence process – (DOD) The process by which information is converted into intelligence and made available to users, consisting of the six interrelated intelligence operations: planning and direction, collection, processing and exploitation, analysis and production, dissemination and integration, and evaluation and feedback. (JP 2-01) See ATP 3-55.6.

intelligence reach – The activity by which intelligence organizations proactively and rapidly access information from, receive support from, and conduct direct collaboration and information sharing with other units and agencies, both within and outside the area of operations, unconstrained by geographic proximity, echelon, or command. (ADRP 2-0)

intelligence requirement – (DOD) 1. Any subject, general or specific, upon which there is a need for the collection of information, or the production of intelligence. 2. A requirement for intelligence to fill a gap in the command's knowledge or understanding of the operational environment or threat forces. (JP 2-0) See ATP 2-01.3, ATP 3-05.20.

intelligence, surveillance, and reconnaissance – (DOD) An activity that synchronizes and integrates the planning and operation of sensors, assets, and processing, exploitation, and dissemination systems in direct support of current and future operations. This is an integrated intelligence and operations function. Also called ISR. (JP 2-01) See ADRP 2-0, FM 2-0, FM 3-16, FM 6-05, ATP 3-04.64, FM 3-16, ATP 3-55.3, ATP 3-55.6, ATP 3-60.2.

intelligence synchronization – The "art" of integrating information collection and intelligence analysis with operations to effectively and efficiently support decisionmaking. (ADRP 2-0)

intelligence system – (DOD) Any formal or informal system to manage data gathering, to obtain and process the data, to interpret the data, and to provide reasoned judgments to decision makers as a basis for action. (JP 2-01) See ATP 2-91.7.

intelligence warfighting function – The related tasks and systems that facilitate understanding the enemy, terrain, and civil considerations. (ADRP 3-0) See also **warfighting function**.

interagency – (DOD) Of or pertaining to United States Government agencies and departments, including the Department of Defense. (JP 3-08) See FM 3-07, FM 3-53, ATP 1-06.2, ATP 3-57.20, ATP 3-57.60.

interagency coordination – (DOD) Within the context of Department of Defense involvement, the coordination that occurs between elements of Department of Defense, and engaged US Government agencies and departments for the purpose of achieving an objective. (JP 3-0) See ADRP 3-0, FM 3-07, FM 3-53, ATP 2-91.7, ATP 3-05.2, ATP 3-57.20, ATP 3-57.60.

interdict – A tactical mission task where the commander prevents, disrupts, or delays the enemy's use of an area or route. (FM 3-90-1) See also **delay, disrupt, tactical mission task**.

interdiction – (DOD) 1. An action to divert, disrupt, delay, or destroy the enemy's military surface capability before it can be used effectively against friendly forces, or to otherwise achieve objectives. (JP 3-03) See ATP 3-91.1.

intergovernmental organization – (DOD) An organization created by a formal agreement between two or more governments on a global, regional, or functional basis to protect and promote national interests shared by member states. Also called IGO. (JP 3-08) See ADRP 3-0, FM 3-07, FM 3-24, FM 3-50, ATP 3-05.2, ATP 3-07.31, ATP 3-57.20, ATP 3-57.60.

interior lines – Lines on which a force operates when its operations diverge from a central point. (ADRP 3-0)

intermediate staging base – (DOD) A tailorable, temporary location used for staging forces, sustainment and/or extraction into and out of an operational area. Also called ISB. (JP 3-35) See ADRP 3-0, FM 3-99, ATP 3-35.

intermodal – (DOD) Type of international freight system that permits transshipping among sea, highway, rail, and air modes of transportation through use of American National Standards Institute and International Organization for Standardization containers, linehaul assets, and handling equipment. (JP 4-09) See FM 4-01, ATP 4-13.

intermodal operations – The process of using multimodal capabilities (air, sea, highway, rail) and conveyances (truck, barge, containers, pallets) to move troops, supplies and equipment through expeditionary entry points and the network of specialized transportation nodes to sustain land forces. (ATP 4-13)

internal defense and development – (DOD) The full range of measures taken by a nation to promote its growth and to protect itself from subversion, lawlessness, insurgency, terrorism, and other threats to its security. Also called IDAD. (JP 3-22) See ADRP 3-07, FM 3-24, FM 3-57, ATP 3-05.2, ATP 3-57.20, ATP 3-57.30, ATP 3-57.80, ATP 3-91.

internally displaced person – (DOD) Any person who has been forced or obliged to flee or to leave their home or place of habitual residence, in particular as a result of or in order to avoid the effects of armed conflict, situations of generalized violence, violations of human rights or natural or human-made disasters, and who have not crossed an internationally recognized state border. Also called IDP. (JP 3-29) See FM 3-57.

internal trust – Reliance on the character, competence, and commitment of Army professionals to live by and uphold the Army Ethic. (ADRP 1)

international military education and training – (DOD) Formal or informal instruction provided to foreign military students, units, and forces on a nonreimbursable (grant) basis by offices or employees of the United States, contract technicians, and contractors. Instruction may include correspondence courses; technical, educational, or informational publications; and media of all kinds. Also called IMET. (JP 3-22) See ATP 3-57.30.

international organization – (NATO) An intergovernmental, regional or global organization governed by international law and established by a group of states, with international juridical personality given by international agreement, however characterized, creating enforceable rights and obligations for the purpose of fulfilling a given function and pursuing common aims. (STANAG 3680/AAP-6) See ATP 3-07.31.

interoperability – (DOD) 1.The ability to operate in synergy in the execution of assigned tasks. (JP 3-0) FM-3-16, FM 4-01, FM 6-05. 2. The condition achieved among communications-electronics systems or items of communications-electronics equipment when information or services can be exchanged directly and satisfactorily between them and/or their users. The degree of interoperability should be defined when referring to specific cases. (JP 6-0) See FM 6-05, ATP 3-07.31.

interorganizational coordination – (DOD) The interaction that occurs among elements of the Department of Defense; engaged United States Government agencies; state, territorial, local, and tribal agencies; foreign military forces and government agencies; intergovernmental organizations; nongovernmental organizations; and the private sector. (JP 3-08) See ADRP 3-0, FM 3-07, ATP 2-91.7.

intertheater airlift – (DOD) The common-user airlift linking theaters to the continental United States and to other theaters as well as the airlift within the continental United States. (JP 3-17) See FM 4-01, ATP 4-48.

interzonal operations – Operations which cross area of operation boundaries of a specific transportation organization and operate under the area control of more than one headquarters or command. (ATP 4-11)

in-transit visibility – (DOD) The ability to track the identity, status, and location of Department of Defense units, and non-unit cargo (excluding bulk petroleum, oils, and lubricants) and passengers; patients, and personal property from origin to consignee or destination across the range of military operations. Also called ITV. (JP 4-01.2) See ADP 4-0, FM 4-01, FM 4-40, FM 4-95, ATP 4-0.1, ATP 4-12.

intratheater airlift – (DOD) Airlift conducted within a theater with assets assigned to a geographic combatant commander or attached to a subordinate joint force commander. (JP 3-17) See FM 4-01, ATP 4-48.

intrazonal operations – Operations confined within a specific transportation organization's area of operation. (ATP 4-11)

inventory control – (DOD) That phase of military logistics that includes managing, cataloging, requirements determinations, procurement, distribution, overhaul, and disposal of materiel. Also called inventory management; materiel control; materiel management; supply management. (JP 4-09) See ATP 4-42.2.

irregular warfare – (DOD) A violent struggle among state and non-state actors for legitimacy and influence over the relevant population(s). Also called IW. (JP 1) See ADRP 3-05, FM 3-05, FM 3-24, FM 3-53, FM 6-05, ATP 3-07.5.

isolate – A tactical mission task that requires a unit to seal off—both physically and psychologically—an enemy from sources of support, deny the enemy freedom of movement, and prevent the isolated enemy force from having contact with other enemy forces. (FM 3-90-1) See also **encirclement, tactical mission task**.

isolated personnel report – (DOD) A Department of Defense form (DD 1833) containing information designed to facilitate the identification and authentication of an isolated person by a recovery force. Also called ISOPREP. (JP 3-50) See ATP 3-50.3, ATP 3-53.1.

—J—

joint – (DOD) Connotes activities, operations, organizations, etc., in which elements of two or more Military Departments participate. (JP 1) See ATP 3-05.20.

joint air attack team – (DOD) A combination of attack and/or scout rotary-wing aircraft and fixed-wing close air support aircraft operating together to locate and attack high priority targets and other targets of opportunity. Also called JAAT. (JP 3-09.3) See FM 3-09.

joint air operations center – (DOD) A jointly staffed facility established for planning, directing, and executing joint air operations in support of the joint force commander's operation or campaign objectives. Also called JAOC. (JP 3-30) ATP 3-52.3, ATP 3-60.2.

joint combined exchange training – (DOD) A program conducted overseas to fulfill United States forces training requirements and at the same time exchange the sharing of skills between United States forces and host nation counterparts. Also called JCET. (JP 3-05) See ATP 3-53.1.

joint deployment and distribution enterprise – (DOD) The complex of equipment, procedures, doctrine, leaders, technical connectivity, information, shared knowledge, organizations, facilities, training, and materiel necessary to conduct joint distribution operations. Also called JDDE. (JP 4-0) See ATP 4-13.

joint deployment and distribution operations center – (DOD) A combatant command movement control organization designed to synchronize and optimize national and theater multimodal resources for deployment, distribution, and sustainment, Also called JDDOC. (JP 4-09) See FM 4-01.

joint doctrine – (DOD) Fundamental principles that guide the employment of United States military forces in coordinated action toward a common objective and may include terms, tactics, techniques, and procedures. (CJCSI 5120.02) See ADP 1-01.

joint engagement zone – (DOD) In air defense, that airspace of defined dimensions within which multiple air defense systems (surface-to-air missiles and aircraft) are simultaneously employed to engage air threats. Also called JEZ. (JP 3-01) See ATP 3-01.15.

joint field office – (DOD) A temporary multiagency coordination center established at the incident site to provide a central location for coordination of federal, state, local, tribal, nongovernmental, and private-sector organizations with primary responsibility for incident oversight, direction, or assistance to effectively coordinate protection, prevention, preparedness, response, and recovery actions. Also called JFO. (JP 3-28) See ATP 2-91.7.

joint fire support – (DOD) The joint fires that assist air, land, maritime, and special operations forces to move, maneuver, and control territory, populations, airspace, and key waters. (JP 3-0) See ADRP 3-09, FM 6-05, ATP 3-52.2.

joint fires – (DOD) Fires delivered during the employment of forces from two or more components in coordinated action to produce desired effects in support of a common objective. (JP 3-0) See ADP 3-09, ADRP 3-09, FM 3-09, FM 6-05, FM 3-52, ATP 3-52.2, ATP 3-60.2.

joint fires element – (DOD) An optional staff element that provides recommendations to the operations directorate to accomplish fires planning and synchronization. Also called JFE. (JP 3-60) See ATP 3-60.2.

joint fires observer – (DOD) A trained and certified Service member who can request, adjust, and control surface-to-surface fires, provide targeting information in support of Type 2 and 3 close air support terminal attack controls, and perform autonomous terminal guidance operations. (JP 1-02) See ADRP 3-09, FM 3-09, FM 3-52.

joint fire support – (DOD) Joint fires that assist air, land, maritime, and special operations forces to move, maneuver, and control territory, populations, airspace, and key waters. (JP 3-0) See FM 3-09.

joint force – (DOD) A general term applied to a force composed of significant elements, assigned or attached, of two or more Military Departments operating under a single joint force commander. (JP 3-0) See ATP 3-05.20, ATP 3-52.2.

joint force air component commander – (DOD) The commander within a unified command, subordinate unified command, or joint task force responsible to the establishing commander for recommending the proper employment of assigned, attached, and/or made available for tasking air forces; planning and coordinating air operations; or accomplishing such operational missions as may be assigned. Also called JFACC. (JP 3-0) See ATP 3-34.84, ATP 3-52.2, ATP 3-60.2.

joint force commander – (DOD) A general term applied to a combatant commander, subunified commander, or joint task force commander authorized to exercise combatant command (command authority) or operational control over a joint force. Also called JFC. See also joint force. (JP 1) See FM 4-40, ATP 3-34.84, ATP 3-52.2, ATP 3-52.3, ATP 3-60.2.

joint force land component commander – (DOD) The commander within a unified command, subordinate unified command, or joint task force responsible to the establishing commander for recommending the proper employment of assigned, attached, and/or made available for tasking land forces; planning and coordinating land operations; or accomplishing such operational missions as may be assigned. Also called JFLCC. (JP 3-0) See ATP 3-52.2, ATP 3-60.2, ATP 4-43.

joint force maritime component commander – (DOD) The commander within a unified command, subordinate unified command, or joint task force responsible to the establishing commander for recommending the proper employment of assigned, attached, and/or made available for tasking maritime forces and assets; planning and coordinating maritime operations; or accomplishing such operational missions as may be assigned. Also called JFMCC. (JP 3-0) See ATP 3-52.2, ATP 3-60.2.

joint force special operations component commander – (DOD) The commander within a unified command, subordinate unified command, or joint task force responsible to the establishing commander for recommending the proper employment of assigned, attached, and/or made available for tasking special operations forces and assets; planning and coordinating special operations; or accomplishing such operational missions as may be assigned. Also called JFSOCC. (JP 3-05) See ADRP 3-05, ATP 3-52.2, ATP 3-60.2.

joint intelligence preparation of the operational environment – (DOD) The analytical process used by joint intelligence organizations to produce intelligence estimates and other intelligence products in support of the joint force commander's decision-making process. It is a continuous process that includes defining the operational environment; describing the impact of the operational environment; evaluating the adversary; and determining adversary courses of action. Also called JIPOE. (JP 2-01.3) See ATP 3-05.1, ATP 3-05.20.

joint integrated prioritized target list – (DOD) A prioritized list of targets approved and maintained by the joint force commander. Also called JIPTL. (JP 3-60) See ATP 3-09.34.

joint logistics over-the-shore operations – (DOD) Operations in which Navy and Army logistics over-the-shore forces conduct logistics over-the-shore operations together under a joint force commander. Also called JLOTS. (JP 4-01.6) See FM 4-01, ATP 3-34.84, ATP 4-15, ATP 4-43.

joint operation planning process – (DOD) An orderly, analytical process that consists of a logical set of steps to analyze a mission, select the best course of action, and produce a joint operation plan or order. Also called JOPP. (JP 5-0) See ATP 3-57.60.

joint operations – (DOD) A general term to describe military actions conducted by joint forces and those Service forces employed in specified command relationships with each other, which of themselves, do not establish joint forces. (JP 3-0) See ATP 3-05.20, ATP 3-09.13, ATP 3-52.2.

joint operations area – (DOD) An area of land, sea, and airspace, defined by a geographic combatant commander or subordinate unified commander, in which a joint force commander (normally a joint task force commander) conducts military operations to accomplish a specific mission. Also called JOA. (JP 3-0) See FM 4-40, FM 6-05, ATP 3-52.2, ATP 3-60.2.

joint personnel recovery center – (DOD) The primary joint force organization responsible for planning and coordinating personnel recovery for military operations within the assigned operational area. Also called JPRC. (JP 3-50) See FM 3-50.

joint reception, staging, onward movement, and integration – (DOD) A phase of joint force projection occurring in the operational area during which arriving personnel, equipment, and materiel transition into forces capable of meeting operational requirements. Also called JRSOI. (JP 3-35) See ATP 3-35.

joint special operations air component commander – (DOD) The commander within a force special operations command responsible for planning and executing joint special operations air activities. Also called JSOACC. (JP 3-05) See ADP 3-05.

joint special operations area – (DOD) An area of land, sea, and airspace assigned by a joint force commander to the commander of a joint special operations force to conduct special operations activities. Also called JSOA. (JP 3-0) See ADRP 3-05, FM 3-05, FM 6-05, ATP 3-18.4, ATP 3-60.2.

joint special operations task force – (DOD) A joint task force composed of special operations units from more than one Service, formed to carry out a specific special operation or prosecute special operations in support of a theater campaign or other operations. Also called JSOTF. (JP 3-05) See ADRP 3-05, FM 3-05, FM 3-18, FM 6-05, ATP 3-05.11, ATP 3-34.84, ATP 3-60.2, ATP 3-75.

joint targeting coordination board – (DOD) A group formed by the joint force commander to accomplish broad targeting oversight functions that may include but are not limited to coordinating targeting information, providing targeting guidance, synchronization, and priorities, and refining the joint integrated prioritized target list. Also called JTCB. (JP 3-60) See ATP 3-09.34.

joint task force – (DOD) A joint force that is constituted and so designated by the Secretary of Defense, a combatant commander, subunified commander, or an existing joint task force commander. Also called JTF. (JP 1) See FM 3-57, ATP 3-07.31, ATP 3-52.3, ATP 3-57.10, ATP 3-57.70.

joint terminal attack controller – (DOD) A qualified (certified) Service member who, from a forward position, directs the action of combat aircraft engaged in close air support and other offensive air operations. Also called JTAC. (JP 3-09.3) See FM 3-09, ATP 3-04.64, ATP 3-52.2, ATP 4-01.45.

jumpmaster – (DOD) The assigned airborne qualified individual who controls paratroops from the time they enter the aircraft until they exit. (JP 3-17) See ATP 3-18.11.

—K—

key communicator – An individual to whom the target audience turns most often for an analysis or interpretation of information and events. (FM 3-53)

key leader engagement – Planned meeting(s) with an influential leader with the intent of building a relationship that facilitates communication and cooperation across a wider population. (FM 3-53)

key tasks – Those activities the force must perform as a whole to achieve the desired end state. (ADRP 5-0)

key terrain – (DOD) Any locality, or area, the seizure or retention of which affords a marked advantage to either combatant. (JP 2-01.3) See FM 3-90-1, FM 6-0, ATP 2-01.3, ATP 2-19.4.

kill box – (DOD) A three-dimensional area used to facilitate the integration of joint fires. (JP 3-09) See FM 3-09, ATP 3-52.2, ATP 3-60.2.

kill box coordinator – The aircraft assigned responsibility to de-conflict aircraft and manage/direct effective target engagement in a kill box. Also called KBC. (ATP 3-09.34)

kill zone – That part of an ambush site where fire is concentrated to isolate, fix, and destroy the enemy. See also **ambush, destroy, fix, isolate**. (FM 3-90-1)

knowledge management – The process of enabling knowledge flow to enhance shared understanding, learning, and decisionmaking. (ADRP 6-0)

—L—

land domain – (DOD) The area of the Earth's surface ending at the high water mark and overlapping with the maritime domain in the landward segment of the littorals. (JP 3-31) See ADP 1-01.

landing area – (DOD) 1. That part of the operational area within which are conducted the landing operations of an amphibious force. 2. In airborne operations, the general area used for landing troops and materiel either by airdrop or air landing. 3. Any specially prepared or selected surface of land, water, or deck designated or used for takeoff and landing of aircraft. (JP 3-02) See ATP 3-17.2.

landing zone – (DOD) Any specific zone used for the landing of aircraft. Also called LZ. (JP 3-17) See ATP 3-17.2, ATP 3-60.1.

land mine – A munition on or near the ground or other surface area that is designed to be exploded by the presence, proximity, or contact of a person or vehicle. (ATP 3-90.8)

landpower – The ability—by threat, force, or occupation—to gain, sustain, and exploit control over land, resources, and people. (ADRP 3-0)

LandWarNet – The Army's portion of the Department of Defense information networks. A technical network that encompasses all Army information management systems and information systems that collect, process, store, display, disseminate, and protect information worldwide. (FM 6-02)

lane – A route through, over, or around an enemy or friendly obstacle that provides safe passage of a passing force. The route may be reduced and proofed as part of a breaching operation, constructed as part of the obstacle, or marked as a bypass. (ATTP 3-90.4)

lateral boundary – A boundary that extends from the rear boundary to the unit's forward boundary. (FM 3-90-1)

latest arrival date – (DOD) A day, relative to C-Day, that is specified by the supported combatant commander as the latest date when a unit, a resupply shipment, or replacement personnel can arrive at the port of debarkation and support the concept of operations. Also called LAD. (JP 5-0) See FM 4-01.

latest time information is of value – The time by which an intelligence organization or staff must deliver information to the requestor in order to provide decisionmakers with timely intelligence. This must include the time anticipated for processing and disseminating that information as well as for making the decision. Also called LTIOV. (ATP 2-01)

law enforcement interrogation – The systematic effort by law enforcement investigators to prove, disprove, or corroborate information relevant to a criminal investigation using direct questioning in a controlled environment. (ATP 3-39.10)

law of war – (DOD) That part of international law that regulates the conduct of armed hostilities. Also called the law of armed conflict. (JP 1-04) See ATP 3-07.31, ATP 3-60.1. (Army) Also called the law of armed conflict—is that part of international law that regulates the conduct of armed hostilities. (FM 27-10)

lead agency – (DOD) The US Government agency designated to coordinate the interagency oversight of the day-to-day conduct of an ongoing operation. (JP 3-08) See FM 3-16.

lead Service or agency for common-user logistics – (DOD) A Service component or Department of Defense agency that is responsible for execution of common-user item or service support in a specific combatant command or multinational operation as defined in the combatant or subordinate joint force commander's operation plan, operation order, and/or directives. (JP 4-0) See FM 4-95.

leadership – The process of influencing people by providing purpose, direction, and motivation to accomplish the mission and improve the organization. (ADP 6-22)

letter of authorization – (DOD) A document issued by the procuring contracting officer or designee that authorizes contractor personnel authorized to accompany the force to travel to, from, and within the operational area; and, outlines government furnished support authorizations within the operational area. Also called LOA. (JP 4-10) See ATTP 4-10.

level I threat – A small enemy force that can be defeated by those units normally operating in the echelon support area or by the perimeter defenses established by friendly bases and base clusters. (ATP 3-91)

level II threat – An enemy force or activities that can be defeated by a base or base cluster's defensive capabilities when augmented by a response force. (ATP 3-91)

level III threat – An enemy force or activities beyond the defensive capability of both the base and base cluster and any local reserve or response force. (ATP 3-91)

level of detail – (DOD) Within the current joint planning and execution system, movement characteristics for both personnel and cargo are described at six distinct levels of detail. Levels I, V, and VI describe personnel and Levels I through IV and VI for cargo. Levels I through IV are coded and visible in the Joint Operation Planning and Execution System automated data processing. Levels V and VI are used by Joint Operation Planning and Execution System automated data processing feeder systems. a. level I - personnel: expressed as total number of passengers by unit line number. Cargo: expressed in total short tons, total measurement tons, total square feet, and total thousands of barrels by unit line number. Petroleum, oils, and lubricants is expressed by thousands of barrels by unit line number. b. level II - cargo: expressed by short tons and measurement tons of bulk, oversize, outsize, and non-air transportable cargo by unit line number. Also square feet for vehicles and non self-deployable aircraft and boats by unit line number. c. level III - cargo: detail by cargo category code expressed as short tons and measurement tons as well as square feet associated to that cargo category code for an individual unit line number. d. level IV - cargo: detail for individual dimensional data expressed in length, width, and height in number of inches, and weight/volume in short tons/measurement tons, along with a cargo description. Each cargo item is associated with a cargo category code and a unit line number). e. level V - personnel: any general summarization/aggregation of level VI detail in distribution and deployment. f. level VI - personnel: detail expressed by name, Service, military occupational specialty and unique identification number. Cargo: detail expressed by association to a transportation control number or single tracking number or item of equipment to include federal stock number/national stock number and/or requisition number. Nested cargo, cargo that is contained within another equipment item, may similarly be identified. Also called JOPES level of detail. (JP 1-02) See FM 4-01, ATP 3-35.

levels of warfare – A framework for defining and clarifying the relationship among national objectives, the operational approach, and tactical tasks. (ADP 1-01)

L-hour – (DOD) The specific hour on C-day at which a deployment operation commences or is to commence. (JP 5-0) FM 6-0.

liaison – (DOD) That contact or intercommunication maintained between elements of military forces or other agencies to ensure mutual understanding and unity of purpose and action. (JP 3-08) See ADRP 5-0, FM 6-0, ATP 2-91.7.

lighterage – (DOD) The process in which small craft are used to transport cargo or personnel from ship-to-shore using amphibians, landing craft, discharge lighters, causeways, and barges. (JP 4-01.6) See FM 4-01, ATP 4-13, ATP 4-15.

limited depositary account – A checking account in a United States or foreign commercial bank that is designated by the Treasury Department to receive deposits from Disbursing Officers for credit to their official limited depositary checking accounts. (FM 1-06)

limit of advance – A phase line used to control forward progress of the attack. The attacking unit does not advance any of its elements or assets beyond the limit of advance, but the attacking unit can push its security forces to that limit. Also called LOA. (ADRP 3-90)

line formation – When a unit's subordinate ground maneuver elements move abreast of each other. (FM 3-90-1)

line haul – An operation in which vehicles cannot make more than one round trip per day due to distance, terrain restrictions, or transit time. (ATP 4-11)

line of communications – (DOD) A route, either land, water, and/or air, that connects an operating military force with a base of operations and along which supplies and military forces move. Also called LOC. (JP 2-01.3) See FM 3-90-1, FM 4-01, FM 4-40, ATP 2-01.3, ATP 3-60.2.

line of communications bridging – Bridges used to establish semipermanent or permanent support to road networks. (ATTP 3-90.4)

line of contact – A general trace delineating the locations where friendly and enemy forces are engaged. (FM 3-90-1) See also **forward edge of the battle area, forward line of own troops, line of departure**.

line of demarcation – (DOD) A line defining the boundary of a buffer zone used to establish the forward limits of disputing or belligerent forces after each phase of disengagement or withdrawal has been completed. (JP 3-07.3) See ATP 3-07.31.

line of departure – (Army) A phase line crossed at a prescribed time by troops initiating an offensive operation. Also called LD. (ADRP 3-90) See also **line of contact, phase line**.

line of effort – (DOD) In the context of joint operation planning, using the purpose (cause and effect) to focus efforts toward establishing operational and strategic conditions by linking multiple tasks and missions. (JP 5-0) See FM 3-24. (Army) A line that links multiple tasks using the logic of purpose rather than geographical reference to focus efforts toward establishing operational and strategic conditions. Also called LOE. (ADRP 3-0)

line of fire – 1. As it relates to the principle of the reciprocal laying of field artillery weapons, any line parallel to the azimuth of fire. 2. The direction of the line established by the tube or any line parallel to that line in the firing battery. (ATP 3-09.50)

line of operation(s) – (DOD) A line that defines the interior or exterior orientation of the force in relation to the enemy or that connects actions on nodes and/or decisive points related in time and space to an objective(s). Also called LOO. (JP 5-0) See FM 3-24. (Army) A line that defines the directional orientation of a force in time and space in relation to the enemy and links the force with its base of operations and objectives. (ADRP 3-0)

line of sight – (Army, Marine Corps) The unobstructed path from a Soldier's/*Marine's* weapon, weapon sight, electronic sending and receiving antennas, or piece of reconnaissance equipment from one point to another. (ATP 2-01.3/MCRP-2-3A)

lines of patient drift – Natural routes along which wounded Soldiers may be expected to go back for medical care from a combat position. (ATP 4-02.2)

link – (DOD) 1. A behavioral, physical, or functional relationship between nodes. See FM 3-57. 2. In communications, a general term used to indicate the existence of communications facilities between two points. 3. A maritime route, other than a coastal or transit route, which links any two or more routes. (JP 3-0) See ATP 3-57.70.

linkup – A meeting of friendly ground forces, which occurs in a variety of circumstances. (ADRP 3-90)

linkup point – The point where two infiltrating elements in the same or different infiltration lanes are scheduled to meet to consolidate before proceeding on with their missions. (FM 3-90-1)

local haul – An operation in which vehicles can make two or more round trips per day based on distance and transit time. (ATP 4-11)

local security – A security task that includes low-level security activities conducted near a unit to prevent surprise by the enemy. (ADRP 3-90) See also **security operations**.

lodgment – (DOD) A designated area in a hostile or potentially hostile operational area that, when seized and held, makes the continuous landing of troops and materiel possible and provides maneuver space for subsequent operations. (JP 3-18) See ADRP 3-0, FM 3-99, ATP 3-91.

logistics – (DOD) Planning and executing the movement and support of forces. (JP 4-0) See ATP 4-48. (Army) Planning and executing the movement and support of forces. It includes those aspects of military operations that deal with: design and development, acquisition, storage, movement, distribution, maintenance, evacuation and disposition of materiel, acquisition or construction, maintenance, operation, and disposition of facilities, and acquisition or furnishing of services. (ADP 4-0)

logistics package – A grouping of multiple classes of supply and supply vehicles under the control of a single convoy commander. Also called LOGPAC. (FM 3-90-1) See also **classes of supply**.

logistics over-the-shore operations – (DOD) The loading and unloading of ships without the benefit of deep draft-capable, fixed port facilities; or as a means of moving forces closer to tactical assembly areas dependent on threat force capabilities. Also called LOTS operations. (JP 4-01.6) See ATP 3-34.84, ATP 4-15.

low visibility operations – (DOD) Sensitive operations wherein the political-military restrictions inherent in covert and clandestine operations are either not necessary or not feasible; actions are taken as required to limit exposure of those involved and/or their activities. Execution of these operations is undertaken with the knowledge that the action and/or sponsorship of the operation may preclude plausible denial by the initiating power. (JP 3-05) See ATP 3-05.1.

—M—

main battle area – The area where the commander intends to deploy the bulk of the unit's combat power and conduct decisive operations to defeat an attacking enemy. Also called MBA. (ADRP 3-90) See also **combat power, defensive operations, forward edge of the battle area, handover line**.

main body – The principal part of a tactical command or formation. It does not include detached elements of the command, such as advance guards, flank guards, and covering forces. (ADRP 3-90) See also **covering force, flank guard**.

main command post – A facility containing the majority of the staff designed to control current operations, conduct detailed analysis, and plan future operations. (FM 6-0)

main effort – A designated subordinate unit whose mission at a given point in time is most critical to overall mission success. (ADRP 3-0)

main supply route – (DOD) The route or routes designated within an operational area upon which the bulk of traffic flows in support of military operations. Also called MSR. (JP 4-01.5) See FM 3-90-1, FM 4-01.

main track – Track that extends through yards and between stations. (ATP 4-14)

major operation – (DOD) 1. A series of tactical actions (battles, engagements, strikes) conducted by combat forces of a single or several Services, coordinated in time and place, to achieve strategic or operational objectives in an operational area. 2. For noncombat operations, a reference to the relative size and scope of a military operation. (JP 3-0) See ADRP 3-0, ATP 3-07.5.

manned unmanned teaming – The integrated maneuver of Army Aviation rotary wing and unmanned aircraft system to conduct movement to contact, attack, reconnaissance, and security tasks. Also called MUM-T. (FM 3-04)

man the force – Manning combines anticipation, movement, and skillful positioning of personnel so that the commander has the personnel required to accomplish the mission. Manning the force involves human resources functions of personnel readiness management, personnel accountability, strength reporting, retention, and personnel information management. Man the force ensures the right person is in the right positions with the right skills and training at the right time. (ATP 1-0.2)

maneuver – (DOD) 1. A movement to place ships, aircraft, or land forces in a position of advantage over the enemy. See FM 3-07. 2. A tactical exercise carried out at sea, in the air, on the ground, or on a map in imitation of war. See FM 3-07. 3. The operation of a ship, aircraft, or vehicle, to cause it to perform desired movements. See FM 3-07, ATP 3-18.14. 4. Employment of forces in the operational area through movement in combination with fires to achieve a position of advantage in respect to the enemy. (JP 3-0) See ADP 3-90, ADRP 3-90, FM 3-07.

maneuver support operations – Integrate the complementary and reinforcing capabilities of mobility, countermobility, protection, and sustainment tasks to enhance decisive action. (FM 3-81)

march column – A march column consists of all elements using the same route for a single movement under control of a single commander. (FM 3-90-2) See also **march serial, march unit**.

march serial – A major subdivision of a march column that is organized under one commander who plans, regulates, and controls the serial. (FM 3-90-2) See also **march column, march unit.**

march unit – A subdivision of a march serial. It moves and halts under the control of a single commander who uses voice and visual signals. (FM 3-90-2) See also **march column, march serial.**

Marine air command and control system – (DOD) A system that provides the aviation combat element commander with the means to command, coordinate, and control all air operations within an assigned sector and to coordinate air operations with other Services. It is composed of command and control agencies with communications-electronics equipment that incorporates a capability from manual through semiautomatic control. Also called MACCS. (JP 3-09.3) See ATP 3-60.2.

marking obscuration – Obscuration effects that are employed to mark targets for destruction by lethal fires, identify friendly positions and locations, and provide a form of prearranged area of operations communications. (ATP 3-11.50)

marshalling – (DOD) 1. The process by which units participating in an amphibious or airborne operation group together or assemble when feasible or move to temporary camps in the vicinity of embarkation points, complete preparations for combat, or prepare for loading. (JP 3-17) See FM 3-99.

marshalling area – (DOD) A location in the vicinity of a reception terminal or pre-positioned equipment storage site where arriving unit personnel, equipment, materiel, and accompanying supplies are reassembled, returned to the control of the unit commander, and prepared for onward movement. See also marshalling. (JP 3-35) See FM 4-01, ATP 3-35.

mass atrocity response operations – (DOD) Military activities conducted to prevent or halt mass atrocities. Also called MARO. (JP 3-07.3) See ATP 3-07.31.

mass casualty – (DOD) Any large number of casualties produced in a relatively short period of time, usually as the result of a single incident such as a military aircraft accident, hurricane, flood, earthquake, or armed attack that exceeds local logistic support capabilities. Also called MASCAL. (JP 4-02) See FM 4-02, ATP 4-02.2.

massed fire – (DOD) 1. The fire of the batteries of two or more ships directed against a single target. 2. Fire from a number of weapons directed at a single point or small area. (JP 3-02) See FM 3-09.

materiel – (DOD) All items necessary to equip, operate, maintain, and support military activities without distinction as to its application for administrative or combat purposes. (JP 4-0) See ATP 3-22.40, ATP 3-90.15, ATP 4-42.2, ATP 4-90.

M-day – (DOD) Mobilization day; unnamed day on which mobilization of forces begins. (JP 4-06) See FM 6-0.

measurement and signature intelligence – (DOD) Intelligence obtained by quantitative and qualitative analysis of data (metric, angle, spatial, wavelength, time dependence, modulation, plasma, and hydromagnetic) derived from specific technical sensors for the purpose of identifying any distinctive features associated with the emitter or sender, and to facilitate subsequent identification and/or measurement of the same. The detected feature may be either reflected or emitted. Also called MASINT. (JP 2-0) See FM 3-16, ATP 3-05.20, ATP 3-13.10.

measurement ton – (DOD) The unit of volumetric measurement of equipment associated with surface-delivered cargo equal to the total cubic feet divided by 40. Also called MTON. (JP 4-01.5) See ATP 4-13.

measure of effectiveness – (DOD) A criterion used to assess changes in system behavior, capability, or operational environment that is tied to measuring the attainment of an end state, achievement of an objective, or creation of an effect. Also called MOE. (JP 3-0) See ADRP 3-07, ADRP 3-37, ADRP 5-0, FM 3-09, FM 3-24, FM 3-57, FM 6-0, ATP 3-07.10, ATP 3-09.24, ATP 3-53.1, ATP 3-53.2, ATP 3-55.3, ATP 3-57.20, ATP 3-57.60, ATP 3-57.80, ATP 6-01.1.

measure of performance – (DOD) A criterion used to assess friendly actions that is tied to measuring task accomplishment. Also called MOP. (JP 3-0) See ADRP 5-0, FM 3-09, FM 3-24, FM 3-57, FM 6-0, ATP 3-09.24, ATP 3-53.1, ATP 3-53.2, ATP 3-55.3, ATP 3-57.20, ATP 3-57.60, ATP 3-57.80, ATP 6-01.1.

media source analysis – The systematic comparison of the content, behavior, patterns, and trends of organic media organizations and sources of a country. (ATP 2-22.9)

medical evacuation – The process of moving any person who is wounded, injured, or ill to and/or between medical treatment facilities while providing en route medical care. (FM 4-02)

medical regulating – (DOD) The actions and coordination necessary to arrange for the movement of patients through the roles of care and to match patients with a medical treatment facility that has the necessary health service support capabilities, and available bed space. (JP 4-02) See FM 4-02, ATP 4-02.2, ATP 4-02.55.

medical treatment facility – (DOD) A facility established for the purpose of furnishing medical and/or dental care to eligible individuals. (JP 4-02) See FM 4-02. (Army) Any facility established for the purpose of providing medical treatment. This includes battalion aid stations, Role 2 facilities, dispensaries, clinics, and hospitals. (FM 4-02)

meeting engagement – A combat action that occurs when a moving force, incompletely deployed for battle, engages an enemy at an unexpected time and place. (FM 3-90-1)

mensuration – (DOD) The process of measurement of a feature or location on the earth to determine an absolute latitude, longitude, and elevation. (JP 3-60) See FM 3-09, ATP 3-91.1.

mentorship – The voluntary developmental relationship that exists between a person of greater experience and a person of lesser experience that is characterized by mutual trust and respect. (AR 600-100) See ADRP 6-22.

message – (DOD) 2. A narrowly focused communication directed at a specific audience to support a specific theme. (JP 3-61) See FM 3-53.

midcourse phase – (DOD) That portion of the flight of a ballistic missile between the boost phase and the terminal phase. (JP 3-01) See ATP 3-27.5.

Military Assistance Program – (DOD) That portion of the US security assistance authorized by the Foreign Assistance Act of 1961, as amended, which provides defense articles and services to recipients on a nonreimbursable (grant) basis. Also called MAP. (JP 3-22) See ATP 3-57.30.

military civic action – (DOD) Programs and projects managed by United States forces but executed primarily by indigenous military or security forces that contribute to the economic and social development of a host nation civil society thereby enhancing the legitimacy and social standing of the host nation government and its military forces. Also called MCA. (JP 3-57) See FM 3-57, ATP 3-57.20, ATP 3-57.30.

military deception – (DOD) Actions executed to deliberately mislead adversary military decision makers as to friendly military capabilities, intentions, and operations, thereby causing the adversary to take specific actions (or inactions) that will contribute to the accomplishment of the friendly mission. Also called MILDEC. (JP 3-13.4) See ADRP 6-0, FM 3-09, FM 3-53, ATP 3-53.1.

military decisionmaking process – An interactive planning methodology to understand the situation and mission, develop a courses of action, and produce an operation plan or order. Also called MDMP. (ADP 5-0) See also **operation order, operation plan**.

military engagement – (DOD) Routine contact and interaction between individuals or elements of the Armed Forces of the United States and those of another nation's armed forces, or foreign and domestic civilian authorities or agencies to build trust and confidence, share information, coordinate mutual activities, and maintain influence. (JP 3-0) See FM 3-07.

military expertise – Ethical design, generation, support, and application of landpower, primarily in unified land operations, and all supporting capabilities essential to accomplish the mission in defense of the American people. (ADRP 1)

military information support operations – (DOD) Planned operations to convey selected information and indicators to foreign audiences to influence their emotions, motives, objective reasoning, and ultimately the behavior of foreign governments, organizations, groups, and individuals in a manner favorable to the originator's objectives. Also called MISO. (JP 3-13.2) See ADP 3-05, ADRP 3-05, FM 3-05, FM 3-18, FM 3-53, ATP 3-07.31, ATP 3-53.2.

Military Sealift Command – (DOD) A major command of the United States Navy reporting to Commander Fleet Forces Command, and the United States Transportation Command's component command responsible for designated common-user sealift transportation services to deploy, employ, sustain, and redeploy United States forces on a global basis. Also called MSC. (JP 4-01.2) See FM 4-01, ATP 4-43.

mine – (DOD) 1. In land mine warfare, an explosive or other material, normally encased, designed to destroy or damage ground vehicles, boats, or aircraft, or designed to wound, kill, or otherwise incapacitate personnel and designed to be detonated by the action of its victim, by the passage of time, or by controlled means. See ATP 4-32.16. 2. In naval mine warfare, an explosive device laid in the water with the intention of damaging or sinking ships or of deterring shipping from entering an area. (JP 3-15) See ATP 4-32.2.

minefield – (DOD) 1. In land warfare, an area of ground containing mines emplaced with or without a pattern. 2. In naval warfare, an area of water containing mines emplaced with or without a pattern. See also mine; mine warfare. (JP 3-15) See ATP 4-32.2.

misinformation – Incorrect information from any source that is released for unknown reasons or to solicit a response or interest from a non-political or nonmilitary target. (FM 3-53)

missile defense – (DOD) Defense measures designed to destroy attacking enemy missiles, or to nullify or reduce the effectiveness of such attack. (JP 3-01) See ATP 3-55.6.

mission – (DOD) 1. The task, together with the purpose, that clearly indicates the action to be taken and the reason therefore. (JP 3-0) See ADP 1-01, ADP 5-0, ADRP 5-0, FM 3-07.

mission command – (Army) The exercise of authority and direction by the commander using mission orders to enable disciplined initiative within the commander's intent to empower agile and adaptive leaders in the conduct of unified land operations. (ADP 6-0) See also **commander's intent, mission orders**. (DOD) The conduct of military operations through decentralized execution based upon mission-type orders. (JP 3-31) See ATP 3-06.1.

mission command system – The arrangement of personnel, networks; information systems, processes and procedures, and facilities and equipment that enable commanders to conduct operations. (ADP 6-0)

mission command warfighting function – The related tasks and systems that develop and integrate those activities enabling a commander to balance the art of command and the science of control in order to integrate the other warfighting functions. (ADRP 3-0)

mission configured load – An ammunition load configured to support specific mission requirements across task forces or organizations. (ATP 4-35)

mission creep – Tangential efforts to assist in areas of concern unrelated to assigned duties that cripple efficient mission accomplishment. (FM 3-16)

mission-essential task – A task a unit could perform based on its design, equipment, manning, and table of organization and equipment/table of distribution and allowances mission. (ADRP 7-0) See also **mission-essential task list**.

mission-essential task list – A compilation of collective mission-essential tasks. Also called METL. (ADRP 7-0) See also **mission-essential task**.

mission orders – Directives that emphasize to subordinates the results to be attained, not how they are achieve them. (ADP 6-0) See also **mission command**.

mission-oriented protective posture – (DOD) A flexible system of protection against chemical, biological, radiological, and nuclear contamination in which personnel are required to wear only that protective clothing and equipment appropriate to the threat level, work rate imposed by the mission, temperature, and humidity. Also called MOPP. (JP 3-11) See ATP 3-05.11.

mission statement – (DOD) A short sentence or paragraph that describes the organization's essential task(s), purpose, and action containing the elements of who, what, when, where, and why. (JP 5-0) See ADP 1-01, FM 6-0, ATP 3-57.60.

mission support site – A preselected area used as a temporary base or stopover point. The mission support site is used to increase the operational range within the joint special operations area. (ATP 3-05.1)

mission variables – The categories of specific information needed to conduct operations. (ADP 1-01)

mixing – Using two or more different assets to collect against the same intelligence requirement. (FM 3-90-2)

mobile defense – A defensive task that concentrates on the destruction or defeat of the enemy through a decisive attack by a striking force. (ADRP 3-90)

mobile training team – (DOD) A team consisting of one or more US military or civilian personnel sent on temporary duty, often to a foreign nation, to give instruction. The mission of the team is to train indigenous personnel to operate, maintain, and employ weapons and support systems, or to develop a self-training capability in a particular skill. The Secretary of Defense may direct a team to train either military or civilian indigenous personnel, depending upon host-nation requests. Also called MTT. (JP 1-02) See ATP 3-07.10, ATP 3-57.30.

mobile security force – (DOD) A dedicated security force designed to defeat Level I and II threats on a base and/or base cluster. Also called MSF. (JP 3-10) See ATP 4-01.45.

mobility – (DOD) A quality or capacity of military forces which permits them to move from place to place while retaining the ability to fulfill their primary mission. (JP 3-17) See ADRP 3-90, FM 3-09, FM 3-90-1.

mobility corridor – (DOD) Areas that are relatively free of obstacles where a force will be canalized due to terrain restrictions allowing military forces to capitalize on the principles of mass and speed. (JP 2-01.3) See ATP 2-01.3, ATP 2-19.4.

mobility operations –Those combined arms activities that mitigate the effects of natural and manmade obstacles to enable freedom of movement and maneuver. (ATTP 3-90.4) See also **breach, countermobility operations**.

mobilization – (DOD) 1. The act of assembling and organizing national resources to support national objectives in time of war or other emergencies. See also industrial mobilization. 2. The process by which the Armed Forces or part of them are brought to a state of readiness for war or other national emergency. This includes activating all or part of the Reserve Component as well as assembling and organizing personnel, supplies, and materiel. Mobilization of the Armed Forces includes but is not limited to the following categories: a. selective mobilization — Expansion of the active Armed Forces resulting from action by Congress and/or the President to mobilize Reserve Component units, Individual Ready Reservists, and the resources needed for their support to meet the requirements of a domestic emergency that is not the result of an enemy attack. b. partial mobilization — Expansion of the active Armed Forces resulting from action by Congress (up to full mobilization) or by the President (not more than 1,000,000 for not more than 24 consecutive months) to mobilize Ready Reserve Component units, individual reservists, and the resources needed for their support to meet the requirements of a war or other national emergency involving an external threat to the national security. c. full mobilization — Expansion of the active Armed Forces resulting from action by Congress and the President to mobilize all Reserve Component units and individuals in the existing approved force structure, as well as all retired military personnel, and the resources needed for their support to meet the requirements of a war or other national emergency involving an external threat to the national security. Reserve personnel can be placed on active duty for the duration of the emergency plus six months. d. total mobilization — Expansion of the active Armed Forces resulting from action by Congress and the President to organize and/or generate additional units or personnel beyond the existing force structure, and the resources needed for their support, to meet the total requirements of a war or other national emergency involving an external threat to the national security. Also called MOB. (JP 4-05) See FM 4-01.

mode operations – The execution of movements using various conveyances (truck, lighterage, railcar, aircraft) to transport cargo. (ADRP 4-0)

modified combined obstacle overlay – (DOD) A joint intelligence preparation of the operational environment product used to portray the militarily significant aspects of the operational environment, such as obstacles restricting military movement, key geography, and military objectives. Also called MCOO. (JP 2-01.3) See ATP 2-01.3.

monitoring – Continuous observation of those conditions relevant to the current operation. (ADRP 5-0)

morale, welfare, recreation, and community support activities – Programs that provide Soldiers, Army civilians, and other authorized personnel with recreational and fitness activities, goods, and services. The moral, welfare, recreation network provides unit recreation and sports programs and rest areas for brigade-sized and larger units. Community support programs include the American Red Cross and family support. (ATP 1-0.2)

motor transportation – A ground support transportation function that includes moving and transferring units, personnel, equipment and supplies by vehicle to support the operations. (ATP 4-11)

movement and maneuver warfighting function – The related tasks and systems that move and employ forces to achieve a position of advantage over the enemy and other threats. (ADRP 3-0) See also **warfighting function**.

movement control – (DOD) The planning, routing, scheduling, and control of personnel and cargo movements over lines of communications, includes maintaining in-transit visibility of forces and material through the deployment and/or redeployment process. (JP 4-01.5) See ADRP 3-90, ATP 4-48. (Army) The dual process of committing allocated transportation assets and regulating movements according to command priorities to synchronize distribution flow over lines of communications to sustain land forces. (ADRP 4-0)

movement control team – (DOD) An Army team used to decentralize the execution of movement responsibilities on an area basis or at key transportation nodes. Also called MCT. (JP 4-09) See FM 4-01.

movement corridor – A designated area established to protect and enable ground movement along a route. (FM 3-81)

movement to contact – (Army) An offensive task designed to develop the situation and establish or regain contact. (ADRP 3-90)

mounted march – The movement of troops and equipment by combat and tactical vehicles. (FM 3-90-2)

multiechelon training – A training technique that allows for the simultaneous training of more than one echelon on different or complementary tasks. (ADRP 7-0)

multimodal –The movement of cargo and personnel using two or more transportation methods (air, highway, rail, sea) from point of origin to destination. (ATP 4-13)

multinational doctrine – (DOD) The agreed upon fundamental principles that guide the employment of forces of two or more nations in coordinated action toward a common objective. (JP 3-16) See ADP 1-01.

multinational logistics – (DOD) Any coordinated logistic activity involving two or more nations supporting a multinational force conducting military operations under the auspices of an alliance or coalition, including those conducted under United Nations mandate. Also called MNL. (JP 4-08) See FM 4-95.

multinational operations – (DOD) A collective term to describe military actions conducted by forces of two or more nations, usually undertaken within the structure of a coalition or alliance. (JP 3-16) See ADRP 3-0, FM 3-07, FM 3-09, FM 3-16, ATP 3-07.31.

multi-Service publication – (DOD) A publication containing principles, terms, tactics, techniques, and procedures used and approved by the forces of two or more Services to perform a common military function consistent with approved joint doctrine. (CJCSM 5120.01) See ADP 1-01.

munition – A complete device charged with explosives, propellants, pyrotechnics, initiating composition or chemical, biological, radiological or nuclear material, for use in operations, including demolitions. (FM 4-30)

music headquarters – The mission command element of a music performance unit. Also called MHQ. (ATP 1-19)

music performance detachment – An organic musical unit of a parent music performance unit at a non-collocated installation. Also called MPD. (ATP 1-19)

music performance team – A compact, modular unit designed to serve as the building block of the music performance unit. Also called MPT. (ATP 1-19)

music performance unit – A modular music unit in Army force structure. Also called MPU. (ATP 1-19)

mutual support – (DOD) That support which units render each other against an enemy, because of their assigned tasks, their position relative to each other and to the enemy, and their inherent capabilities. (JP 3-31) See ADRP 3-0, FM 3-09, FM 6-0.

—N—

named area of interest – (DOD) A geospatial area or systems node or link against which information that will satisfy a specific information requirement can be collected. Named areas of interest are usually selected to capture indications of adversary courses of action, but also may be related to conditions of the operational environment. (JP 2-01.3) See FM 3-09, FM 3-98, ATP 2-01.3, ATP 3-55.3, ATP 3-55.6, ATP 3-60.2.

national defense strategy – (DOD) A document approved by the Secretary of Defense for applying the Armed Forces of the United States in coordination with Department of Defense agencies and other instruments of national power to achieve national security strategy objectives. Also called NDS. (JP 1) See ADRP 3-28.

National Incident Management System – (DOD) A national crisis response system that provides a consistent, nationwide approach for Federal, state, local, and tribal governments; the private sector; and nongovernmental organizations to work effectively and efficiently together to prepare for, respond to, and recover from domestic incidents, regardless of cause, size or complexity. Also called NIMS. (JP 3-41) See ADP 3-28, ATP 2-91.7, ATP 3-28.1.

national military strategy – (DOD) A document approved by the Chairman of the Joint Chiefs of Staff for distributing and applying military power to attain national security strategy and national defense strategy objectives. Also called NMS. (JP 1) See ADRP 3-28.

national security strategy – (DOD) A document approved by the President of the United States for developing, applying, and coordinating the instruments of national power to achieve objectives that contribute to national security. Also called NSS. (JP 1) See ADRP 3-28.

nation assistance – (DOD) Assistance rendered to a nation by foreign forces within that nation's territory based on agreements mutually concluded between nations. (JP 3-0) See FM 3-24, FM 3-57, ATP 3-57.20, ATP 3-57.30, ATP 3-57.70, ATP 3-57.80.

natural disaster – (DOD) An emergency situation posing significant danger to life and property that results from a natural cause. See also domestic emergencies. (JP 3-29) See FM 3-53.

national special security event – (DOD) A designated event that, by virtue of its political, economic, social, or religious significance, may be the target of terrorism or other criminal activity. Also called NSSE. (JP 3-28) See ATP 2-91.7.

naval surface fire support – (DOD) Fire provided by Navy surface gun and missile systems in support of a unit or units. Also called NSFS. (JP 3-09.3) See FM 3-09.

N-day – Day an active duty unit is notified for deployment or redeployment. (JP 1-02) See FM 6-0.

nerve agent – (DOD) A potentially lethal chemical agent that interferes with the transmission of nerve impulses. (JP 3-11) See ATP 3-05.11.

nested concept – A planning technique to achieve unity of purpose whereby each succeeding echelon's concept of operations is aligned by purpose with the higher echelon's concept of operations. (ADRP 5-0) See also **concept of operations**.

net control station – A communications station designated to control traffic and enforce circuit discipline within a given net. Also called NCS. (ADRP 1-02)

net division tonnage – The tonnage in short tons, or payload, which can be moved over a railway division each day. Also called NDT. (ATP 4-14)

net trainload – The payload carried by a train. Also called NTL. (ATP 4-14)

networked munitions – (DOD) Remotely controlled, interconnected, weapons system designed to provide rapidly emplaced ground-based countermobility and protection capability through scalable application of lethal and nonlethal means. (JP 3-15) See ATP 3-90.8.

network operations – (DOD) Activities conducted to operate and defend the Global Information Grid. Also called NETOPS. (JP 6-0) See FM 6-02.

network transport – A system of systems including the people, equipment, and facilities that provide end-to-end communications connectivity for network components. (FM 6-02)

neutral – (DOD) In combat and combat support operations, an identity applied to a track whose characteristics, behavior, origin, or nationality indicate that it is neither supporting nor opposing friendly forces (JP 3-0) See FM 3-07. (Army) A party identified as neither supporting nor opposing friendly or enemy forces. (ADRP 3-0)

neutralization – In the context of the computed effects of field artillery fires renders a target ineffective for a short period of time, producing 10-percent casualties or materiel damage. (FM 3-09)

neutralization fire – Fire delivered to render the target ineffective or unusable. (FM 3-09)

neutralize – (Army) A tactical mission task that results in rendering enemy personnel or materiel incapable of interfering with a particular operation. (FM 3-90-1)

N-hour – The time a unit is notified to assemble its personnel and begin the deployment sequence. (FM 3-99)

N-hour sequence – Starts the reverse planning necessary after notification to have the first assault aircraft en route to the objective area for commencement of the parachute assault according to the order for execution. (FM 3-99)

night vision device – (DOD) Any electro-optical device that is used to detect visible and near-infrared energy, and provide a visible image. Night vision goggles, forward-looking infrared, thermal sights, and low-light level television are night vision devices. Also called NVD. (JP 3-09.3) See ATP 3-06.1.

night vision goggle – (DOD) An electro optical image intensifying device that detects visible and near-infrared energy, intensifies the energy, and provides a visible image for night viewing. Also called NVG. (JP 3-09.3) See ATP 3-06.1.

node – (DOD) 1. A location in a mobility system where a movement requirement is originated, processed for onward movement, or terminated. (JP 3-17) See FM 4-40, ATP 3-57.60. 3. An element of a system that represents a person, place, or physical thing. (JP 3-0) See FM 4-01, ATP 3-57.60.

no-strike list – (DOD) A list of objectives or entities characterized as protected from the effects of military operations under international law and/or rules of engagement. Also called NSL. (JP 3-60) See ADRP 3-09, ATP 3-57.60, FM 3-09.

no-fire area – (DOD) An area designated by the appropriate commander into which fires or their effects are prohibited. (JP 3-09.3) Also called NFA. See FM 3-09, ATP 3-09.34.

noncombatant evacuation operations – (DOD) Operations directed by the Department of State or other appropriate authority, in conjunction with the Department of Defense, whereby noncombatants are evacuated from foreign countries when their lives are endangered by war, civil unrest, or natural disaster to safe havens as designated by the Department of State. Also called NEO. (JP 3-68) See FM 3-07, ATP 3-05.68, ATP 3-75.

noncombatant evacuees – (DOD) 1. US citizens who may be ordered to evacuate by competent authority include: a. civilian employees of all agencies of the US Government and their dependents, except as noted in 2a below; b. military personnel of the Armed Forces of the United States specifically designated for evacuation as noncombatants; and c. dependents of members of the Armed Forces of the United States. 2. US (and non-US) citizens who may be authorized or assisted (but not necessarily ordered to evacuate) by competent authority include: a. civilian employees of US Government agencies and their dependents, who are residents in the country concerned on their own volition, but express the willingness to be evacuated; b. private US citizens and their dependents; c. military personnel and dependents of members of the Armed Forces of the United States outlined in 1c above, short of an ordered evacuation; and d. designated personnel, including dependents of persons listed in 1a through 1c above, as prescribed by the Department of State. (JP 3-68) See ATP 3-05.68.

noncontiguous area of operations –Where one or more of the commander's subordinate forces' areas of operation do not share a common boundary. See also **area of operations.** (FM 3-90-1)

nonconventional assisted recovery – (DOD) Personnel recovery conducted by indigenous/surrogate personnel that are trained, supported, and led by special operations forces, unconventional warfare ground and maritime forces, or other government agencies' personnel that have been specifically trained and directed to establish and operate indigenous or surrogate infrastructures. Also called NAR. (JP 3-50) See ATP 3-05.1.

nongovernmental organization – (DOD) A private, self-governing, not-for-profit organization dedicated to alleviating human suffering; and/or promoting education, health care, economic development, environmental protection, human rights, and conflict resolution; and/or encouraging the establishment of democratic institutions and civil society. Also called NGO (JP 3-08) See ADRP 3-0, FM 3-07, FM 3-24, FM 3-50, ATP 3-07.31, ATP 3-57.20, ATP 3-57.60.

nonlethal weapon – (DOD) A weapon that is explicitly designed and primarily employed so as to incapacitate personnel or materiel, while minimizing fatalities, permanent injury to personnel, and undesired damage to property and the environment. Also called NLW. (JP 3-28) See ATP 3-07.31.

nonpersistent agent – (DOD) A chemical agent that when released dissipates and/or loses its ability to cause casualties after 10 to 15 minutes. (JP 3-11) See ATP 3-05.11.

nonstandard bridging – Bridging that is purposely designed for a particular gap and typically built using commercial off-the-shelf or locally available materials. (ATTP 3-90.4)

nontransportable patient – A patient whose medical condition is such that he could not survive further evacuation to the rear without surgical intervention to stabilize his medical condition. (FM 4-02)

—O—

objective – (DOD) The clearly defined, decisive, and attainable goal toward which every operation is directed. (JP 5-0) See ADRP 5-0. (Army) A location on the ground used to orient operations, phase operations, facilitate changes of direction, and provide for unity of effort. (ADRP 3-90)

objective rally point – A rally point established on an easily identifiable point on the ground where all elements of the infiltrating unit assemble and prepare to attack the objective. (ADRP 3-90)

obscurant – Material that decreases the level of energy available for the functions of seekers, trackers, and vision enhancement devices. (ATP 3-11.50)

obscuration – The employment of materials into the environment that degrade optical and/or electro-optical capabilities within select portions of the electromagnetic spectrum in order to deny acquisition by or deceive an enemy or adversary. (ATP 3-11.50)

obscuration blanket – A dense horizontal concentration of smoke covering an area of ground with visibility inside the concentration less than 50 meters. (ATP 3-11.50)

obscuration curtain – A vertical development of smoke that reduces the enemy's ability to clearly identify what is occurring on the other side of the cloud. (ATP 3-11.50)

obscuration haze – Obscuration placed over friendly areas to restrict adversary observations and fire, but not dense enough to disrupt friendly operations within the screen. (ATP 3-11.50)

observation – The condition of weather and terrain that permits a force to see the friendly, enemy, and neutral personnel and systems, and key aspects of the environment. (ADRP 1-02)

observation post – A position from which military observations are made, or fire directed and adjusted, and which possesses appropriate communications. While aerial observers and sensors systems are extremely useful, those systems do not constitute aerial observation posts. Also called OP. (FM 3-90-2)

obstacle – (DOD) Any natural or man-made obstruction designed or employed to disrupt, fix, turn, or block the movement of an opposing force, and to impose additional losses in personnel, time, and equipment on the opposing force. (JP 3-15) See FM 3-90-1, FM 3-99, FM 6-0, ATP 2-01.3, ATP 2-19.4.

obstacle belt – (DOD) A brigade-level command and control measure, normally given graphically, to show where within an obstacle zone the ground tactical commander plans to limit friendly obstacle employment and focus the defense. (JP 3-15) See FM 3-90-1.

obstacle control measures – Specific measures that simplify the granting of obstacle-emplacing authority while providing obstacle control. (FM 3-90-1)

obstacle groups – One or more individual obstacles grouped to provide a specific obstacle effect. (FM 3-90-1)

obstacle line – A conceptual control measure used at battalion or brigade level to show placement intent without specifying a particular type of linear obstacle. (ADRP 1-02)

obstacle restricted areas – (DOD) A command and control measure used to limit the type or number of obstacles within an area. (JP 3-15) See also **obstacle.** See ATP 3-90.8.

obstacle zone – (DOD) A division-level command and control measure, normally done graphically, to designate specific land areas where lower echelons are allowed to employ tactical obstacles. (JP 3-15) See FM 3-90-1.

occupy – A tactical mission task that involves a force moving a friendly force into an area so that it can control that area. Both the force's movement to and occupation of the area occur without enemy opposition. (FM 3-90-1)

offensive counterair – (DOD) Offensive operations to destroy, disrupt, or neutralize enemy aircraft, missiles, launch platforms, and their supporting structures and systems both before and after launch, and as close to their source as possible. Also called OCA. (JP 3-01) See FM 3-01, FM 3-09, ATP 3-14.5, ATP 3-55.6.

offensive cyberspace operations – (DOD) Cyberspace operations intended to project power by the application of force in or through cyberspace. Also called OCO. (JP 1-02) See FM 3-38, FM 6-02.

offensive fires – Fires that preempt enemy actions. (FM 3-09)

offensive task – A task conducted to defeat and destroy enemy forces and seize terrain, resources, and population centers. (ADRP 3-0)

offshore petroleum discharge system – (DOD) Provides bulk transfer of petroleum directly from an offshore tanker to a beach termination unit located immediately inland from the high watermark. Bulk petroleum then is either transported inland or stored in the beach support area. Also called OPDS. (JP 4-03) See ATP 4-43.

on-call target – (DOD) Planned target upon which fires or other actions are determined using deliberate targeting and triggered, when detected or located, using dynamic targeting. (JP 3-60) See ATP 3-60.1.

on-order mission – A mission to be executed at an unspecified time. (FM 6-0)

open source – Any person or group that provides information without the expectations of privacy – the information, the relationship, or both is not protected against public disclosure. (ATP 2-22.9)

open-source intelligence – (DOD) Information of potential intelligence value that is available to the general public. Also called OSINT. (JP 2-0) See FM 3-16, ATP 2-22.9, ATP 3-05.20.

operation – (DOD) 1. A sequence of tactical actions with a common purpose or unifying theme. (JP 1) See FM 3-09. 2. A military action or the carrying out of a strategic, operational, tactical, service, training, or administrative military mission. (JP 3-0) See ADP 1-01, ADRP 3-90, FM 3-09, FM 3-90-1.

operational approach – (DOD) A description of broad actions the force must take to transform current conditions into those desired at end state. (JP 5-0) See ADP 1-01, ADP 5-0, ADRP 3-0, ADRP 5-0, FM 3-24, FM 6-0, ATP 5-0.1.

operational area – (DOD) An overarching term encompassing more descriptive terms (such as area of responsibility and joint operations area) for geographic areas in which military operations are conducted. Also called OA. (JP 3-0) See FM 3-07, FM 6-05, ATP 3-09.24, ATP 3-60.2.

operational area security – A form of security operations conducted to protect friendly forces, installations, routes, and actions within an area of operations. (ADRP 3-37)

operational art – (DOD) The cognitive approach by commanders and staffs—supported by their skill, knowledge, experience, creativity, and judgment—to develop strategies, campaigns, and operations to organize and employ military forces by integrating ends, ways, and means. (JP 3-0) See ADP 1-01, ADP 3-0, ADRP 3-0, ADRP 5-0, ATP 5-0.1.

operational concept – A fundamental statement that frames how Army forces, operating as part of a joint force, conduct operations. (ADP 1-01)

operational contract support – (DOD) The process of planning for and obtaining supplies, services, and construction from commercial sources in support of operations along with the associated contractor management functions. Also called OCS. (JP 4-10) See ATTP 4-10.

operational control – (DOD) The authority to perform those functions of command over subordinate forces involving organizing and employing commands and forces, assigning tasks, designating objectives, and giving authoritative direction necessary to accomplish the mission. Also called OPCON. (JP 1) See ADRP 5-0, FM 3-09, FM 4-30, FM 6-0, FM 6-05, ATP 3-04.64, ATP 3-53.1, ATP 4-32.16, ATP 4-43.

operational environment – (DOD) A composite of the conditions, circumstances, and influences that affect the employment of capabilities and bear on the decisions of the commander. Also called OE. (JP 3-0) See ADP 1-01, ADP 5-0, ADRP 3-0, ADRP 6-0, FM 3-07, FM 3-13, FM 3-24, FM 3-38, FM 3-52, FM 3-53, FM 3-57, FM 4-30, FM 4-95, FM 6-05, ATP 2-01.3, ATP 2-19.4, ATP 3-05.1, ATP 3-05.2, ATP 3-52.2, ATP 3-53.2, ATP 3-55.6, ATP 3-57.10, ATP 3-57.20, ATP 3-57.60, ATP 3-57.70, ATP 3-57.80, ATP 4-43, ATP 4-48, ATP 5-0.1.

operational framework – A cognitive tool used to assist commanders and staffs in clearly visualizing and describing the application of combat power in time, space, purpose, and resources in the concept of operations. (ADP 1-01)

operational initiative – The setting or dictating the terms of action throughout an operation. (ADRP 3-0)

operational reach – (DOD) The distance and duration across which a joint force can successfully employ military capabilities. (JP 3-0) See ADRP 3-0, FM 3-94, FM 4-01, FM 4-95, ATP 4-15.

operational training domain – The training activities organizations undertake while at home station, at maneuver combat training centers, during joint exercises, at mobility centers, and while operationally deployed. (ADP 7-0)

operational variables – A comprehensive set of information categories used to define an operational environment. (ADP 1-01)

operation order – (DOD) A directive issued by a commander to subordinate commanders for the purpose of effecting the coordinated execution of an operation. Also called OPORD. (JP 5-0) See ADP 1-01, FM 6-0, ATP 3-57.20, ATP 3-57.60.

operation plan – (DOD) 1. Any plan for the conduct of military operations prepared in response to actual and potential contingencies. See ATP 4-43. 2. A complete and detailed joint plan containing a full description of the concept of operations, all annexes applicable to the plan, and a time phased force and deployment data. Also called OPLAN. (JP 5-0) See FM 6-0, ATP 3-57.20, ATP 3-57.60.

operations data report – An annotated chronology of the unit's operations that will be fully supported by an indexed set of key historical documents. (ATP 1-20)

operations process – The major mission command activities performed during operations: planning, preparing, executing, and continuously assessing the operation. (ADP 5-0)

operations security – (DOD) A process of identifying critical information and subsequently analyzing friendly actions attendant to military operations and other activities. Also called OPSEC. (JP 3-13.3) See ADRP 3-37, ATP 3-05.20, ATP 3-07.31.

ordnance – (DOD) Explosives, chemicals, pyrotechnics, and similar stores, e.g., bombs, guns and ammunition, flares, smoke, or napalm. (JP 3-15) See ATP 4-32.2, ATP 4-32.16.

organic – (DOD) Assigned to and forming an essential part of military organization. Organic parts of a unit are those listed in its table of organization for the Army, Air Force, and Marine Corps, and are assigned to the administrative organizations of the operating forces for the Navy. (JP 1-02) See ADRP 5-0, FM 3-09, FM 4-30, FM 6-0.

orienting angle – A horizontal, clockwise angle measured from the line of fire to the orienting line. (ATP 3-09.50)

orienting line – A line of known direction in the firing unit's area that serves as a basis for laying the firing unit for direction. (ATP 3-09.50)

orienting station – 1.A point established on the ground that has directional control. 2. An orienting device, such as an aiming circle or gun laying and positioning system, set up over a point to lay the weapons by the orienting angle method. (ATP 3-09.50)

outpatient – A person receiving medical/dental examination and/or treatment from medical personnel and in a status other than being admitted to a hospital. Included in this category is the person who is treated and retained (held) in a medical treatment facility (such as a Role 2 facility) other than a hospital. (FM 4-02)

overbridging – A method used to reinforce, provide emergency repair, or augment existing bridges or bridge spans using standard bridging. In close combat this is typically provided through the employment of tactical bridging. (ATTP 3-90.4)

overhead persistent infrared – (DOD) Those systems originally developed to detect and track foreign intercontinental ballistic missile systems. Also called OPIR. (JP 3-14) See ATP 3-14.5, ATP 3-27.5.

overt – (DOD) Activities that are openly acknowledged by or are readily attributable to the United States Government, including those designated to acquire information through authorized and open means without concealment. Overt information may be collected by observation, elicitation, or from knowledgeable human sources. (JP 2-01.2) See ATP 3-53.1.

overt operation – (DOD) An operation conducted openly, without concealment. (JP 2-01.2) See ATP 3-05.2.

—P—

parallel planning – Two or more echelons planning for the same operation and sharing information sequentially through warning orders from the higher headquarters prior to the headquarters publishing its operation plan or operation order. (ADRP 5-0)

partner nation – (DOD) Those nations that the United States works with to disrupt the production, transportation, distribution, and sale of illicit drugs, as well as the money involved with this illicit activity. Also called PN. (JP 3-07.4) See FM 3-53, ATP 3-57.20.

passage lane – A lane through an enemy or friendly obstacle that provides a safe passage for a passing force. (FM 3-90-2)

passage of lines – (DOD) An operation in which a force moves forward or reward through another force's combat positions with the intention of moving into or out of contact with the enemy. (JP 3-18) See ADRP 3-90, FM 3-09, FM 3-90-1, FM 3-90-2.

passage point – A specifically designated place where the passing units will pass through the stationary unit. Also called PP. (FM 3-90-2) See also **passage of lines**.

passive air defense – (DOD) All measures other than active air defense, taken to minimize the effectiveness of hostile air and missile threats against friendly forces and assets. (JP 3-01) See ADRP 3-09, ADRP 3-90, FM 3-90-1, ATP 3-27.5.

patient – A sick, injured or wounded Soldier who receives medical care or treatment from medically trained personnel. (FM 4-02)

patient decontamination – The removal and/or the neutralization of hazardous levels of chemical, biological, radiological and nuclear contamination from patients at a medical treatment facility. Patient decontamination is performed under the supervision of medical personnel to prevent further injury to the patient and to maintain the patient's health status during the decontamination process. Patient decontamination serves multiple purposes; it protects the patient from further injury, it prevents exposing medical personnel to the contamination, and it prevents contamination of the medical treatment facility. (FM 4-02.7)

patient estimates – Estimates derived from the casualty estimate prepared by the personnel staff officer/assistant chief of staff, personnel. The patient medical workload is determined by the Army Health System support planner. Patient estimate only encompasses medical casualty. (FM 4-02)

patient movement – The act of moving a sick, injured, wounded, or other person to obtain medical and/or dental care or treatment. Functions include medical regulating, patient evacuation, and en route medical care. (FM 4-02)

peace building – (DOD) Stability actions, predominately diplomatic and economic, that strengthen and rebuild governmental infrastructure and institutions in order to avoid a relapse into conflict. Also called PB. (JP 3-07.3) See ADRP 3-07, ATP 3-07.31, ATP 3-07.5.

peace enforcement – (DOD) Application of military force, or threat of its use, normally pursuant to international authorization, to compel compliance with resolutions or sanctions designed to maintain or restore peace and order. (JP 3-07.3) See ADRP 3-07, ATP 3-07.31, ATP 3-07.5.

peacekeeping – (DOD) Military operations undertaken with the consent of all major parties to a dispute, designed to monitor and facilitate implementation of an agreement (cease fire, truce, or other such agreement) and support diplomatic efforts to reach a long-term political settlement. (JP 3-07.3) See ADRP 3-07, FM 3-53, ATP 3-07.31, ATP 3-07.5.

peacemaking – (DOD) The process of diplomacy, mediation, negotiation, or other forms of peaceful settlements that arranges an end to a dispute and resolves issues that led to it. (JP 3-07.3) See ADRP 3-07, FM 3-07, ATP 3-07.5, ATP 3-07.31.

peace operations – (DOD) A broad term that encompasses multiagency and multinational crisis response and limited contingency operations involving all instruments of national power with military missions to contain conflict, redress the peace, and shape the environment to support reconciliation and rebuilding and facilitate the transition to legitimate governance. Also called PO. (JP 3-07.3) See ADRP 3-07, FM 3-07, FM 3-53, ATP 3-07.31, ATP 3-07.5.

penetration – A form of maneuver in which an attacking force seeks to rupture enemy defenses on a narrow front to disrupt the defensive system. (FM 3-90-1)

performance work statement – (DOD) A statement of work for performance based acquisitions that describe the results in clear, specific, and objective terms with measurable outcomes. Also called PWS (JP 4-10) See ATP 1-06.2.

permissive environment – (DOD) Operational environment in which host country military and law enforcement agencies have control as well as the intent and capability to assist operations that a unit intends to conduct. (JP 3-0) See FM 3-57, ATP 3-57.10.

persistent agent – (DOD) A chemical agent that, when released, remains able to cause casualties for more than 24 hours to several days or weeks. (JP 3-11) See ATP 3-05.11.

personal protective equipment – (DOD) The equipment provided to shield or isolate a person from the chemical, physical, and thermal hazards that can be encountered at a hazardous materials incident. Personal protective equipment includes both personal protective clothing and respiratory protection. Also called PPE. (JP 3-11) See ATP 4-02.84.

personnel accountability – **(Army)** The by-name recording of specific data on individuals as they arrive and depart from units or theater transition points. Data includes information such as duty status, changes in duty status, changes in location, and grade changes. (ATP 1-0.2)

personnel readiness management – Personnel readiness management is the distribution of Soldiers and Army civilians to command and organizations based on documented manpower requirements, authorizations, and predictive analysis in support of the commander's plans and priorities need to accomplish its mission. (ATP 1-0.2)

personnel recovery – (DOD) The sum of military, diplomatic, and civil efforts to prepare for and execute the recovery and reintegration of isolated personnel. Also called PR. (JP 3-50) See ATP 3-04.64, ATP 3-05.20, ATP 3-55.6.

personnel services – Sustainment functions that man and fund the force, maintain Soldier and family readiness, promote the moral and ethical values of the nation, and enable the fighting qualities of the Army. (ADP 4-0)

personnel support – Personnel support encompasses the functions of morale, welfare, recreation, command interest programs, and band operations and contributes to unit readiness by promoting fitness, building morale and cohesion, enhancing quality of life, and by providing recreational, social, and other support services for Soldiers, Department of Defense civilians, and other personnel authorized to accompany the force. (ATP 1-0.2)

phase – (Army) A planning and execution tool used to divide an operation in duration or activity. (ADRP 3-0)

phase line – (DOD) line utilized for control and coordination of military operations, usually an easily identified feature in the operational area. Also called PL. (JP 3-09) See FM 3-09, FM 3-90-1, ATP 3-20.15.

P-hour – The specific hour on D-day at which a parachute assault commences with the exit of the first Soldier from an aircraft over a designated drop zone. P-hour may or may not coincide with H-hour. (FM 6-0)

physical security – (DOD) 1. That part of security concerned with physical measures designed to safeguard personnel; to prevent unauthorized access to equipment, installations, material, and documents; and to safeguard them against espionage, sabotage, damage, and theft. (JP 3-0) See FM 3-13.

physical security inspection – A formal, recorded assessment of the physical protective measures and security procedures that are implemented to protect unit and activity assets. (ATP 3-39.32)

physical security survey – A formal recorded assessment of an installation's overall physical security program, including electronic security measures. (ATP 3-39.32)

piecemeal commitment – The immediate employment of units in combat as they become available instead of waiting for larger aggregations of units to ensure mass, or The unsynchronized employment of available forces so that their combat power is not employed effectively. (ADRP 3-90) See also **combat power**.

pilot team – A deliberately structured composite organization comprised of Special Forces operational detachment members, with likely augmentation by interagency or other skilled personnel, designed to infiltrate a designated area to conduct sensitive preparation of the environment activities and assess the potential to conduct unconventional warfare in support of U.S. objectives. (ATP 3-05.1)

planned target – (DOD) Target that is known to exist in the operational environment, upon which actions are planned using deliberate targeting, creating effects which support commander's objectives. There are two subcategories of planned targets: scheduled and on-call. (JP 3-60) See ATP 3-60.1, FM 3-09, FM 3-99.

planning – The art and science of understanding a situation, envisioning a desired future, and laying out effective ways of bringing that future about. (ADP 5-0)

planning horizon – A point in time commanders use to focus the organization's planning efforts to shape future events. (ADRP 5-0) See also **planning**.

plan requirements and assess collection – The task of analyzing requirements, evaluating available assets (internal and external), recommending to the operations staff taskings for information collection assets, submiting requests for information for adjacent and higher collection support, and assessing the effectiveness of the information collection plan. (ATP 2-01)

platoon – A subdivision of a company or troop consisting of two or more squads or sections. (ADRP 3-90)

point of breach – The location at an obstacle where the creation of a lane is being attempted. (ATTP 3-90.4)

point of departure – The point where the unit crosses the line of departure and begins moving along a direction of attack. (ADRP 3-90) See also **line of departure**.

point of penetration – The location, identified on the ground, where the commander concentrates his efforts at the enemy's weakest point to seize a foothold on the farside objective. (ATTP 3-90.4)

police information – Available information concerning known and potential enemy and criminal threats and vulnerabilities collected during police activities, operations, and investigations. (FM 3-39)

police intelligence - The application of systems, technologies, and processes that analyze applicable data and information necessary for situational understanding and focusing policing activities to achieve social order. (FM 3-39)

policing – The application of control measures within an area of operations to maintain law and order, safety, and other matters affecting the general welfare of the population. (FM 3-39)

populace and resources control – Operations which provide security for the populace, deny personnel and materiel to the enemy, mobilize population and materiel resources, and detect and reduce the effectiveness of enemy agents. Populace control measures include curfews, movement restrictions, travel permits, registration cards, and resettlement of civilians. Resource control measures include licensing, regulations or guidelines, checkpoints (for example, road blocks), ration controls, amnesty programs, and inspection of facilities. Most military operations employ some type of populace and resources control measures. Also called PRC. (FM 3-57)

port of debarkation – (DOD) The geographic point at which cargo or personnel are discharged. Also called POD. (JP 4-0) See FM 4-01, ATP 4-13.

port of embarkation – (DOD) The geographic point in a routing scheme from which cargo or personnel depart. Also called POE. See also port of debarkation. (JP 4-01.2) See FM 4-01, ATP 4-13.

port opening – The ability to establish, initially operate and facilitate throughput for ports of debarkation to support unified land operations. (ADRP 4-0)

port support activity – (DOD) A tailorable support organization composed of mobilization station assets that ensures the equipment of the deploying units is ready to load. Also called PSA. See also support. (JP 3-35) See FM 4-01, ATP 3-35.

position area for artillery – An area assigned to an artillery unit where individual artillery systems can maneuver to increase their survivability. A position area for artillery is not an area of operations for the artillery unit occupying it. Also called PAA. (FM 3-90-1)

positive control – (DOD) A method of airspace control that relies on positive identification, tracking, and direction of aircraft within an airspace, conducted with electronic means by an agency having the authority and responsibility therein. (JP 3-52) See FM 3-52, ATP 3-06.1.

positive identification – (DOD) An identification derived from observation and analysis of target characteristics including visual recognition, electronic support systems, non-cooperative target recognition techniques, identification friend or foe systems, or other physics-based identification techniques. (JP 3-01) See ATP 3-60.2.

postal operations – Postal operations provide a network to process mail and provide postal service to Soldiers and other personnel authorized to receive postal entitlements. Processing mail involves receiving, separating, sorting, dispatching, and redirecting ordinary and accountable mail; completing international mail exchange; handling casualty and enemy prisoner of war mail; and screening for contaminated or suspicious mail. Postal services involve selling stamps, cashing and selling money orders, providing registered (including classified up to secret), insured and certified mail services, and processing postal claims and inquiries. (ATP 1-0.2)

precision-guided munition – (DOD) A guided weapon intended to destroy a point target and minimize collateral damage. Also called PGM, (JP 3-03) See FM 3-09, FM 3-98, ATP 3-09.50.

precision munition – A munition that corrects for ballistic conditions using guidance and control up to the aimpoint or submunitions dispense with terminal accuracy less than the lethal radius of effects. (FM 3-09)

precision smart munition – A munition or submunition that autonomously searches for, detects, classifies, selects, and engages a target or targets. A precision smart munition has a limited target discrimination capability. (FM 3-09)

preparation – Those activities performed by units and Soldiers to improve their ability to execute an operation. (ADP 5-0)

preparation fire – Normally a high volume of fires delivered over a short period of time to maximize surprise and shock effect. Preparation fire include electronic attack and should be synchronized with other electronic warfare activities. (FM 3-09)

preparation of the environment – (DOD) An umbrella term for operations and activities conducted by selectively trained special operations forces to develop an environment for potential future special operations. Also called PE. (JP 3-05) See ADP 3-05, ADRP 3-05, FM 3-18, ATP 3-05.1, ATP 3-05.2, ATP 3-53.1.

presumptive identification – The employment of technologies with limited specificity and sensitivity by general-purpose forces in a field environment to determine the presence of a chemical, biological, radiological, and/or nuclear hazard with a low level of confidence and degree of certainty necessary to support immediate tactical decisions. (ATP 3-11.37)

preventive medicine – The anticipation, prediction, identification, prevention, and control of communicable diseases (including vector-, food-, and waterborne diseases), illnesses, injuries, and diseases due to exposure to occupational and environmental threats, including nonbattle injury threats, combat stress responses, and other threats to the health and readiness of military personnel and military units. (FM 4-02)

primary position – The position that covers the enemy's most likely avenue of approach into the area of operations. (ADRP 3-90) See also **alternate position, area of operations, avenue of approach, battle position, subsequent position, supplementary position.**

principle – A comprehensive and fundamental rule or an assumption of central importance that guides how an organization or function approaches and thinks about the conduct of operations. (ADP 1-01)

principles of joint operations – Time-tested general characteristics of successful operations that serve as guides for the conduct of future operations. (ADP 1-01)

priority intelligence requirement – (DOD) An intelligence requirement, stated as a priority for intelligence support, that the commander and staff need to understand the adversary or other aspects of the operational environment. Also called PIR. (JP 2-01) See ADRP 2-0, ADRP 5-0, FM 3-57, FM 6-0, ATP 2-01, ATP 2-01.3, ATP 2-19.4, ATP 2-91.7, ATP 3-55.3, ATP 3-55.6, ATP 3-57.60.

priority of fires – The commander's guidance to his staff, subordinate commanders, fire support planners, and supporting agencies to organize and employ fire support in accordance with the relative importance of the unit's mission. (ADRP 3-09) See also **fire support**.

priority of support – A priority set by the commander to ensure a subordinate unit has support in accordance with its relative importance to accomplishing the mission. (ADRP 5-0)

priority target – A target, based on either time or importance, on which the delivery of fires takes precedence over all the fires for the designated firing unit or element. (FM 3-09)

private information – Data, facts, instructions, or other material intended for or restricted to a particular person, group, or organization. (ATP 2-22.9)

private sector – (DOD) An umbrella term that may be applied in the United States and in foreign countries to any or all of the nonpublic or commercial individuals and businesses, specified nonprofit organizations, most of academia and other scholastic institutions, and selected nongovernmental organizations. (JP 3-57) See ATP 3-57.20.

probable line of deployment – A phase line that designates the location where the commander intends to deploy the unit into assault formation before beginning the assault. Also called PLD. (ADRP 3-90) See also **phase line**.

procedures – (DOD) Standard, detailed steps that prescribe how to perform specific tasks. (CJCSM 5120.01) See ADP 1-01, ATP 3-90.90.

procedural control – (DOD) A method of airspace control which relies on a combination of previously agreed and promulgated orders and procedures. (JP 3-52) See FM 3-52.

Profession of Arms – A Community within the Army Profession composed of Soldiers of the Regular Army, Army National Guard, and Army Reserve. (ADRP 1)

program of targets – A number of planned targets of a similar nature that are planned for sequential attack. (FM 3-09)

progressive yard – A multifunctional yard structured to move cars in a fluid and rapid manner, containing receiving, classification, and departure yards. (ATP 4-14)

projected obscuration – An obscurant produced by artillery or mortar munitions, naval gunfire, helicopter-delivered rockets or, potentially, weapon grenades. (ATP 3-11.50)

proof – The verification that a lane is free of mines or explosive hazards and that the width and trafficability at the point of breach are suitable for the passing force. (ATTP 3-90.4)

propaganda – (DOD) Any form of adversary communication, especially of a biased or misleading nature, designed to influence the opinions, emotions, attitudes, or behavior of any group in order to benefit the sponsor, either directly or indirectly. (JP 3-13.2) See FM 3-53.

protection – (DOD) 1. Preservation of the effectiveness and survivability of mission-related military and nonmilitary personnel, equipment, facilities, information, and infrastructure deployed or located within or outside the boundaries of a given operational area. (JP 3-0) See ADRP 3-37, ADRP 5-0, FM 1-04, and FM 3-07. 2. In space usage, active and passive defensive measures to ensure the United States and friendly space systems perform as designed by seeking to overcome an adversary's attempts to negate them and to minimize damage if negation is attempted. (JP 3-14) See FM 3-07, ATP 3-05.11.

protection obscuration – Obscuration effects placed within the area of operations that contribute to the increased protection of United States forces and their interests by defeating on degrading adversary detection, observation, and engagement capabilities. (ATP 3-11.50)

protection warfighting function – The related tasks and systems that preserve the force so the commander can apply maximum combat power to accomplish the mission. (ADRP 3-0) See also **warfighting function**.

prudent risk – A deliberate exposure to potential injury or loss when the commander judges the outcome in terms of mission accomplishment as worth the cost. (ADP 6-0)

purple kill box – A fire support and airspace coordination measure used to facilitate the attack of surface targets with subsurface-, surface-to-surface, and air-to-surface munitions without further coordination with the establishing headquarters. Also called PKB. (ATP 3-09.34)

psychological action – Lethal and nonlethal actions planned, coordinated, and conducted to produce a psychological effect in a foreign individual, group, or population. (FM 3-53)

psychological objective – A statement of a measurable response that reflects the desired attitude or behavior change of a selected foreign target audience as a result of Military Information Support Operations. (FM 3-53)

public affairs – (DOD) Those public information, command information, and community engagement activities directed toward both the external and internal publics with interest in the Department of Defense. Also called PA. (JP 3-61) See FM 3-24.

public information – (DOD) Within public affairs, that information of a military nature, the dissemination of which is consistent with security and approved for release. (JP 3-61) See FM 3-53.

public key infrastructure – (DOD) An enterprise-wide service that supports digital signatures and other public key-based security mechanisms for Department of Defense functional enterprise programs, including generation, production, distribution, control, and accounting of public key certificates. Also called PKI. (JP 2-03) See ATP 6-02.75.

publicly available information – Data, facts, instructions, or other material published or broadcast for general public consumption; available on request to a member of the general public; lawfully seen or heard by any casual observer; or made available to a meeting open to the general public. (ATP 2-22.9)

pursuit – An offensive operation designed to catch or cut off a hostile force attempting to escape, with the aim of destroying it. (ADRP 3-90)

—Q—

quartering party – A group of unit representatives dispatched to a probable new site of operations in advance of the main body to secure, reconnoiter, and organize an area prior to the main body's arrival and occupation. (FM 3-90-2)

quay – (DOD) A structure of solid construction along a shore or bank that provides berthing and generally provides cargo-handling facilities. (JP 4-01.5) See ATP 4-13.

quick response force – A dedicated force on a base with adequate tactical mobility and fire support designated to defeat Level I and Level II threats and shape Level III threats until they can be defeated by a tactical combat force or other available response forces. (ATP 3-37.10)

—R—

radio frequency countermeasures – (DOD) Any device or technique employing radio frequency materials or technology that is intended to impair the effectiveness of enemy activity, particularly with respect to precision guided weapons and sensor systems. Also called RF CM. (JP 3-13.1) See FM 3-38.

raid – (DOD) An operation to temporarily seize an area in order to secure information, confuse an adversary, capture personnel or equipment, or to destroy a capability culminating with a planned withdrawal. (JP 3-0) See FM 3-90-1, ATP 3-05.1, ATP 3-75.

rally point – 1. An easily identifiable point on the ground at which aircrews and passengers can assemble and reorganize following an incident requiring a forced landing. 2. An easily identifiable point on the ground at which units can reassemble and reorganize if they become dispersed. Also called RP. (ADRP 1-02)

Rangers – (DOD) Rapidly deployable airborne light infantry organized and trained to conduct highly complex joint direct action operations in coordination with or in support of other special operations units of all Services. (JP 3-05) See ADP 3-05, ADRP 3-05, FM 3-05, ATP 3-75.

Rapid Engineer Deployable Heavy Operational Repair Squadron Engineer – (DOD) Air Force units wartime-structured to provide a heavy engineer capability that are mobile, rapidly deployable, and largely self-sufficient for limited periods of time. Also called RED HORSE. (JP 3-34) See ATP 4-32.2.

R-day – (DOD) Redeployment day. (JP 1-02) See FM 6-0.

reachback – (DOD) The process of obtaining products, services, and applications, or forces, or equipment, or material from organizations that are not forward deployed. (JP 3-30) See FM 3-57, ATP 3-52.2, ATP 3-57.60.

ready-to-load date – (DOD) The date when a unit will be ready to move from the origin, i.e., mobilization station. Also called RLD. (JP 5-0) See FM 4-01.

rear boundary – A boundary that defines the rearward limits of a unit's area. It usually also defines the start of the next echelon's support area. (FM 3-90-1) See also **area of operations, boundary**.

rear echelon – The echelon containing those elements of the force that are not required in the objective area. (FM 3-99)

rearward extension of the line of fire – An imaginary line in the exact opposite direction of the line of fire that extends through the center axis of the tube when looking down through the muzzle to the breech of the weapon. (ATP 3-09.50)

rearward passage of lines – Occurs when a unit passes through another unit's positions while moving away from the enemy. (ADRP 3-90) See also **forward passage of lines, passage of lines**.

reattack recommendation – (DOD) An assessment, derived from the results of battle damage assessment and munitions effectiveness assessment, providing the commander systematic advice on reattack of a target. Also called RR. (JP 3-60) See FM 3-09.

receiving yard – Yard where trains are cleared promptly on arrival to prevent main line congestion. (ATP 4-14)

reception – (DOD) 3. The process of receiving, off-loading, marshalling, accounting for, and transporting of personnel, equipment, and materiel from the strategic and/or intratheater deployment phase to a sea, air, or surface transportation point of debarkation to the marshalling area. (JP 3-35) See FM 4-01, ATP 3-35.

reconnaissance – (DOD) A mission undertaken to obtain, by visual observation or other detection methods, information about the activities and resources of an enemy or adversary, or to secure data concerning the meteorological, hydrographic or geographic characteristics of a particular area. Also called RECON. (JP 2-0) See ADRP 3-90, FM 2-0, FM 3-90-2, FM 3-98, FM 3-99, ATP 3-04.64, ATP 3-07.31, ATP 3-55.3, ATP 3-55.6, ATP 3-91.

reconnaissance by fire – A technique in which a unit fires on a suspected enemy position to cause the enemy forces to disclose their presence by movement or return fire. (FM 3-90-2)

reconnaissance handover – The action that occurs between two elements in order to coordinate the transfer of information and/or responsibility for observation of potential threat contact, or the transfer of an assigned area from one element to another. (FM 3-98)

reconnaissance handover line – A designated phase line on the ground where reconnaissance responsibility transitions from one element to another. (FM 3-98)

reconnaissance in force – A deliberate combat operation designed to discover or test the enemy's strength, dispositions, and reactions or to obtain other information. (ADRP 3-90)

reconnaissance objective – A terrain feature, geographical area, enemy force, adversary, or other mission or operational variable, such as specific civil considerations, about which the commander wants to obtain additional information. (ADRP 3-90)

reconnaissance-pull – Reconnaissance that determines which routes are suitable for maneuver, where the enemy is strong and weak, and where gaps exist, thus pulling the main body toward and along the path of least resistance. This facilitates the commander's initiative and agility. (FM 3-90-2)

reconnaissance-push – Reconnaissance that refines the common operational picture, enabling the commander to finalize the plan and support shaping and decisive operations. It is normally used once the commander commits to a scheme of maneuver or course of action. (FM 3-90-2)

reconnaissance squadron – A unit consisting of two or more company, battery, or troopsize units and a headquarters. (ADRP 3-90)

reconstitution – Actions that commanders plan and implement to restore units to a desired level of combat effectiveness commensurate with mission requirements and available resources. (ADRP 1-02)

recovered explosive ordnance – Devices that are retrieved in the operational environment, from field storage sites and licensed storage areas that contain explosives, propellants, pyrotechnics, initiating composition, or nuclear, biological or chemical material for use in operations, including demolitions which when salvaged have not been primed for use and may or may not be in their primary or logistic packaging. (FM 4-30)

recovery – (DOD) 4. Actions taken to extricate damaged or disabled equipment for return to friendly control or repair at another location. (JP 3-34) See ATP 4-32.16.

recovery force – (DOD) In personnel recovery, an organization consisting of personnel and equipment with a mission of locating, supporting, and recovering isolated personnel, and returning them to friendly control. (JP 3-50) See FM 3-50.

recovery vehicle – (DOD) In personnel recovery, the vehicle on which isolated personnel are boarded and transported from the recovery site. (JP 3-50) See ATP 3-18.14.

redeployment – (DOD) The transfer or rotation of forces and materiel to support another joint force commander's operational requirements, or to return personnel, equipment, and materiel to the home and/or demobilization stations for reintegration and/or outprocessing. (JP 3-35) See FM 4-01, FM 4-95, ATP 3-35. (Army) The transfer of forces and materiel to home and/or demobilization stations for reintegration and/or out-processing. (ATP 3-35)

red team – (DOD) An organizational element comprised of trained and educated members that provide an independent capability to fully explore alternatives in plans and operations in the context of the operational environment and from the perspective of adversaries and others. (JP 2-0) See FM 3-38, ATP 2-19.3.

reduce – 1. A tactical mission task that involves the destruction of an encircled or bypassed enemy force. (FM 3-90-1) 2. A mobility task to create and mark lanes through, over, or around an obstacle to allow the attacking force to accomplish its mission. (ATTP 3-90.4) See also **assault, bypass, destroy, neutralize, obscure, secure, suppress, tactical mission task**.

reduction area – A number of adjacent points of breach that are under the control of the breaching commander. (ATTP 3-90.4)

redundancy – Using two or more like assets to collect against the same intelligence requirement. (FM 3-90-2)

refer – To measure, using the panoramic telescope, the deflection to a given aiming point without moving the tube of the weapon. (ATP 3-09.50)

referred deflection – The deflection measured to an aiming point without moving the tube of the weapon. (ATP 3-09.50)

refugee – (DOD) A person who, owing to a well-founded fear of being persecuted for reasons of race, religion, nationality, membership of a particular social group or political opinion, is outside the country of his or her nationality and is unable or, owing to such fear, is unwilling to avail himself or herself of the protection of that country. (JP 3-29) See FM 3-07, FM 3-57.

regional air defense commander – (DOD) Commander, subordinate to the area air defense commander, who is responsible for air and missile defense in the assigned region and exercises authorities as delegated by the area air defense commander. Also called RADC. (JP 3-01) See ATP 3-27.5.

regionally aligned forces – Those forces that provide a combatant commander with up to joint task force capable headquarters with scalable, tailorable capabilities to enable the combatant commander to shape the environment. They are those Army units assigned to combatant commands, those Army units allocated to a combatant command, and those Army capabilities distributed and prepared by the Army for combatant command regional missions. (FM 3-22)

regional mechanism – The primary method through which friendly forces affect indigenous populations, host nations, or the enemy to establish the conditions needed to safeguard our interests and those of our allies. (ADP 3-05)

registering piece – The howitzer designated by the fire direction center to conduct a registration fire mission. (ATP 3-09.50)

rehearsal – A session in which a staff or unit practices expected actions to improve performance during execution. (ADRP 5-0)

reinforcing – A support relationship requiring a force to support another supporting unit. (ADRP 5-0)

reintegration – The process through which former combatants, belligerents, and displaced civilians receive amnesty, reenter civil society, gain sustainable employment, and become contributing members of the local populace. (ADRP 3-07)

relay – A single transport mission completed in one trip and utilizes multiple vehicles without transferring the load. (ATP 4-11)

release point – A location on a route where marching elements are released from centralized control. Also called RP. (FM 3-90-2) See also **lane, march column, march serial, march unit, start point**.

relief in place – (DOD) An operation in which, by direction of higher authority, all or part of a unit is replaced in an area by the incoming unit and the responsibilities of the replaced elements for the mission and the assigned zone of operations are transferred to the incoming unit. (JP 3-07.3) See FM 3-90-2.

religious advisement – (DOD) The practice of informing the commander on the impact of religion on joint operations to include, but not limited to: worship, rituals, customs and practices of US military personnel, international forces, and the indigenous population; as well as the impact of military operations on the religious and humanitarian dynamics in the operational area. (JP 1-05) See ATP 1-05.03.

render safe procedures – (DOD) The portion of the explosive ordnance disposal procedures involving the application of special explosive ordnance disposal methods and tools to provide for the interruption of functions or separation of essential components of unexploded explosive ordnance to prevent an unacceptable detonation. (JP 3-15.1) See ATP 4-32.2.

reorganization – All measures taken by the commander to maintain unit combat effectiveness or return it to a specified level of combat capability. (FM 3-90-1)

repatriation – (DOD) 2. The release and return of enemy prisoners of war to their own country in accordance with the 1949 Geneva Convention Relative to the Treatment of Prisoners of War. (JP 1-0) See ATP 3-07.31.

request for information – (DOD) 1. Any specific time-sensitive ad hoc requirement for intelligence information or products to support an ongoing crisis or operation not necessarily related to standing requirements or scheduled intelligence production. Also called RFI. See also intelligence. (JP 2-0) See ATP-2-01.3.

required delivery date – (DOD) The date that a force must arrive at the destination and complete unloading. Also called RDD. (JP 5-0) See FM 4-01.

required supply rate – The amount of ammunition expressed in terms of rounds per weapon per day for ammunition items fired by weapons, in terms of other units of measure per day for bulk allotment, and other items estimated to be required to sustain operations of any designated force without restriction for a specified period. (ATP 3-09.23)

requiring activity – (DOD) A military or other designated supported organization that identifies and receives contracted support during military operations. See also supported unit. (JP 4-10) See ATTP 4-10.

rerailer – Cast iron devices used in simple derailments to retrack cars and locomotives. (ATP 4-14)

reserve – (Army) That portion of a body of troops which is withheld from action at the beginning of an engagement, in order to be available for a decisive movement. (ADRP 3-90)

reserved obstacle – (Army/Marine Corps) Obstacles of any type, for which the commander restricts execution authority. (ATP 3-90.8)

resistance movement – (DOD) An organized effort by some portion of the civil population of a country to resist the legally established government or an occupying power and to disrupt civil order and stability. (JP 3-05) See ADRP 3-05, ATP 3-05.1, ATP 3-05.2.

resource management operations – The execution of the resource management mission includes analyze resource requirements, ensure commanders are aware of existing resource implications in order for them to make resource informed decisions, and then obtain the necessary funding that allows the commander to accomplish the overall unit mission. (FM 1-06)

resources – (DOD) The forces, materiel, and other assets or capabilities apportioned or allocated to the commander of a unified or specified command. (JP 1-02) See ATP 3-07.20.

responsiveness – The ability to react to changing requirements and respond to meet the needs to maintain support. (ADP 4-0)

restricted target – (DOD) A valid target that has specific restrictions placed on the actions authorized against it due to operational considerations. (JP 3-60) See ADRP 3-09, FM 3-09.

restricted target list – (DOD) A list of restricted targets nominated by elements of the joint force and approved by the joint force commander or directed by higher authorities. Also called RTL. (JP 3-60) See ADRP 3-09, FM 3-09.

restrictive fire area – (DOD) An area in which specific restrictions are imposed and into which fires that exceed those restrictions will not be delivered without coordination with the establishing headquarters. Also called RFA. (JP 3-09) See FM 3-09, FM 3-90-1, ATP 3-09.34.

restrictive fire line – (DOD) A line established between converging friendly surface forces that prohibits fires or their effects across that line. Also called RFL. See FM 3-09.

resuscitative care – Advanced trauma management care and surgery limited to the minimum required to stabilize a patient for transportation to a higher role of care. (FM 4-02)

retain – A tactical mission task in which the commander ensures that a terrain feature controlled by a friendly force remains free of enemy occupation or use. (FM 3-90-1) See also **tactical mission task**.

retention – An Army program that ensures all Soldiers, regardless of the type of military operation, have access to career counseling and retention processing. Retention improves readiness of the force, assists in force alignment, and contributes to maintaining the Army end strength. (ATP 1-0.2)

retirement – A form of retrograde in which a force out of contact moves away from the enemy. (ADRP 3-90)

retrograde – (DOD) The process for the movement of non-unit equipment and materiel from a forward location to a reset (replenishment, repair, or recapitalization) program or to another directed area of operations to replenish unit stocks, or to satisfy stock requirements. (JP 4-09) See ATP 4-48. (Army) A defensive task that involves organized movement away from the enemy. (ADRP 3-90)

retrograde movement – Any movement of a command to the rear, or away from the enemy. It may be focused by the enemy or may be made voluntarily. Such movements may be classified as a withdrawal, retirement, or delaying action. (FM 3-90-1)

retrograde of material – An Army logistics function of returning materiel from the owning or using unit back through the distribution system to the source of supply, directed ship to location, or point of disposal. (ATP 4-0.1)

return to duty – A patient disposition which, after medical evaluation and treatment when necessary, returns a Soldier for duty in his unit. (FM 4-02)

riot control agent – (DOD) Any chemical, not listed in a schedule of the Convention on the Prohibition of the Development, Production, Stockpiling and Use of Chemical Weapons and on their Destruction that can produce rapidly in humans sensory irritation or disabling physical effects that disappear within a short time following termination of exposure. Also called RCA. (JP 3-11) See ATP 3-05.11.

risk – (DOD) Probability and severity of loss linked to hazards. (JP 5-0) See ATP 5-19.

risk assessment – (DOD) The identification and assessment of hazards (first two steps of risk management process). Also called RA. (JP 3-07.2) See ATP 3-60.1, ATP 5-19.

risk management – (DOD) The process of identifying, assessing, and controlling risks arising from operational factors and making decisions that balance risk cost with mission benefits. Also called RM. (JP 3-0) See ADRP 5-0, FM 3-57, FM 6-0, ATP 2-19.4, ATP 3-57.60, ATP 3-60.1, ATP 6-01.1.

role – The broad and enduring purpose for which the organization or branch is established. (ADP 1-01)

route – The prescribed course to be traveled from a specific point of origin to a specific destination. (FM 3-90-1)

route classification – (DOD) Classification assigned to a route using factors of minimum width, worst route type, least bridge, raft, or culvert military load classification, and obstructions to traffic flow. (JP 3-34) See ATP 4-01.45.

route reconnaissance – A directed effort to obtain detailed information of a specified route and all terrain from which the enemy could influence movement along that route. (ADRP 3-90)

rolling resistance – The force components acting on a train in a direction parallel with the track, which tend to hold or retard the train's movement. Also called RR. (ATP 4-14)

rule of law – A principle under which all persons, institutions, and entities, public and private, including the state itself, are accountable to laws that are publicly promulgated, equally enforced, and independently adjudicated, and that are consistent with international human rights principles. (FM 3-07)

rules of engagement – (DOD) Directives issued by competent military authority that delineate the circumstances and limitations under which United States forces will initiate and/or continue combat engagement with other forces encountered. Also called ROE. (JP 1-04) See ADRP 3-0, ADRP 3-05, ADRP 3-09, FM 1-04, FM 3-07, FM 3-24, ATP 3-01.15, ATP 3-06.1, ATP 3-07.31, ATP 3-09.34, ATP 3-60.1, ATP 3-60.2, ATP 4-01.45.

running estimate – The continuous assessment of the current situation used to determine if the current operation is proceeding according to the commander's intent and if planned future operations are supportable. (ADP 5-0)

running track – Tracks that extend the entire length of the yard and provide a route of travel to any point in the yard independent of the switching leads and classification tracks. (ATP 4-14)

ruse – (DOD) In military deception, a trick of war designed to deceive the adversary, usually involving the deliberate exposure of false information to the adversary's intelligence collection system. (JP 3-13.4) See FM 6-0.

—S—

safe area – (DOD) A designated area in hostile territory that offers the evader or escapee a reasonable chance of avoiding capture and of surviving until he or she can be evacuated. (JP 3-50) See ADRP 3-05.

safety factor – The ratio of the strength of the rope to the working load. (ATP 4-14)

salvo – (DOD) 1. In naval gunfire support, a method of fire in which a number of weapons are fired at the same target simultaneously. 2. In close air support or air interdiction operations, a method of delivery in which the release mechanisms are operated to release or fire all ordnance of a specific type simultaneously. (JP 1-02) See ATP 3-01.15.

scheme of fires – The detailed, logical sequence of targets and fire support events to find and engage targets to accomplish the supported commander's intent. (FM 3-09)

science of control – The systems and procedures used to improve the commander's understanding and support accomplishing missions. (ADP 6-0)

science of tactics – Encompasses the understanding of those military aspects of tactics—capabilities, techniques and procedures—that can be measured and codified. (ADRP 3-90)

scientific and technical intelligence – (DOD) The product resulting from the collection, evaluation, analysis, and interpretation of foreign scientific and technical information that covers: a. foreign developments in basic and applied research and in applied engineering techniques; and b. scientific and technical characteristics, capabilities, and limitations of all foreign military systems, weapons, weapon systems, and materiel; the research and development related thereto; and the production methods employed for their manufacture. Also called S&TI. (JP 2-01) See ATP 2-22.4, ATP 3-13.10.

screen – A security task that primarily provides early warning to the protected force. (ADRP 3-90) See also **flank guard, guard, security operations, sensor, surveillance.**

seaport – (DOD) A land facility designated for reception of personnel or materiel moved by sea, and that serves as an authorized port of entrance into or departure from the country in which located. See also port of debarkation; port of embarkation. (JP 4-01.2) See FM 4-01, ATP 4-13.

search and attack – A technique for conducting a movement to contact that shares many of the characteristics of an area security mission. (FM 3-90-1)

section – An Army unit smaller than a platoon and larger than a squad. (ADRP 3-90)

sector air defense commander – (DOD) Commander subordinate to an area/regional air defense commander, who is responsible for air and missile defense in the assigned sector and exercises authorities delegated by the area/regional air defense commander. Also called SADC. (JP 3-01) See ATP 3-27.5.

sector of fire – That area assigned to a unit, crew-served weapon, or an individual weapon within which it will engage targets as they appear in accordance with established engagement priorities. (FM 3-90-1)

secure – A tactical mission task that involves preventing a unit, facility, or geographical location from being damaged or destroyed as a result of enemy action. (FM 3-90-1) See also **assault, breach, denial measure, destroy, reduce, suppress, tactical mission task**.

security area – That area that begins at the forward area of the battlefield and extends as far to the front and flanks as security forces are deployed. Forces in the security area furnish information on the enemy and delay, deceive, and disrupt the enemy and conduct counterreconnaissance. (ADRP 3-90) See also **counterreconnaissance, delay, disrupt**.

security assistance – (DOD) Group of programs authorized by the Foreign Assistance Act of 1961, as amended, and the Arms Export Control Act of 1976, as amended, or other related statutes by which the United States provides defense articles, military training, and other defense-related services, by grant, loan, credit, or cash sales in furtherance of national policies and objectives. Security assistance is an element of security cooperation funded and authorized by Department of State to be administered by Department of Defense/Defense Security Cooperation Agency. Also called SA. (JP 3-22) See FM 3-57, ATP 3-05.2, ATP 3-07.10, ATP 3-57.30, ATP 3-57.80.

security classification – (DOD) A category to which national security information and material is assigned to denote the degree of damage that unauthorized disclosure would cause to national defense or foreign relations of the United States and to denote the degree of protection required. There are three such categories: top secret, secret, and confidential. (JP 1-02) See ATP 3-35.

security cooperation – (DOD) All Department of Defense interactions with foreign defense establishments to build defense relationships that promote specific US security interests, develop allied and friendly military capabilities for self-defense and multinational operations, and provide US forces with peacetime and contingency access to a host nation. Also called SC. (JP 3-22) See ADRP 3-0, ADRP 3-07, FM 3-07, FM 3-24, ATP 3-05.2, ATP 3-07.10, ATP 3-93.

security cooperation organization – (DOD) All Department of Defense elements located in a foreign country with assigned responsibilities for carrying out security assistance/cooperation management functions. It includes military assistance advisory groups, military missions and groups, offices of defense and military cooperation, liaison groups, and defense attaché personnel designated to perform security assistance/cooperation functions. Also called SCO. (JP 3-22) See ATP 3-05.2.

security force assistance – (DOD) The Department of Defense activities that contribute to unified action by the U.S. Government to support the development of the capacity and capability of foreign security forces and their supporting institutions. Also called SFA. (JP 3-22) See ADRP 3-05, ADRP 3-07, FM 3-07, FM 3-24, ATP 3-05.2, ATP 3-07.5, ATP 3-07.10, ATP 3-53.1, ATP 3-93. (Army) The unified action to generate, employ, and sustain local, host nation, or regional security forces in support of legitimate authority. (FM 3-07)

security forces – (DOD) Duly constituted military, paramilitary, police, and constabulary forces of a state. (JP 3-22) See ATP 3-05.2.

security operations – Those operations undertaken by a commander to provide early and accurate warning of enemy operations, to provide the force being protected with time and maneuver space within which to react to the enemy, and to develop the situation to allow the commander to effectively use the protected force. (ADRP 3-90) See also **cover, guard, screen**.

security sector reform – (DOD) A comprehensive set of programs and activities undertaken to improve the way a host nation provides safety, security, and justice. Also called SSR. (JP 3-07) See FM 3-07.

seize – (Army) A tactical mission task that involves taking possession of a designated area using overwhelming force. (FM 3-90-1) See also **contain, tactical mission task**.

self-defense obscuration – Obscuration employed to protect United States forces at the vehicle level. Self-defense obscuration is used to defeat or degrade adversary area of operations weapon system guidance links. (ATP 3-11.50)

self-development training domain – Planned, goal-oriented learning that reinforces and expands the depth and breadth of an individual's knowledge base, self-awareness, and situational awareness; complements institutional and operational learning; enhances professional competence; and meets personal objectives. (ADP 7-0)

senior airfield authority – (DOD) An individual designated by the joint force commander to be responsible for the control, operation, and maintenance of an airfield to include the runways, associated taxiways, parking ramps, land, and facilities whose proximity directly affects airfield operations. Also called SAA. (JP 3-17) See ATP 3-17.2, ATP 3-52.3.

sensitive site – (DOD) A geographically limited area that contains, but is not limited to, adversary information systems, war crimes sites, critical government facilities, and areas suspected of containing high value targets. (JP 3-31) See ATP 3-90.15.

sensitive-site assessment – Determination of whether threats or hazards associated with a sensitive site warrant exploitation. Also called SSA. (ATP 3-11.23)

sequel – (DOD) The subsequent major operation or phase based on the possible outcomes (success, stalemate, or defeat) of the current major operation or phase. (JP 5-0) See FM 3-07, FM 6-0, ATP 2-01.

Service component command – (DOD) A command consisting of the Service component commander and all those Service forces, such as individuals, units, detachments, organizations, and installations under that command, including the support forces that have been assigned to a combatant command or further assigned to a subordinate unified command or joint task force. (JP 1) See ATP 3-52.3.

Service doctrine – Those publications approved by a single Service for use within that Service. (ADP 1-01)

shadow government – Governmental elements and activities performed by the irregular organization that will eventually take the place of the existing government. Members of the shadow government can be in any element of the irregular organization (underground, auxiliary, or guerrilla force). (ATP 3-05.1)

shaping operation – An operation that establishes conditions for the decisive operation through effects on the enemy, other actors, and the terrain. (ADRP 3-0) See also **decisive operation, sustaining operation**.

sheaf – The lateral distribution of the bursts of two or more pieces fired together. (ATP 3-09.50)

shifting fire – (DOD) Fire delivered at constant range at varying deflections; used to cover the width of a target that is too great to be covered by an open sheaf. (JP 1-02) See FM 3-09.

short ton – Equivalent of 2,000 pounds (0.907 metric ton) of weight. (ATP 4-35)

shuttle – A single transport mission completed in repeated trips by the same vehicles between two points. (ATP 4-11)

signal operating instructions – (DOD) A series of orders issued for technical control and coordination of the signal communication activities of a command. (JP 6-0) See FM 6-02.53.

signals intelligence – (DOD) 1. A category of intelligence comprising either individually or in combination all communications intelligence, electronic intelligence, and foreign instrumentation signals intelligence, however transmitted. (JP 2-0) See FM 2-0, FM 3-16. 2. Intelligence derived from communications, electronic, and foreign instrumentation signals. Also called SIGINT. (JP 2-0) See FM 2-0, ATP 3-05.20.

simplicity – Relates to processes and procedures to minimize the complexity of sustainment. (ADP 4-0)

single envelopment – A form of maneuver that results from maneuvering around one assailable flank of a designated enemy force. (FM 3-90-1)

single port manager – (DOD) The transportation component, designated by the Department of Defense through the US Transportation Command, responsible for management of all common-user aerial and seaports worldwide. Also called SPM. (JP 4-01.5) See ATP 4-13.

site characterization – A complete description and inventory of all personnel, equipment, material, and information discovered during exploitation. (ATP 3-11.23)

site exploitation – (DOD) A series of activities to recognize, collect, process, preserve, and analyze information, personnel, and/or materiel found during the conduct of operations. Also called SE. (JP 3-31) See FM 3-24, FM 3-98, ATP 3-60. (Army) The synchronized and integrated application of scientific and technological capabilities and enablers to answer information requirements, facilitate subsequent operations, and support host-nation rule of law. (ATP 3-90.15)

situation template – (DOD) A depiction of assumed adversary dispositions, based on that adversary's preferred method of operations and the impact of the operational environment if the adversary should adopt a particular course of action. (JP 2-01.3) See ATP 2-01.3.

situational obstacle – An obstacle that a unit plans and possibly prepares prior to starting an operation, but does not execute unless specific criteria are met. (ATP 3-90.8)

situational understanding – The product of applying analysis and judgment to relevant information to determine the relationship among the operational and mission variables to facilitate decisionmaking. (ADP 5-0)

sociocultural factors – (DOD) The social, cultural, and behavioral factors characterizing the relationships and activities of the population or a specific region or operational environment. (JP 2-01.3) See ADRP 3-05, ATP-2-01.3.

Soldier and leader engagement – Interpersonal interactions by Soldiers and leaders with audiences in an area of operations. (FM 3-13)

space coordinating authority – (DOD) A commander or individual assigned responsibility for planning, integrating, and coordinating space operations support in the operational area. Also called SCA. (JP 3-14) See ATP 3-52.2.

special air operation – An air operation conducted in support of special operations and other clandestine, covert, and military information support activities. (ATP 3-76)

special forces – (DOD) US Army forces organized, trained, and equipped to conduct special operations with an emphasis on unconventional warfare capabilities. Also called SF. (JP 3-05) See ADP 3-05, FM 3-05, FM 3-18, ATP 3-75.

special operations – (DOD) Operations requiring unique modes of employment, tactical techniques, equipment, and training often conducted in hostile, denied, or politically sensitive environments and characterized by one or more of the following: time sensitive, clandestine, low visibility, conducted with and/or through indigenous forces, requiring regional expertise, and/or a high degree of risk. Also called SO. (JP 3-05) See ADP 3-05, ADRP 3-05, FM 1-04, FM 3-05, FM 3-57, ATP 3-05.2, ATP 3-09.13, ATP 3-18.4, ATP 3-52.2, ATP 3-57.50, ATP 3-57.70, ATP 3-75.

special operations command – (DOD) A subordinate unified or other joint command established by a joint force commander to plan, coordinate, conduct, and support joint special operations within the joint force commander's assigned operational area. See also **special operations**. (JP 3-05) See ADP 3-05, ATP 3-05.11.

special operations command and control element – (DOD) A special operations element that is the focal point for the synchronization of special operations forces activities with conventional forces. Also called SOCCE. (JP 3-05) See ADRP 3-05.

special operations forces – (DOD) Those Active and Reserve Component forces of the Military Service designated by the Secretary of Defense and specifically organized, trained, and equipped to conduct and support special operations. Also called SOF. (JP 3-05) See ADP 3-05, ADRP 3-05, FM 3-53, FM 6-05, ATP 3-09.13, ATP 3-52.2, ATP 3-57.20, ATP 3-60.2.

special operations liaison element – (DOD) A special operations liaison team provided by the joint force special operations component commander to the joint force air component commander (if designated) or appropriate Service component air command and control organization, to coordinate, deconflict, and integrate special operations air, surface, and subsurface operations with conventional air operations. Also called SOLE. (JP 3-05) See ADRP 3-05.

special operations-peculiar – (DOD) Equipment, material, supplies, and services required for special operations missions for which there is no Service-common requirement. Also called SO-peculiar. (JP 3-05) See ADRP 3-05, FM 3-53.

special operations task force – A temporary or semipermanent grouping of Army special operations forces units under one commander and formed to carry out a specific operation or a continuing mission. Also called SOTF. (ADRP 3-05)

special reconnaissance – (DOD) Reconnaissance and surveillance actions conducted as a special operation in hostile, denied, or politically sensitive environments to collect or verify information of strategic or operational significance, employing military capabilities not normally found in conventional forces. Also called SR. (JP 3-05) See ADP 3-05, ADRP 3-05, ADRP 3-90, FM 3-05, FM 3-90-2, FM 3-98, ATP 3-18.4.

special tactics team – (DOD) An Air Force task-organized element of special tactics that may include combat control, pararescue, tactical air control party, and special operations weather personnel. Also called STT. (JP 3-05) See ATP 3-18.11.

special warfare – The execution of activities that involve a combination of lethal and nonlethal actions taken by a specially trained and educated force that has a deep understanding of cultures and foreign language, proficiency in small-unit tactics, and the ability to build and fight alongside indigenous combat formations in a permissive, uncertain, or hostile environment. (ADP 3-05)

specified task – (Army) A task specifically assigned to a unit by its higher headquarters. (FM 6-0)

spectrum management operations – The interrelated functions of spectrum management, frequency assignment, host nation coordination, and policy that together enable the planning, management, and execution of operations within the electromagnetic operational environment during all phases of military operations. Also called SMO. (FM 6-02)

spoiling attack – A tactical maneuver employed to seriously impair a hostile attack while the enemy is in the process of forming or assembling for an attack. (FM 3-90-1)

spoke – (DOD) The portion of the hub and spoke distribution system that refers to transportation mode operators responsible for scheduled delivery to a customer of the "hub". (JP 4-09) See FM 4-01. (Army) A portion of the distribution system that refers to the transportation mode operator's responsibility for scheduled delivery to a receiving unit. (ATP 4-11)

spot report – (DOD) A concise narrative report of essential information covering events or conditions that may have an immediate and significant effect on current planning and operations that is afforded the most expeditious means of transmission consistent with requisite security. Also called SPOTREP. (Note: In reconnaissance and surveillance usage, spot report is not to be used.) (JP 3-09.3) See ATP 3-04.64, ATP 4-32.2.

squad – A small military unit typically containing two or more fire teams. (ADRP 3-90)

stability mechanism – The primary method through which friendly forces affect civilians in order to attain conditions that support establishing a lasting, stable peace. (ADRP 3-0)

stability operations– (DOD) An overarching term encompassing various military missions, tasks, and activities conducted outside the United States in coordination with other instruments of national power to maintain or reestablish a safe and secure environment, provide essential governmental services, emergency infrastructure reconstruction, and humanitarian relief. (JP 3-0) See ADP 3-05, ADRP 3-37, FM 1-04, FM 3-07, FM 3-57, ATP 3-57.70, FM 3-94, FM 4-30, FM 4-95, ATP 2-19.3, ATP 2-22.4, ATP 2-33.4, ATP 3-57.10.

stability tasks – Tasks conducted as part of operations outside the United States in coordination with other instruments of national power to maintain or reestablish a safe and secure environment and provide essential governmental services, emergency infrastructure reconstruction, and humanitarian relief. (ADRP 3-07)

stabilization – The process by which underlying tensions that might lead to resurgence in violence and a breakdown in the law and order are managed and reduced, while efforts are made to support preconditions for successful long-term development. (FM 3-07)

stabilized patient – (1) Patient may require emergency intervention, but not surgery, within the next 24 hours. The patient's condition is characterized by a secure airway, control or absence of hemorrhage, shock adequately treated, vital signs stable, and major fractures immobilized. Stabilization is a precondition of extended duration evacuation (up to 24 hours). This includes, but is not limited to: (a) Ventilator. (b) Physiologic monitors. (c) Skull free of air or functioning drains in place. (d) Chest tube functional or x-ray free of pneumothorax. (e) Oxygen requirement is acceptable. (f) Functioning nasogastric tube or absence of ileus. (g) Bone fixator is acceptable. (h) Plaster bi-valved. (i) Pulses present after vascular repair. Despite these definitive example characteristics, there are patients who do not fit these descriptions, and yet may be considered stabilized—as always, this clinical decision is decided on between the originating and receiving physicians. (2) Patient whose condition may require emergency interventions within the next 24 hours. The patient's condition is characterized by a minimum of a secured airway, control or absence of hemorrhage, treated shock, and immobilized fractures. Stabilization is a necessary precondition for further evacuation. (3) A patient whose airway is secured, hemorrhage is controlled, shock is treated, and fractures are immobilized. (FM 4-02)

staff section – A group of staff members by area of expertise under a coordination, special, or personal staff officer. (FM 6-0)

staging – (DOD) Assembling, holding, and organizing arriving personnel, equipment, and sustaining materiel in preparation for onward movement. See also staging area. (JP 3-35) See ATP 3-35, ATP 4-13.

staging area – (DOD) 2. Other movements – a general locality established for the concentration of troop units and transient personnel between movements over the lines of communications. Also called SA. (JP 3-35) See ATP 3-35, FM 4-01.

standard bridging – Any bridging derived from manufactured bridge systems and components that are designed to be transportable, easily constructed, and reused. (ATTP 3-90.4)

standardization – (DOD) The process by which the Department of Defense achieves the closest practicable cooperation among the Services and Department of Defense agencies for the most efficient use of research, development, and production resources, and agrees to adopt on the broadest possible basis the use of: a. common or compatible operational, administrative, and logistic procedures; b. common or compatible technical procedures and criteria; c. common, compatible, or interchangeable supplies, components, weapons, or equipment; and d. common or compatible tactical doctrine with corresponding organizational compatibility. (JP 4-02) See FM 3-16, ATP 4-12.

standard operating procedure – (DOD) A set of instructions covering those features of operations which lend themselves to a definite or standardized procedure without loss of effectiveness. The procedure is applicable unless ordered otherwise. Also called SOP. (JP 1-02) See ATP 2-19.4, ATP 3-57.60, ATP 3-90.90.

start point – A location on a route where the march elements fall under the control of a designated march commander. Also called SP. (FM 3-90-2)

starting tractive effort – The power exerted by a locomotive to move itself and its load from a dead stop. Also called STE. (ATP 4-14)

status-of-forces agreement – (DOD) A bilateral or multilateral agreement that defines the legal position of a visiting military force deployed in the territory of a friendly state. Also called SOFA. (JP 3-16) See ATP 3-07.31.

stay behind operation – An operation in which the commander leaves a unit in position to conduct a specified mission while the remainder of the forces withdraw or retire from an area. (FM 3-90-1)

strength reporting – The numerical end product of the personnel accountability process to reflect the combat power of a unit. It is based on fill versus authorizations and drives Army readiness and personnel readiness management. Strength reporting is used to monitor unit strength, prioritize replacements, execute strength distribution, and make tactical and human resources support decisions. Strength reporting is conducted at each mission command level. (ATP 1-0.2)

stewardship – The responsibility of Army professionals to strengthen the Army as a profession and to care for the people and other resources entrusted to them by the American people. (ADRP 1)

strategic communication – (DOD) Focused United States Government efforts to understand and engage key audiences to create, strengthen, or preserve conditions favorable for the advancement of United States Government interests, policies, and objectives through the use of coordinated programs, plans, themes, messages, and products synchronized with the actions of all interments of national power. Also called SC. (JP 5-0) See FM 3-13, FM 3-53, ATP 3-55.12.

strike – (DOD) An attack to damage or destroy an objective or a capability. (JP 3-0) See FM 3-09, FM 3-24.

strike coordination and reconnaissance – A mission flown for the purpose of detecting targets and coordinating or performing attack or reconnaissance on those targets. Also called SCAR. (JP 3-03) See ATP 3-09.34, ATP 3-55.6, ATP 3-60.1, ATP 3-60.2.

striking force – A dedicated counterattack force in a mobile defense constituted with the bulk of available combat power. (ADRP 3-90) See also **mobile defense**.

strong point – A heavily fortified battle position tied to a natural or reinforcing obstacle to create an anchor for the defense or to deny the enemy decisive or key terrain. (ADRP 3-90) Also called SP. See also **battle position, mobile defense**.

subsequent position – A position that a unit expects to move to during the course of battle. (ADRP 3-90)

subversion – (DOD) Actions designated to undermine the military, economic, psychological, or political strength or morale of a governing authority. (JP 3-24) See ADRP 3-05, ATP 3-05.1, ATP 3-05.2.

subversive political action – (DOD) A planned series of activities designed to accomplish political objectives by influencing, dominating, or displacing individuals or groups who are so placed as to affect the decisions and actions of another government. (JP 1-02) See ATP 3-05.1.

supplementary position – A defensive position located within a unit's assigned area of operations that provides the best sectors of fire and defensive terrain along an avenue of approach that is not the primary avenue along where the enemy is expected to attack. (ADRP 3-90) See also **alternate position, area of operations, avenue of approach, battle position, sector of fire**.

supply – (DOD) The procurement, distribution, maintenance while in storage, and salvage of supplies, including the determination of kind and quantity of supplies. a. **producer phase**—That phase of military supply that extends from determination of procurement schedules to acceptance of finished supplies by the Services. b. **consumer phase**— That phase of military supply that extends from receipt of finished supplies by the Services through issue for use or consumption. (JP 4-0) See ATP 4-48. (Army) The process of providing all items necessary to equip, maintain, and operate a military command. (ADRP 1-02) See also **classes of supply**.

supply chain – (DOD) The linked activities associated with providing materiel from a raw materiel stage to an end user as a finished product. (JP 4-09) See ATP 4-0.1.

supply discipline – Command responsibility to identify and redistribute excess materials, observe senior commander's priorities, and ensure subordinates operate within the legal boundaries of the logistics system. (ADRP 1-02)

support – (DOD) 1. The action of a force that aids, protects, complements, or sustains another force in accordance with the directive requiring such action. 2. A unit that helps another unit in battle. 3. An element of a command that assists, protects, or supplies other forces in combat. (JP 1) See FM 3-07, FM 6-05.

support area – In contiguous areas of operations, an area for any commander that extends from its rear boundary forward to the rear boundary of the next lower level of command. (ADRP 3-0)

support bridging – Bridges used to establish semipermanent or permanent support to planned movements and road networks. Normally used to replace tactical bridging when necessary. (ATTP 3-90.4)

support by fire – A tactical mission task in which a maneuver force moves to a position where it can engage the enemy by direct fire in support of another maneuvering force. (FM 3-90-1) See also **attack by fire, overwatch, tactical mission task**.

support by fire position – The general position from which a unit conducts the tactical mission task of support by fire. (ADRP 3-90)

support to civil administration – Assistance given by U.S. armed forces to stabilize or to continue the operations of the governing body or civil structure of a foreign country, whether by assisting an established government or by establishing military authority over an occupied population. Also called SCA. (FM 3-57)

supported commander – (DOD) 1. The commander having primary responsibility for all aspects of a task assigned by the Joint Strategic Capabilities Plan or other joint operation planning authority. 2. In the context of joint operation planning, the commander who prepares operation plans or operation orders in response to requirements of the Chairman of the Joint Chiefs of Staff. 3. In the context of a support command relationship, the commander who receives assistance from another commander's force or capabilities, and who is responsible for ensuring that the supporting commander understands the assistance required. (JP 3-0) See FM 6-05, ATP 3-60.2.

supported unit – As related to contracted support, a supported unit is the organization that is the recipient, but not necessarily the requester of, contractor-provided support. (ATTP 4-10)

supporting arms coordination center – (DOD) A single location on board an amphibious command ship in which all communication facilities incident to the coordination of fire support of the artillery, air, and naval gunfire are centralized. This is the naval counterpart to the fire support coordination center utilized by the landing force. Also called SACC. (JP 3-09.3) See ATP 3-60.2.

supporting commander – (DOD) 1.A commander who provides augmentation forces or other support to a supported commander or who develops a supporting plan. 2. In the context of a support command relationship, the commander who aids, protects, complements, or sustains another commander's force, and who is responsible for providing the assistance required by the supported commander. See also support, supported commander. (JP 3-0) See FM 6-05.

supporting distance – The distance between two units that can be traveled in time for one to come to the aid of the other and prevent its defeat by an enemy or ensure it regains control of a civil situation. (ADRP 3-0) See also **supporting range**.

supporting effort – A designated subordinate unit with a mission that supports the success of the main effort. (ADRP 3-0)

supporting plan – (DOD) An operation plan prepared by a supporting commander, a subordinate commander, or an agency to satisfy the requests or requirements of the supported commander's plan. (JP 5-0) See FM 6-0, ATP 3-57.60.

supporting range – The distance one unit may be geographically separated from a second unit yet remain within the maximum range of the second unit's weapons systems. (ADRP 3-0) See also **supporting distance**.

suppress – A tactical mission task that results in temporary degradation of the performance of a force or weapons system below the level needed to accomplish the mission. (FM 3-90-1) See also **tactical mission task**.

suppression – In the context of the computed effects of field artillery fires, renders a target ineffective for a short period of time producing at least 3-percent casualties or materiel damage. (FM 3-09)

suppression of enemy air defenses – (DOD) Activity that neutralizes, destroys, or temporarily degrades surface-based enemy air defenses by destructive and/or disruptive means. Also called SEAD. (JP 3-01) See FM 3-09, ATP 3-55.6.

suppressive fire – Fires on or about a weapons system to degrade its performance below the level needed to fulfill its mission objectives during the conduct of the fires. (FM 3-09)

surgical strike – The execution of activities in a precise manner that employ special operations forces in hostile, denied, or politically sensitive environments to seize, destroy, capture, exploit, recover, or damage designated targets, or influence threats. (ADP 3-05)

surveillance – (DOD) The systematic observation of aerospace, surface or subsurface areas, places, persons, or things by visual, aural, electronic, photographic, or other means. (JP 3-0) See FM 2-0, FM 3-55, FM 3-99, ATP 3-04.64, ATP 3-13.10, ATP 3-55.3, ATP 3-55.6.

survivability – (DOD) All aspects of protecting personnel, weapons, and supplies while simultaneously deceiving the enemy. (JP 3-34) See ADRP 3-90, FM 3-90-1, FM 4-01, FM 4-95, ATP 3-91. (Army, Marine Corps) A quality or capability of military forces which permits them to avoid or withstand hostile actions or environmental conditions while retaining the ability to fulfill their primary mission. (ATP 3-37.34)

survivability move – A move that involves rapidly displacing a unit, command post, or facility in response to direct and indirect fires, the approach of an enemy unit, a natural phenomenon, or as a proactive measure based on intelligence, meteorological data, and risk analysis of enemy capabilities and intentions (including weapons of mass destruction). (ADRP 3-90)

survivability operations – (Army, Marine Corps) Those military activities that alter the physical environment to provide or improve cover, concealment, and camouflage. (ATP 3-37.34)

sustaining operation – An operation at any echelon that enables the decisive operation or shaping operations by generating and maintaining combat power. (ADRP 3-0) See also **decisive operation, shaping operation**.

sustainment – (DOD) The provision of logistics and personnel services required to maintain and prolong operations until successful mission accomplishment. (JP 3-0) See FM 4-40. (Army) The provision of logistics, personnel services, and health service support necessary to maintain operations until successful mission completion. (ADP 4-0)

sustainment maintenance – Off-system component repair and/or end item repair and return to the supply system or by exception to the owning unit, performed by national level maintenance providers. (FM 4-30)

sustainment preparation of the operational environment – The analysis to determine infrastructure, environmental factors, and resources in the operational environment that will optimize or adversely impact friendly forces means for supporting and sustaining the commander's operations plan. (ADP 4-0)

sustainment warfighting function – The related tasks and systems that provide support and services to ensure freedom of action, extend operational reach, and prolong endurance. (ADRP 3-0) See also **warfighting function.**

subballast – Gravel, sand, or cinders used to provide a level surface for the ballast and other track components. (ATP 4-14)

switch engines – The type of motive power used for receiving cars, classifying, and reassembling them for delivery or forward movement. (ATP 4-14)

switch stand – The mechanism which controls the operation of the switch and shows its position. (ATP 4-14)

switch tie – Specially cut and formed hardwood crossties, designed to support switches, switch stands, and the moveable rails of a switch. (ATP 4-14)

synchronization – (DOD) 1. The arrangement of military actions in time, space, and purpose to produce maximum relative combat power at a decisive place and time (JP 2-0) See ADP 3-0, ADRP 3-0, FM 3-09, FM 3-38, FM 6-05. 2. In the intelligence context, application of intelligence sources and methods in concert with the operation plan to ensure intelligence requirements are answered in time to influence the decisions they support. (JP 2-0) See FM 3-07, FM 6-05, ATP 3-07.20.

system – (DOD) A functionally, physically, and/or behaviorally related group of regularly interacting or interdependent elements; that group of elements forming a unified whole. (JP 3-0) See FM 3-07.

—T—

tacit knowledge - What individuals know; a unique, personal store of knowledge gained from life experiences, training, and networks of friends, acquaintances, and professional colleagues. (ATP 6-01.1)

tactical air command center – (DOD) The principal US Marine Corps air command and control agency from which air operations and air defense warning functions are directed. It is the senior agency of the US Marine air command and control system that serves as the operational command post of the aviation combat element commander. It provides the facility from which the aviation combat element commander and his battle staff plan, supervise, coordinate, and execute all current and future air operations in support of the Marine air-ground task force. The tactical air command center can provide integration, coordination, and direction of joint and combined air operations. Also called TACC (USMC). (JP 3-09.3) See ATP 3-01.15, ATP 3-04.64, ATP 3-52.2, ATP 3-60.2.

tactical air control center – (DOD) The principal air operations installation (ship-based) from which all aircraft and air warning functions of tactical air operations are controlled. Also called TACC (USN). (JP 3-09.3) See ATP 3-01.15, ATP 3-52.2, ATP 3-60.2.

tactical air control party – (DOD) A subordinate operational component of a tactical air control system designed to provide air liaison to land forces and for the control of aircraft. Also called TACP. (JP 3-09.3) See FM 3-09, FM 6-05, ATP 3-52.2.

tactical air coordinator (airborne) – (DOD) An officer who coordinates, from an aircraft, the actions of other aircraft engaged in air support of ground or sea forces. Also called TAC(A). (JP 3-09.3) See FM 3-09.

tactical assembly area – (DOD) An area that is generally out of the reach of light artillery and the location where units make final preparations (pre-combat checks and inspections) and rest, prior to moving to the line of departure. See also line of departure. Also called TAA. (JP 3-35) See ATP 3-35.

tactical bridging – Bridges used for the immediate mobility support of combat maneuver forces in close combat. They are very often employed under the threat of direct or indirect fire and are intended to be used multiple times for short periods. (ATTP 3-90.4)

tactical combat force – (DOD) A combat unit, with appropriate combat support and combat service support assets, that is assigned the mission of defeating Level III threats. Also called TCF. (JP 3-10) See ATP 3-91.

tactical command post – A facility containing a tailored portion of a unit headquarters designed to control portions of an operation for a limited time. (FM 6-0)

tactical control – (DOD) The authority over forces that is limited to the detailed direction and control of movements or maneuvers within the operational area necessary to accomplish missions or tasks assigned. Also called TACON. (JP 1) See ADRP 5-0, FM 3-09, FM 6-0, FM 6-05, ATP 4-32.16, ATP 3-53.1.

tactical level of war – (DOD) The level of war at which battles and engagements are planned and executed to achieve military objectives assigned to tactical units or task forces. (JP 3-0) See ADP 3-90.

tactical mission task – The specific activity performed by a unit while executing a form of tactical operation or form of maneuver. It may be expressed in terms of either actions by a friendly force or effects on an enemy force. See also **mission statement, operation order**. (FM 3-90-1)

tactical questioning – (DOD) The field-expedient initial questioning for information of immediate tactical value of a captured or detained person at or near the point of capture and before the individual is placed in a detention facility. Also called TQ. (JP 3-63) See FM 1-04, FM 2-91.6.

tactical road march – A rapid movement used to relocate units within an area of operations to prepare for combat operations. (ADRP 3-90) See also **area of operations**.

tactics – (DOD) The employment and ordered arrangement of forces in relation to each other. (CJCSM 5120.01) See ADP 1-01, ADP 3-90, ADRP 3-90, FM 3-09.

tailgate medical support – An economy of force device employed primarily to retain maximum mobility during movement halts or to avoid the time and effort required to set up a formal, operational treatment facility (for example, during rapid advance and retrograde operations). (FM 4-02)

target – (DOD) 1. An entity or object considered for possible engagement or other action. See ADP 3-09, ADRP 3-09, ATP 2-01.3. 2. In intelligence usage, a country, area, installation, agency, or person against which intelligence operations are directed. See FM 3-09, ATP 2-01.3, ATP 2-19.4. 3. An area designated and numbered for future firing. See FM 3-90-1, ATP 2-01.3. 4. In gunfire support usage, an impact burst that hits the target. (JP 3-60) See ATP 3-06.1, ATP 3-07.20, ATP 3-60.

target acquisition – (DOD) The detection, identification, and location of a target in sufficient detail to permit the effective employment of weapons. Also called TA. (JP 3-60) See ADRP 3-09, FM 3-09, ATP 3-04.64, ATP 3-09.12, ATP 3-55.6.

target analysis – (DOD) An examination of potential targets to determine military importance, priority of attack, and weapons required to obtain a desired level of damage or casualties. (JP 3-60) See ATP 3-55.6.

target area of interest – (DOD) The geographical area where high-value targets can be acquired and engaged by friendly forces. Also called TAI. (JP 2-01.3) See FM 3-09, FM 3-98, ATP 3-55.6, ATP 3-60.2. See also **high-payoff target, high-value target**.

target audience – (DOD) An individual or group selected for influence. Also called TA (JP 3-13) See FM 3-24, FM 3-53, ATP 3-07.20, ATP 3-53.2.

target development – (DOD) The systematic examination of potential target systems - and their components, individual targets, and even elements of targets - to determine the necessary type and duration of the action that must be exerted on each target to create an effect that is consistent with the commander's specific objectives. (JP 3-60) See ATP 2-19.3, ATP 2.19.4.

target identification – The accurate and timely characterization of a detected object on the battlefield as friend, neutral, or enemy. (ADRP 1-02)

targeting – (DOD) The process of selecting and prioritizing targets and matching the appropriate response to them, considering operational requirements and capabilities. (JP 3-0) See ADP 3-09, ADRP 3-09, ADRP 5-0, FM 3-09, FM 3-16, FM 3-98, ATP 2-01.3, ATP 2-19.3, ATP 2-22.4, ATP 3-52.2, ATP 3-53.2, ATP 3-55.6, ATP 3-60.

target intelligence – (DOD) Intelligence that portrays and locates the components of a target or target complex and indicates its vulnerability and relative importance. (JP 3-60) See ATP 2-19.3.

target of opportunity – (DOD) 1. A target identified too late, or not selected for action in time, to be included in deliberate targeting that, when detected or located, meets criteria specific to achieving objectives and is processed using dynamic targeting. See ATP 3-53.2. 2. A target visible to a surface or air sensor or observer, which is within range of available weapons and against which fire has not been scheduled or requested. (JP 3-60) See ATP 3-60.1.

target reference point – An easily recognizable point on the ground (either natural or man-made) used to initiate, distribute, and control fires. Also called TRP. (ADRP 1-02) See also **engagement area, sector of fire**.

task – (DOD) A clearly defined action or activity specifically assigned to an individual or organization that must be done as it is imposed by an appropriate authority. (JP 1) See ADP 1-01, ATP 3-93.

task organization – (Army) A temporary grouping of forces designed to accomplish a particular mission. (ADRP 5-0)

task-organizing – (DOD) An organization that assigns to responsible commanders the means with which to accomplish their assigned tasks in any planned action. (JP 3-33) See FM 3-98. (Army) The act of designing an operating force, support staff, or sustainment package of specific size and composition to meet a unique task or mission. (ADRP 3-0)

technical control – The supervision of human intelligence, counterintelligence, and signals intelligence collection tactics, techniques, and procedures. Technical control does not interfere with tasking organic human intelligence, counterintelligence, and signals intelligence collection assets; it ensures adherence to existing policies or regulations by providing technical guidance for human intelligence, counterintelligence, and signals intelligence tasks within the information collection plan. (ATP 2-01)

technical intelligence – (DOD) Intelligence derived from the collection, processing, analysis, and exploitation of data and information pertaining to foreign equipment and materiel for the purposes of preventing technological surprise, assessing foreign scientific and technical capabilities, and developing countermeasures designed to neutralize an adversary's technological advantages. Also called TECHINT. (JP 2-0) See FM 3-16, ATP 2-22.4, ATP 3-05.20.

techniques – (DOD) Non-prescriptive ways or methods used to perform missions, functions, or tasks. (CJCSM 5120.01) See ADP 1-01, ADRP 3-0, ADRP 5-0, ADRP 6-0, ATP 3-09.50.

tempo – The relative speed and rhythm of military operations over time with respect to the enemy. (ADRP 3-0)

tenets of operations – Desirable attributes that should be built into all plans and operations and are directly related to the Army's operational concept. (ADP 1-01)

terminal – (DOD) A facility designed to transfer cargo from one means of conveyance to another. (JP 4-01.6) See FM 4-01, ATP 4-13.

terminal attack control – (DOD) The authority to control the maneuver of and grant weapons release clearance to attacking aircraft. (JP 3-09.3) See FM 3-09.

terminal guidance operations – (DOD) Those actions that provide electronic, mechanical, voice or visual communications that provide approaching aircraft and/or weapons additional information regarding a specific target location. Also called TGO. (JP 3-09)

terminal operations – (DOD) The reception, processing, and staging of passengers; the receipt, transit storage and marshalling of cargo; the loading and unloading of modes of transport conveyances; and the manifesting and forwarding of cargo and passengers to a destination. (JP 4-01.5) See FM 4-01, ATP 4-11, ATP 4-13.

terminal phase – (DOD) That portion of the flight of a ballistic missile that begins when the warhead or payload reenters the atmosphere and ends when the warhead or payload detonates, release its submunitions, or impacts. (JP 3-01) See ATP 3-27.5.

terrain analysis – DOD) The collection, analysis, evaluation, and interpretation of geographic information on the natural and man-made features of the terrain, combined with other relevant factors, to predict the effect of the terrain on military operations. (JP 2-03) See ATP 2-01.3. (Army) The study of the terrain's properties and how they change over time, with use, and under varying weather conditions. (ATP 3-34.80)

terrain management – The process of allocating terrain by establishing areas of operation, designating assembly areas, and specifying locations for units and activities to deconflict activities that might interfere with each other. (ADRP 5-0) See also **area of operations**.

terrorism – (DOD) The unlawful use of violence or threat of violence, often motivated by religious, political, or other ideological beliefs, to instill fear and coerce governments or societies in the pursuit of goals that are usually political. (JP 3-07.2) See FM 3-57, ATP 3-05.2, ATP 3-07.31, ATP 3-57.80.

theater – (DOD) The geographical area for which a commander of a geographic combatant command has been assigned responsibility. (JP 1) See ATP 3-52.2, ATP 3-57.20, ATP 3-57.60.

theater closing – The process of redeploying Army forces and equipment from a theater, the drawdown and removal or disposition of Army non-unit equipment and materiel, and the transition of materiel and facilities back to host nation or civil authorities. (ADP 4-0)

theater container management – The supervision and control of containers as they move through the distribution system to ensure they are delivered, discharged and returned in accordance to the combatant commander's concept of operations. (ATP 4-12)

theater distribution – (DOD) The flow of equipment, personnel, and material within theater to meet the geographic combatant commander's mission. (JP 4-09) See ADRP 4-0, ATP 4-0.1, ATP 4-48.

theater distribution system – (DOD) A distribution system comprised of four independent and mutually supported networks within theater to meet the geographic combatant commander's requirements: the physical network; the financial network; the information network; and the communications network. (JP 4-01) See FM 4-01.

theater evacuation policy – A command decision indicating the length in days of the maximum period of noneffectiveness that patients may be held within the command for treatment. Patients that, in the opinion of a responsible medical officer, cannot be returned to duty status within the period prescribed are evacuated by the first available means, provided the travel involved will not aggravate their disabilities. (FM 4-02)

theater event system – (DOD) Architecture for reporting ballistic missile events, composed of three independent processing and reporting elements: the joint tactical ground stations, tactical detection and reporting, and the space-based infrared system mission control station. Also called TES. (JP 3-14) See ATP 3-14.5.

theater of operations – (DOD) An operational area defined by the geographic combatant commander for the conduct or support of specific military operations. Also called TO. (JP 3-0) See ATP 3-52.2, ATP 4-43.

theater opening – The ability to establish and operate ports of debarkation (air, sea, and rail), to establish a distribution system, and to facilitate throughput for the reception, staging, and onward movement of forces within a theater of operations. (ADP 4-0)

theater special operations command – (DOD) A subordinate unified command established by a combatant commander to plan, coordinate, conduct, and support joint special operations. Also called TSOC. (JP 3-05) See ADP 3-05, ADRP 3-05, FM 3-53, ATP 3-05.2.

theater validation identification – The employment of multiple independent, established protocols and technologies by scientific experts in the controlled environment of a fixed or mobile/transportable laboratory to characterize a chemical, biological, radiological, and/or nuclear hazard with a high level of confidence and degree of certainty necessary to support operational-level decisions. (ATP 3-11.37)

threat – Any combination of actors, entities, or forces that have the capability and intent to harm United States forces, United States national interests, or the homeland. (ADRP 3-0)

threat assessment – (DOD) In antiterrorism, examining the capabilities, intentions, and activities, past and present, of terrorist organizations as well as the security environment within which friendly forces operate to determine the level of threat. Also called TA. (JP 3-07.2) See FM 3-16.

throughput – (DOD) 1. In transportation, the average quantity of cargo and passengers that can pass through a port on a daily basis from arrival at the port to loading onto a ship or plane, or from the discharge from a ship or plane to the exit (clearance) from the port complex. (JP 4-01.5) See FM 4-01, ATP 3-01.91, ATP 4-0.1, ATP 4-13. 2. In patient movement and care, the maximum number of patients (stable or stabilized) by category, that can be received at the airport, staged, transported, and received at the proper hospital within any 24-hour period. (JP 4-02) See FM 4-01.

throughput capacity – (DOD) The estimated capacity of a port or an anchorage to clear cargo and/or passengers in 24 hours usually expressed in tons for cargo, but may be expressed in any agreed upon unit of measurement. See also clearance capacity. (JP 4-01.5) See FM 4-01.

throughput distribution – A method of distribution which bypasses one or more intermediate supply echelons in the supply system to avoid multiple handling. (ATP 4-11)

time of attack – The moment the leading elements of the main body cross the line of departure or, in a night attack, the point of departure. (ADRP 3-90) See also **line of departure, point of departure**.

time on target – (DOD) The actual time at which munitions impact the target. Also called TOT. (JP 3-09.3) See FM 100-30, ATP 3-60.2.

time-phased force and deployment data – (DOD) The time-phased force data, non-unit cargo and personnel data, and movement data for the operation plan or operation order or ongoing rotation of forces. Also called TPFDD. See also time-phased force and deployment list. (JP 5-0) See FM 4-01.

time-phased force and deployment list – (DOD) Appendix 1 to Annex A of the operation plan. It identifies types and/or actual units required to support the operation plan and indicates origin and ports of debarkation or ocean area. It may also be generated as a computer listing from the time-phased force and deployment data. Also called TPFDL. See also time-phased force and deployment data. (JP 4-05) See FM 4-01.

times – (DOD) The Chairman of the Joint Chiefs of Staff coordinates the proposed dates and times with the commanders of the appropriate unified and specified commands, as well as any recommended changes to when specified operations are to occur (C-, D-, M-days end at 2400 hours Universal Time [Zulu time] and are assumed to be 24 hours long for planning). (JP 5-0) See FM 6-0.

time-sensitive target – (DOD) A joint force commander validated target or set of targets requiring immediate response because it is a highly lucrative, fleeting target of opportunity or it poses (or will soon pose) a danger to friendly forces. Also called TST. (JP 3-60) See FM 6-05, ATP 2-19.3, ATP 3-53.2, ATP 3-60, ATP 3-60.1.

token – An electronic identification method used within a multi-node configured command and control, battle management, and communications suite to identify the lead server for transmission of track data. The token may be transferred between suites to maintain positive integrity of track data. The suite where the token resides is the only suite that may make changes to the AN/TPY-2 system configuration. The token methodology also applies within a single node command and control, battle management, and communications suite, but the token remains within the single node. (ATP 3-27.5)

toxic industrial biological – (DOD) Any biological material manufactured, used, transported, or stored by industrial, medical, or commercial processes which could pose an infectious or toxic threat. Also called TIB. (JP 3-11) See ATP 3-05.11.

toxic industrial chemical – (DOD) A chemical developed or manufactured for use in industrial operations or research by industry, government, or academia that poses a hazard. Also called TIC. (JP 3-11) See ATP 3-05.11.

toxic industrial material – (DOD) A generic term for toxic, chemical, biological, or radioactive substances in solid, liquid, aerosolized, or gaseous form that may be used, or stored for use, for industrial, commercial, medical, military, or domestic purposes. Also called TIM. (JP 3-11) See ATP 3-05.11.

track – (DOD) 1. A series of related contacts displayed on a data display console or other display device. (JP 3-01) See ATP 3-27.5. 2. To display or record the successive positions of a moving object. (JP 3-01) See ATP 3-27.5. 3, ATP 3-55.6. To lock onto a point of radiation and obtain guidance therefrom. 4. To keep a gun properly aimed, or to point continuously a target-locating instrument at a moving target. 5. The actual path of an aircraft above or a ship on the surface of the Earth. 6. One of the two endless belts of which a full-track or half-track vehicle runs. 7. A metal part forming a path of a moving object such as the track around the inside of a vehicle for moving a mounted machine gun. (JP 3-01) See ATP 3-27.5.

track alignment – The horizontal dimension of a track; for example, curves. (ATP 4-14)

track profile – The vertical dimensions of the track caused by terrain features such as hills or valleys. (ATP 4-14)

tractive effort – A measure of the potential power of a locomotive expressed in pounds. (ATP 4-14)

traffic control post – A manned post that is used to preclude the interruption of traffic flow or movement along a designated route. (FM 3-39)

trail party – The last march unit in a march column and normally consists of primarily maintenance elements in a mounted march. (FM 3-90-2) See also **march serial, march unit**.

train density – The number of trains that may be operated safely over a division in each direction during a 24-hour period. Also called TD. (ATP 4-14)

train dispatcher – Responsible for main-line movement of passenger and freight trains on a division. (ATP 4-14)

training and evaluation outline – A summary document that provides information on collective training objectives, related individual training objectives, resource requirements, and applicable evaluation procedures for a type of organization. (ADRP 7-0)

training objective – A statement that describes the desired outcome of a training activity in the unit. (ADRP 7-0)

transitional military authority – A temporary military government exercising the functions of civil administration in the absence of a legitimate civil authority. (FM 3-07)

transportation – A logistics function that includes movement control and associated activities to incorporate military, commercial, and multinational motor, rail, air, and water mode assets in the movement of units, personnel, equipment, and supplies in support the concept of operations. (ADRP 1-02)

transportation component command – (DOD) A major command of its parent Service under United States Transportation Command, which includes Air Force Air Mobility Command, Navy Military Sealift Command, and Army Military Surface Deployment and Distribution Command. Also called TCC. (JP 4-01.6) See FM 4-01.

transportation feasibility – (DOD) A determination that the capability exists to move forces, equipment, and supplies from the point of origin to the final destination within the time required. (JP 4-09) See FM 4-01.

transportation priorities – (DOD) Indicators assigned to eligible traffic that establish its movement precedence. (JP 4-09) See FM 4-01.

transportation system – (DOD) All the land, water, and air routes and transportation assets engaged in the movement of United States forces and their supplies during military operations, involving both mature and immature theaters and at the strategic, operational, and tactical levels of war. (JP 4-01) See FM 4-01.

traveling overwatch – A movement technique used when contact with enemy forces is possible. The lead element and trailing element are separated by a short distance which varies with the terrain. The trailing element moves at variable speeds and may pause for short periods to overwatch the lead element. It keys its movement to terrain and the lead element. The trailing element over-watches at such a distance that, should the enemy engage the lead element, it will not prevent the trailing element from firing or moving to support the lead element. (FM 3-90-2)

triage – The medical sorting of patients. The categories are: MINIMAL (OR AMBULATORY)— those who require limited treatment and can be returned to duty; IMMEDIATE—patients requiring immediate care to save life or limb; DELAYED—patients who, after emergency treatment, incur little additional risk by delay or further treatment; and EXPECTANT—patients so critically injured that only complicated and prolonged treatment will improve life expectancy. (FM 4-02)

trigger line – A phase line located on identifiable terrain that crosses the engagement area that is used to initiate and mass fires into an engagement area at a predetermined range for all or like weapon systems. (ADRP 1-02) See also **engagement area, phase line**.

troop – A company-size unit in a cavalry organization. (ADRP 3-90)

troop leading procedures – A dynamic process used by small-unit leaders to analyze a mission, develop a plan, and prepare for an operation. (ADP 5-0)

troop movement – The movement of troops from one place to another by any available means. (ADRP 3-90)

turn – 1. A tactical mission task that involves forcing an enemy force from one avenue of approach or mobility corridor to another. 2. A tactical obstacle effect that integrates fire planning and obstacle effort to divert an enemy formation from one avenue of approach to an adjacent avenue of approach or into an engagement area. (FM 3-90-1) See also **avenue of approach, tactical mission task**.

turning movement – (Army) A form of maneuver in which the attacking force seeks to avoid the enemy's principle defensive positions by seizing objectives behind the enemy's current positions thereby causing the enemy force to move out of their current positions or divert major forces to meet the threat. (FM 3-90-1)

two-level maintenance – Tiered maintenance system comprised of field and sustainment maintenance. (FM 4-30)

—U—

unanticipated target – (DOD) A target of opportunity that was unknown or not expected to exist in the operational environment. (JP 3-60) See ATP 3-60.1.

unauthorized commitment – (DOD) An agreement that is not binding solely because the United States Government representative who made it lacked the authority to enter into that agreement on behalf of the United States Government. (JP 4-10) See FM 1-04.

uncertain environment – (DOD) Operational environment in which host government forces, whether opposed to or receptive to operations that a unit intends to conduct, do not have totally effective control of the territory and population in the intended operational area. (JP 3-0) See FM 3-57, ATP 3-57.10.

uncommitted force – A force that is not in contact with an enemy and is not already deployed on a specific mission or course of action. (ADRP 3-90)

unconventional assisted recovery – (DOD) Nonconventional assisted recovery conducted by special operations forces. Also called UAR. (JP 3-50) See ATP 3-05.1.

unconventional assisted recovery coordination cell – (DOD) A compartmented special operations forces facility, established by the joint force special operations component commander, staffed on a continuous basis by supervisory personnel and tactical planners to coordinate, synchronize, and de-conflict nonconventional assisted recovery operations within the operational area assigned to the joint force commander. Also called UARCC. (JP 3-50) See ATP 3-05.1.

unconventional warfare – (DOD) Activities conducted to enable a resistance movement or insurgency to coerce, disrupt, or overthrow a government or occupying power by operating through or with an underground, auxiliary, and guerilla force in a denied area. Also called UW. (JP 3-05) See ADP 3-05, ADRP 3-05, FM 3-05, FM 3-53, FM 6-05, ATP 3-05.1, ATP 4-14.

underframe – The structure of a railcar under the deck that supports the weight of the load. (ATP 4-14)

underground – A cellular covert element within unconventional warfare that is compartmentalized and conducts covert or clandestine activities in areas normally denied to the auxiliary and the guerrilla force. (ADRP 3-05)

unexploded explosive ordnance – (DOD) Explosive ordnance which has been primed, fused, armed or otherwise prepared for action, and which has been fired, dropped, launched, projected, or placed in such a manner as to constitute a hazard to operations, installations, personnel, or material and remains unexploded either by malfunction or design or for any other cause. Also called UXO. (JP 3-15) See ATP 3-07.31, ATP 4-32, ATP 4-32.2, ATP 4-32.16.

unified action – (DOD) The synchronization, coordination, and/or integration of the activities of governmental and nongovernmental entities with military operations to achieve unity of effort. (JP 1) See ADP 6-0, ADRP 3-0, ADRP 3-07, ADRP 3-28, ADRP 6-0, FM 3-24, FM 3-52, ATP 2-01, ATP 2-19.3, ATP 2-19.4.

unified action partners – Those military forces, governmental and nongovernmental organizations, and elements of the private sector with whom Army forces plan, coordinate, synchronize, and integrate during the conduct of operations. (ADRP 3-0)

unified command – (DOD) A command with a broad continuing mission under a single commander and composed of significant assigned components of two or more Military Departments that is established and so designated by the President, through the Secretary of Defense with the advice and assistance of the Chairman of the Joint Chiefs of Staff. Also called unified combatant command. (JP 1) See FM 3-57, ATP 4-43.

Unified Command Plan – (DOD) The document, approved by the President, that sets forth basic guidance to all unified combatant commanders; establishes their missions, responsibilities, and force structure; delineates the general geographical area of responsibility for geographic combatant commanders; and specifies functional responsibilities for functional combatant commanders. Also called UCP. See also combatant command; combatant commander. (JP 1) See FM 3-53, ATP 3-27.5.

unified land operations – How the Army seizes, retains, and exploits the initiative to gain and maintain a position of relative advantage in sustained land operations through simultaneous offensive, defensive, and stability operations in order to prevent or deter conflict, prevail in war, and create the conditions for favorable conflict resolution. (ADP 3-0)

unit – (DOD) Any military element whose structure is prescribed by competent authority, such as a table of organization and equipment; specifically, part of an organization. (JP 3-33) See chapter 4 of this publication.

unit distribution – A method of distributing supplies by which the receiving unit is issued supplies in its own area, with transportation furnished by the issuing agency. (FM 4-40)

unit historical officer – An individual, military or civilian, who is designated as the unit historian and is responsible for military history activities. (ATP 1-20)

unit history – An informal narrative that covers the entire history of a specific unit, written in an easy-to-read manner for the benefit of the Soldiers. (ATP 1-20)

unit line number – (DOD) A seven-character alphanumeric code that describes a unique increment of a unit deployment, i.e., advance party, main body, equipment by sea and air, reception team, or trail party, in the time-phased force and deployment data. Also called ULN. (JP 3-35) See FM 4-01, ATP 3-35.

unit movement data – (DOD) A unit equipment and/or supply listing containing corresponding transportability data. Tailored unit movement data has been modified to reflect a specific movement requirement. Also called UMD. (JP 3-35) See FM 4-01, ATP 3-35.

unity of command – (DOD) The operation of all forces under a single responsible commander who has the requisite authority to direct and employ those forces in pursuit of a common purpose. (JP 3-0) See ADP 6-0, FM 3-24.

unity of effort – (DOD) Coordination, and cooperation toward common objectives, even if the participants are not necessarily part of the same command or organization—the product of successful unified action. (JP 1) See ADP 3-0, ADP 6-0, ADRP 3-0, ADRP 3-28, ADRP 6-0, FM 3-16, FM 3-24, FM 3-98, FM 6-05, ATP 3-07.5.

unmanned aircraft – (DOD) An aircraft that does not carry a human operator and is capable of flight with or without human remote control. Also called UA. (JP 3-30) See ATP 3-04.64, ATP 3-60.2.

unmanned aircraft system – (DOD) That system whose components include the necessary equipment, network, and personnel to control an unmanned aircraft. Also called UAS. (JP 3-30) See FM 3-16, ATP 3-01.15, ATP 3-04.64.

unplanned target – (DOD) A target of opportunity that is known to exist in the operational environment. (JP 3-60) See ATP 3-60.1.

unexploded explosive ordnance – (DOD) Explosive ordnance which has been primed, fused, armed or otherwise prepared for action, and which has been fired, dropped, launched, projected, or placed in such a manner as to constitute a hazard to operations, installations, personnel, or material and remains unexploded either by malfunction or design or for any other cause. Also called UXO. (JP 3-15) See ATP 4-32.

urban operations – Operations across the range of military operations planned and conducted on, or against objectives on a topographical complex and its adjacent natural terrain, where man-made construction or the density of population are the dominant features. (FM 3-06)

U.S. military prisoner – A person sentenced to confinement or death during a court-martial and ordered into confinement by a competent authority, whether or not the convening authority has approved the sentence. (FM 3-39)

—V—

validate – (DOD) Execution procedure used by combatant command components, supporting combatant commanders, and providing organizations to confirm to the supported commander and United States Transportation Command that all the information records in a time-phased force and deployment data not only are error-free for automation purposes, but also accurately reflect the current status, attributes, and availability of units and requirements. (JP 5-0) See FM 4-01.

validation – (DOD) 1. A process associated with the collection and production of intelligence that confirms that an intelligence collection or production requirement is sufficiently important to justify the dedication of intelligence resources, does not duplicate an existing requirement, and has not been previously satisfied. See ATP 3-60.1 2. A part of target development that ensures all vetted targets meet the objectives and criteria outlined in the commander's guidance and ensures compliance with the law of war and rules of engagement. See ATP 3-60.1 3. In computer modeling and simulation, the process of determining the degree to which a model or simulation is an accurate representation of the real world from the perspective of the intended uses of the model or simulation. See ATP 3-60.1 4. Execution procedure whereby all the information records in a time-phased force and deployment data are confirmed error free and accurately reflect the current status, attributes, and availability of units and requirements. (JP 3-35) See ATP 3-35, ATP 3-60.1.

vehicle-borne improvised explosive device – (DOD) A device placed or fabricated in an improvised manner on a vehicle incorporating destructive, lethal, noxious, pyrotechnic, or incendiary chemicals and designed to destroy, incapacitate, harass, or distract. Otherwise known as a car bomb. Also called VBIED. (JP 3-10) See ATP 3-18.14.

vehicle distance – The clearance between vehicles in a column which is measured from the rear of one vehicle to the front of the following vehicle. (ADRP 1-02) See also **march column, march serial, march unit**.

vertical and/or short takeoff and landing – (DOD) Vertical and/or short takeoff and landing capability for aircraft. (JP 1-02) See ATP 3-52.3.

vertical envelopment – (DOD) A tactical maneuver in which troops that are air-dropped, air-land, or inserted via air assault, attack the rear and flanks of a force, in effect cutting off or encircling the force. (JP 3-18) See FM 3-90-1, FM 3-99.

vertical interval – The difference in altitude between the unit or observer and the target or point of burst. (ATP 3-09.50)

vetting – (DOD) A part of target development that assesses the accuracy of the supporting intelligence to targeting. (JP 3-60) See ATP 3-60, ATP 3-60.1.

visual information – (DOD) Various visual media with or without sound. Generally, visual information includes still and motion photography, audio video recording, graphic arts, visual aids, models, displays, and visual presentations. Also called VI. (JP 3-61) See FM 6-02, ATP 3-55.12, ATP 6-02.40.

Voluntary Intermodal Sealift Agreement – (DOD) An agreement that provides the Department of Defense with assured access to United States flag assets, both vessel capacity and intermodal systems, to meet Department of Defense contingency requirements. Also called VISA. (JP 4-01.2) See FM 4-01.

vulnerabilities – Characteristics, motives, or conditions of the target audience that can be used to influence behavior. (FM 3-53)

—W—

waiting area – A location adjacent to the route or axis that may be used for the concealment of vehicles, troops, and equipment while an element is waiting to resume movement. Waiting areas are normally located on both banks (or sides) close to crossing areas. (ATTP 3-90.4)

warfighting function – A group of tasks and systems (people, organizations, information, and processes), united by a common purpose that commanders use to accomplish missions and training objectives. (ADRP 3-0)

warning order – (DOD) 1. A preliminary notice of an order or action that is to follow. (FM 6-0) 2. A planning directive that initiates the development and evaluation of military courses of action by a supported commander and requests that the supported commander submit a commander's estimate. 3. A planning directive that describes the situation, allocates forces and resources, establishes command relationships, provides other initial planning guidance, and initiates subordinate unit mission planning. Also called WARNORD. (JP 5-0) See ATP 3-53.2, ATP 3-57.60.

wartime reserve modes – (DOD) Characteristics and operating procedures of sensor, communications, navigation aids, threat recognition, weapons, and countermeasures systems that will contribute to military effectiveness if unknown to or misunderstood by opposing commanders before they are used, but could be exploited or neutralized if known in advance. Also called WARM. (JP 3-13.1) See FM 3-38, ATP 3-13.10.

waste discharge – The accidental or intentional spilling, leaking, pumping, pouring, emitting, emptying, or dumping of a hazardous waste into or onto any land or water. (ATP 3-34.5)

weaponeering – (DOD) The process of determining the quantity of a specific type of lethal or nonlethal means required to create a desired effect on a given target. (JP 3-60) See FM 3-09, ATP 3-60, ATP 3-91.1.

weapon engagement zone – (DOD) In air defense, airspace of defined dimensions within which the responsibility for engagement of air threats normally rests with a particular weapon system. Also called WEZ. (JP 3-01) See ATP 3-06.1.

weapons of mass destruction – (DOD) Chemical, biological, radiological, or nuclear weapons capable of a high order of destruction or causing mass casualties and exclude the means of transporting or propelling the weapon where such means is a separable and divisible part from the weapon. Also called WMD. (JP 3-40) See ADRP 3-05, FM 3-05, ATP 3-05.2, ATP 4-32, ATP 4-32.2, ATP 4-32.16.

weapons of mass destruction counterforce – Weapons of mass destruction counterforce is a tactical objective to defeat the full range of chemical, biological, radiological, and nuclear threats before they can be employed as weapons. (FM 3-11)

weapons of mass destruction proliferation prevention – The employment of tactical level capabilities to support operational and strategic nonproliferation objectives of combating weapons of mass destruction. (FM 3-11)

weapons technical intelligence – (DOD) A category of intelligence and processes derived from the technical and forensic collection and exploitation of improvised explosive devices, associated components, improvised weapons, and other weapon systems. Also called WTI. (JP 3-15.1) See ATP 2-22.4, ATP 4-32.

wharf – (DOD) A structure built of open rather than solid construction along a shore or a bank that provides cargo-handling facilities. (JP 4-01.5) See ATP 4-13.

wide area security – The application of the elements of combat power in unified action to protect populations, forces, infrastructure, and activities; to deny the enemy positions of advantage; and to consolidate gains in order to retain the initiative. (ADP 3-0)

withdrawal operation – (DOD) A planned retrograde operation in which a force in contact disengages from an enemy force and moves in a direction away from the enemy. (JP 1-02) See ADRP 3-90, FM 3-90-1, ATP 3-91.

working group – (Army) A grouping of predetermined staff representatives who meet to provide analysis, coordinate, and provide recommendations for a particular purpose or function. (FM 6-0)

wreck train – A train specially configured and tailored to conduct wreck recovery operations. (ATP 4-14)

wythe system – A steam and diesel-electric locomotive classification system that groups wheels and uses numerals separated by hyphens to represent the number of wheels in each group. (ATP 4-14)

—X—

X-hour – The unspecified time that commences unit notification for planning and deployment preparation in support of potential contingency operations that do not involve rapid, short notice deployment. (FM 3-99)

X-hour sequence – An extended sequence of events initiated by X-hour that allow a unit to focus on planning for a potential contingency operation, to include preparation for deployment. (FM 3-99).

—Z—

zone of fire – (DOD) An area into which a designated ground unit or fire support ship delivers, or is prepared to deliver, fire support. Fire may or may not be observed. Also called ZF. (JP 3-09) See FM 3-09.

zone reconnaissance – A form of reconnaissance that involves a directed effort to obtain detailed information on all routes, obstacles, terrain, and enemy forces within a zone defined by boundaries. (ADRP 3-90)

This page intentionally left blank.

Chapter 2

Acronyms, Abbreviations, and Country Codes

This chapter presents acronyms, abbreviations, and geographical entity codes.

SECTION I — ACRONYMS AND ABBREVIATIONS

2-1. This section (pages 2-1 to 2-23) lists selected Army and joint acronyms and abbreviations commonly used in Army doctrine.

Bolded entries apply only to the Army. An asterisk (*) marks terms shown in chapter 1.

—A—

AA	***assembly area,** *avenue of approach
AADC	*area air defense commander
AADP	area air defense plan
AAFES	Army Air Force Exchange Service
AAMDC	Army Air and Missile Defense Command
AAP	Allied administrative publication
AAR	after action review
ABCA	American, British, Canadian, Australian, and New Zealand
ABCS	Army Battle Command System
ABCT	**armored brigade combat team**
ABN	airborne
AC	Active Component
ACA	*airspace control authority, *airspace coordination area
ACCE	*air component coordination element
ACH	**advanced combat helmet**
ACM	airspace coordinating measure
ACP	*airspace control plan
ACO	airspace control order
ACS	*airspace control system
ACSA	acquisition and cross-servicing agreement
ACT	*activity
ACU	**Army combat uniform**
AD3E	**assess, decide, develop and detect, deliver, and evaluate**
AD	*air defense
ADA	air defense artillery
A/DACG	arrival/departure airfield control group
ADAM	air defense airspace management

Bolded entries apply only to the Army. An asterisk (*) marks terms shown in chapter 1.	
ADL	*available-to-load date
ADP	**Army doctrine publication**
ADRP	**Army doctrine reference publication**
ADSI	**air defense systems integrator**
AE	**aeromedical evacuation**
AEF	**American Expeditionary Forces**
AEP	**allied engineering publication**
AFATDS	Advanced Field Artillery Tactical Data System
AER	**Army Emergency Relief**
AFB	Air Force base
AFI	Air Force instruction
AFSB	Army field support brigade
AFTTP	Air Force tactics, techniques, and procedures
AFMAN	Air Force manual
AFWA	Air Force Weather Agency
AG	adjutant general
AGL	above ground level
AGM	***attack guidance matrix, air-to-ground missile**
AGO	***air-ground operations**
AGR	**ability group run**
AHS	**Army Health System**
AI	**assistant instructor**
AIS	automated information system
AIT	**Advance Individual Training**, automatic identification technology
AJP	allied joint publication
AKO	**Army Knowledge Online**
ALE	**Army special operations forces liaison element**
ALO	*air liaison officer
ALSA	Air Land Sea Application (Center)
ALT	**alternate**
AM	amplitude modulation
AMC	*airborne mission coordinator, *Air Mobility Command
AMD	air and missile defense
AMDWS	**air and missile defense workstation**
ANDVT	advanced narrowband digital voice terminal
ANG	Air National Guard
ANGLICO	air-naval gunfire liaison company
ANW2	Adaptive Networking Wideband Waveform
AO	area of operations
AOA	amphibious objective area
AOB	*advanced operations base
AOC	air operations center

Bolded entries apply only to the Army. An asterisk (*) marks terms shown in chapter 1.

AOR	*area of responsibility
AOI	area of interest
APC	**armored personnel carrier**
APCO	**association of public safety communication officials**
APEX	*Adaptive Planning and Execution system
APFT	**Army Physical Fitness Test**
APKWS	**advance precision kill weapon system**
APOD	aerial port of debarkation
APOE	aerial port of embarkation
APORT	*aerial port
APP	allied procedural publication
APS	Army pre-positioned stocks
AR	Army regulation
ARCENT	United States Army Central Command
ARFOR	**(Not used as an acronym in Army doctrine. See term in chapter 1.)**
ARFORGEN	**Army force generation**
ARNG	Army National Guard
ARNGUS	Army National Guard of the United States
ARRB	***Army requirements review board**
ARSOAC	**Army Special Operations Aviation Command**
ARSOF	Army special operations forces
ARSTRAT	**U.S. Army Forces Strategic Command**
ASA(ALT)	Assistant Secretary of the Army for Acquisition, Logistics, and Technology
ASAS	All Source Analysis System
ASC	Army Sustainment Command
ASCC	*Army Service component command
ASCOPE	areas, structures, capabilities, organizations, people, and events
ASD (HD&ASA)	Assistant Secretary of Defense (Homeland Defense and Americas' Security Affairs)
ASI	**additional skill identifier**
ASOC	air support operations center
ASP	ammunition supply point
ASR	***alternate supply route**
AT	*antiterrorism, antitank
ATACMS	Army Tactical Missile System
ATGM	antitank guided missile
ATHP	***ammunition transfer holding point**
ATN	**Army Training Network**
ATO	*air tasking order
ATP	**Army techniques publication**
ATS	air traffic service
ATTP	Army tactics, techniques, and procedures

Bolded entries apply only to the Army. An asterisk (*) marks terms shown in chapter 1.

AUTL	**Army Universal Task List**
AV	*asset visibility
AWACS	Airborne Warning and Control System
AWCP	**Army Weight Control Program**
AWOL	absent without leave
AXP	*ambulance exchange point

—B—

BAE	brigade aviation element
BAH	basic allowance for housing
BAS	basic allowance for subsistence
BCS3	**Battle Command Sustainment Support System**
BCD	*battlefield coordination detachment
BCT	*brigade combat team
BDA	*battle damage assessment
BEI	***biometrics enabled intelligence**
BFSB	**battlefield surveillance brigade**
BFT	**blue force tracking**
BHA	**bomb hit assessment**
BHL	***battle handover line**
BHO	**battle handover**
BI	*battle injury
BKB	*blue kill box
BLOS	beyond line-of-sight
BM	*battle management
BMCT	*begin morning civil twilight
BMNT	*begin morning nautical twilight
BN	battalion
BNML	**battalion military liaison**
BP	battle position
BSA	*brigade support area
BSB	**brigade support battalion**
BZ	*buffer zone

—C—

C2	*command and control
CA	*civil administration, *civil affairs, *coordinating altitude
CAAF	contractor personnel authorized to accompany the force
CAB	**combined arms battalion,** combat aviation brigade
CAC	**Combined Arms Center,** common access card
CADD	**Combined Arms Doctrine Directorate**
CAIS	*civil authority information support

Bolded entries apply only to the Army. An asterisk (*) marks terms shown in chapter 1.

CAO	**casualty assistance officer**, civil affairs operations
CAISI	**Combat Service Support Automated Information Systems Interface**
CAL	critical asset list
CALL	**Center for Army Lessons Learned**
CAP	*crisis action planning, *civil augmentation program
CAPT	civil affairs planning team
CARVER	criticality, accessibility, recuperability, vulnerability, effect, and recognizability [a target assessment technique]
CAS	casualty evacuation
CASEVAC	*casualty evacuation
CAT	category, civil affairs team
CATS	**combined arms training strategy**
CBRN	chemical, biological, radiological, and nuclear
CBRNE	chemical, biological, radiological, nuclear, and high-yield explosive
CCA	**close combat attack**
CCD	**charged-coupled device**
CCDR	combatant commander
CCIR	commander's critical information requirement
CCMD	*combatant command
CCO	*container control officer
CCP	casualty collection point
CD	**chaplain detachment**
CDC	Centers for Disease Control and Prevention
CDE	collateral damage estimation
CDRUSSOCOM	Commander, United States Special Operations Command
CED	**captured enemy document**
CEMA	**cyber electromagnetic activities**
CERP	chemical, biological, radiological, nuclear, and high-yield explosives enhanced response force package
CERP	Commanders' Emergency Response Program
CF	*conventional forces
CFL	*coordinated fire line
CFR	Code of Federal Regulations
CFZ	**critical friendly zone**
CGS	common ground station
Chem	**chemical**
CI	*counterintelligence
CIA	Central Intelligence Agency
CID	*combat identification, criminal investigation division
CIM	*civil information management
CIO	chief information officer
CJA	**command judge advocate**

Bolded entries apply only to the Army. An asterisk (*) marks terms shown in chapter 1.	
CJCS	Chairman of the Joint Chiefs of Staff
CJCSI	Chairman of the Joint Chiefs of Staff instruction
CJCSM	Chairman of the Joint Chiefs of Staff manual
CJSOTF	**combined joint special operations task force**
CJTF	combined joint task force (NATO)
CLT	*civil liaison team, **casualty liaison team**
CLS	**combat lifesaver**
CM	**consequence management**
CMD	*command
CME	***civil-military engagement**
CMO	civil-military operations
CMOC	*civil-military operations center
CMSE	***civil-military support element**
CO	commanding officer, cyberspace operations
COA	*course of action
COCOM	*combatant command (command authority)
COG	*center of gravity
COIC	**current operations integration cell**
COIN	*counterinsurgency
COLPRO	*collective protection
COLT	***combat observation and lasing team**
COM	*chief of mission, collection operations management
COMCAM	*combat camera
COMINT	communications intelligence
COMNET	communications network
COMSEC	*communications security
CONOPS	concept of operations
CONPLAN	*concept plan
CONUS	continental United States
COOP	continuity of operations
COP	*common operational picture
COR	*contracting officer representative
COS	chief of staff
COSC	**combat and operational stress control**
COTS	commercial off-the-shelf
CP	*checkpoint, *command post, *contact point, *counterproliferation
CR	***civil reconnaissance, *curve resistance**
CRAF	*Civil Reserve Air Fleet
CRC	control and reporting center
CRSP	centralized receiving and shipping point
CREW	counter radio-controlled improvised explosive device electronic warfare
CRM	collection requirements management

Bolded entries apply only to the Army. An asterisk (*) marks terms shown in chapter 1.

CS	*civil support
CSAR	*combat search and rescue
CSC	convoy support center
CSM	**command sergeant major**
CSPO	***contracting support operations**
CSR	*controlled supply rate
CSSB	combat sustainment support battalion
CT	*counterterrorism
CTE	***continuous tractive effort**
CTP	*common tactical picture
CUL	*common-user logistics
CULT	*common-user land transportation
CW	*chemical warfare

—D—

D3A	decide, detect, deliver, and assess, direct action
DA	Department of the Army
DAADC	deputy area air defense commander
DAFL	*directive authority for logistics
DAL	defended asset list
DA Pam	**Department of the Army pamphlet**
DART	disaster assistance response team
DASC	*direct air support center
DATT	defense attaché
DBP	***drawbar pull**
DC	*dislocated civilian
DCA	*defensive counterair
DCE	*defense coordinating element
DCGS	distributed common ground/surface system
DCGS-A	**Distributed Common Ground System–Army**
DCO	*defensive cyberspace operations, *defense coordinating officer
DCO-RA	*defensive cyberspace operation response action
DD	Department of Defense form
D-day	*unnamed day on which operations commence or are scheduled to commence.
DE	*directed energy
DEERS	Defense Enrollment Eligibility Reporting System
DEPMEDS	deployable medical systems
DFAC	**dining facility**
DFAS	Defense Finance and Accounting Service
DHHS	Department of Health and Human Services
DHS	Department of Homeland Security
DIA	Defense Intelligence Agency

Bolded entries apply only to the Army. An asterisk (*) marks terms shown in chapter 1.	
DIB	*defense industrial base
DIRLAUTH	*direct liaison authorized
DISA	Defense Information Systems Agency
DLIC	*detachment left in contact
DMC	**distribution management center**
DNA	deoxyribonucleic acid
DNBI	*disease and nonbattle injury
DNI	Director of National Intelligence
DOD	Department of Defense
DODD	Department of Defense directive
DOR	date of rank
DOS	Department of State
DOT	Department of Transportation
DOTMLPF	doctrine, organization, training, materiel, leadership and education, personnel, and facilities [the force development domains]
DPICM	dual purpose improved conventional munitions
DRSN	Defense Red Switched Network
DS	*direct support
DSCA	*defense support of civil authorities
DSM	*decision support matrix
DST	*decision support template
DT	**dynamic targeting**
DTG	date-time group
DTMS	**Digital Training Management System**
DTS	*Defense Transportation System
DSN	Defense Switched Network
DVA	Department of Veterans Affairs
DZ	*drop zone

—E—

EA	***engagement area,** *electronic attack, *executive agent
EAB	**echelons above brigade**
EAD	*earliest arrival date
EAP	emergency action plan
EC	*enemy combatant
ECU	environmental control unit
ECC	**expeditionary contracting command**
ECL	**electrochemiluminescence**
EDT	***end delivery tonnage**
EECT	*end evening civil twilight
EEFI	*essential element of friendly information
EENT	*end of evening nautical twilight
EEO	equal employment opportunity

Bolded entries apply only to the Army. An asterisk (*) marks terms shown in chapter 1.

ELINT	*electronic intelligence
EMAC	emergency management assistance compact
EMC	*electromagnetic compatibility
EMCON	*emission control
EMI	*electromagnetic interference
EMOE	*electromagnetic operational environment
EMP	*electromagnetic pulse
EMS	electromagnetic spectrum
EMSO	**electromagnetic spectrum operations**
EO	electro-optical, executive order, equal opportunity
EOC	*emergency operations center
EOD	*explosive ordnance disposal
EO-IR CM	*electro-optical-infrared countermeasure
EP	*electronic protection
EPA	*evasion plan of action
EPLO	*emergency preparedness liaison officer
EPLRS	**enhanced position location and reporting system**
EPW	enemy prisoner of war
ERP	***engineer regulating point**
ES	*electronic warfare support
ESB	**expeditionary signal battalion**
ESC	expeditionary sustainment command
ESF	*emergency support function
EW	*electronic warfare
EXORD	execute order

—F—

1SG	**first sergeant**
F3EAD	find, fix, finish, exploit, analyze, and disseminate
FA	*field artillery
FAA	**Federal Aviation Administration (DOT)**
FAAD	**forward area air defense**
FAC	forward air controller
FAC(A)	*forward air controller (airborne)
FAH	**final attack heading**
FARP	*forward arming and refueling point
FBCB2	**Force XXI Battle Command, brigade and below**
FBI	Federal Bureau of Investigation (DOJ)
FCL	***final coordination line**
FDC	*fire direction center
FEBA	*forward edge of the battle area
FEMA	Federal Emergency Management Agency

Bolded entries apply only to the Army. An asterisk (*) marks terms shown in chapter 1.

FEZ	*fighter engagement zone
FFA	*free-fire area
FFIR	*friendly force information requirement
FHA	*foreign humanitarian assistance
FHP	*force health protection
F-hour	*effective time of announcement by the Secretary of Defense to the Military Departments of a decision to mobilize Reserve units
FID	*foreign internal defense
FISINT	*foreign instrumentation signals intelligence
FIST	fire support team
FLE	*forward logistics element
FLIR	*forward-looking infrared
FLOT	*forward line of own troops
FM	field manual, frequency modulation
FMC	**financial management center, field medical card,** full mission-capable
FMI	field manual-interim
FMT	field maintenance team
FMS	*foreign military sales
FOB	forward operating base
FO	*forward observer
FOO	field ordering officer
FORSCOM	United States Army Forces Command
FOS	*forward operating site
FP	*force protection
FPCON	*force protection condition
FPL	***final protective line**
FRAGORD	*fragmentary order
FS	**fire support**
FSA	*fire support area
FSC	**forward support company**
FSCC	*fire support coordination center
FSCL	*fire support coordination line
FSCM	*fire support coordination measure
FSCOORD	fire support coordinator
FSO	fire support officer
FSS	*fire support station
FTX	**field training exercise**

—G—

G-1	**assistant chief of staff, personnel**
G-2	**assistant chief of staff, intelligence**
G-2X	**counterintelligence and human intelligence staff officer for a general staff**

Bolded entries apply only to the Army. An asterisk (*) marks terms shown in chapter 1.

G-3	assistant chief of staff, operations
G-4	assistant chief of staff, logistics
G-5	assistant chief of staff, plans
G-6	**assistant chief of staff, signal**
G-8	**assistant chief of staff, financial management**
G-9	**assistant chief of staff, civil affairs operations**
GARS	Global Area Reference System
GBMD	*global-based missile defense
GCSS-A	Global Command and Control System-Army
GCSS–Army	**Global Combat Support System–Army**
GCC	geographic combatant commander
GEOINT	*geospatial intelligence
GFM	*global force management
GIG	Global Information Grid
GI&S	*geospatial information and services
GMD	*ground-based midcourse defense
GMTI	ground moving target indicator
GP	general purpose
GPS	Global Positioning System
GPW	Geneva Convention Relative to the Treatment of Prisoners of War
GR	***grade resistance**
GS	*general support
GSB	**group support battalion**
GSR	general support-reinforcing
GSSC	global satellite communications (SATCOM) support center
GTL	*gross trailing load, *gun-target line

—H—

HA	humanitarian assistance
HACC	*humanitarian assistance coordination center
HARM	high-speed antiradiation missile
HAZMAT	hazardous materials
HCA	*humanitarian and civic assistance
HCT	human intelligence (HUMINT) collection team
HD	*homeland defense
HE	high explosive
HEI	high explosives incendiary
HEMTT	heavy expanded mobile tactical truck
HF	high frequency
HHC	headquarters and headquarters company
HLZ	**helicopter landing zone**
HM	hazardous material
HMMWV	high mobility multipurpose wheeled vehicle

Bolded entries apply only to the Army. An asterisk (*) marks terms shown in chapter 1.

HN	*host nation
HNS	*host nation support
HPT	*high-payoff target
HPTL	**high-payoff target list**
HR	human resources
HRF	homeland response force
HRP	*high-risk personnel
HQ	headquarters
HQDA	Headquarters, Department of the Army
HRSC	**human resources sustainment center**
HS	*homeland security
HSS	*health service support
HUMINT	*human intelligence
HVAA	*high-value airborne asset (protection)
HVI	high-value individual
HVT	*high-value target
HW	*hazardous waste

—I—

IAA	*incident awareness and assessment
IAMD	*integrated air and missile defense
IAW	in accordance with
IBCT	infantry brigade combat team
IBS	**integrated broadband system**
IC	intelligence community
ICO	**installation contracting office**
ICS	*incident command system
ICRC	International Committee of the Red Cross
ID	**infantry division,** *identification
IDAD	*internal defense and development
IDN	**initial distribution number**
IDP	*internally displaced person
IE	**information engagement**
IED	*improvised explosive device
IEM	installation emergency management
IFF	identification, friend or foe
IGO	*intergovernmental organization
IMCOM	**United States Army Installation Management Command**
IMET	*international military education and training
IMINT	*imagery intelligence
IMT	**initial military training**
INSCOM	United States Army Intelligence and Security Command

Bolded entries apply only to the Army. An asterisk (*) marks terms shown in chapter 1.

IO	*information operations
IP	initial position
IPB	***intelligence preparation of the battlefield,** *intelligence preparation of the battlespace
IPDS	*inland petroleum distribution system
IPE	*individual protective equipment
IPFU	**individual physical fitness uniform**
IPI	*indigenous populations and institutions
IPERMS	**Interactive Personnel Electronic Records Management System**
IPOE	intelligence preparation of the operational environment
IR	information requirements, infrared
IRC	internet relay chat, *information-related capability
IRR	Individual Ready Reserve
ISB	*intermediate staging base
ISOPREP	*isolated personnel report
ISR	*intelligence, surveillance, and reconnaissance
ITO	installation transportation officer
ITV	*in-transit visibility
IW	*irregular warfare

—J—

J-2	intelligence directorate of a joint staff, intelligence staff section
J-3	operations directorate of a joint staff, operations staff section
J-4	logistics directorate of a joint staff, logistics staff section
J-5	plans directorate of a joint staff
J-6	communications system directorate of a joint staff
JAAT	*joint air attack team
JACCE	joint air component coordination element
JAG	judge advocate general
JAOC	*joint air operations center
JCET	*joint combined exchange training
JCS	Joint Chiefs of Staff
JDDE	*joint deployment and distribution enterprise
JDDOC	*joint deployment and distribution operations center
JCMOTF	joint civil-military operations task force
JDAM	Joint Direct Attack Munition
JEZ	*joint engagement zone
JFACC	*joint force air component commander
JFC	*joint force commander
JFCC-IMD	Joint Functional Component Command for Integrated Missile Defense
JFE	*joint fires element
JFLCC	*joint force land component commander
JFMCC	*joint force maritime component commander

Bolded entries apply only to the Army. An asterisk (*) marks terms shown in chapter 1.

JFO	*joint field office, joint fires observer
JFSOC	joint special operations component
JFSOCC	*joint force special operations component commander
JIC	joint information center
JIPOE	*joint intelligence preparation of the operational environment
JIIM	**joint, interagency, intergovernmental, multinational**
JIPTL	*joint integrated prioritized target list
JOA	*joint operations area
JLOTS	*joint logistics over-the-shore
JOC	joint operations center
JOPES	Joint Operation Planning and Execution System
JOPP	*joint operation planning process
JP	joint publication
JPRC	*joint personnel recovery center
JRSOI	*joint reception, staging, onward movement, and integration
JSCP	Joint Strategic Capabilities Plan
JSOA	*joint special operations area
JSOACC	*joint special operations air component commander
JSOTF	*joint special operations task force
JSTARS	Joint Surveillance Target Attack Radar System
JTAC	*joint terminal attack controller
JTCB	*joint targeting coordination board
JTIDS	Joint Tactical Information Distribution System
JTF	*joint task force
JTT	**joint tactical terminal**
JWICS	Joint Worldwide Intelligence Communications System

—K—

KBC	*kill box coordinator
KIA	**killed in action**

—L—

LAD	*latest arrival date
LAN	local area network
LGB	laser-guided bomb
LD	*line of departure
LNO	liaison officer
LOA	***limit of advance,** *letter of authorization
LOC	*line of communications
LOE	*line of effort
LOGCAP	logistics civil augmentation program
LOGPAC	***logistics package**
LOGSA	**logistics support activity**

Bolded entries apply only to the Army. An asterisk (*) marks terms shown in chapter 1.

LOO	*line of operation
LOS	line of sight
LOTS	*logistics over-the-shore
LTIOV	latest time information is of value
LZ	*landing zone

—M—

MACCS	*Marine air command and control system
MAGTF	Marine air-ground task force
MAP	*Military Assistance Program
MARO	*mass atrocity response operations
MASCAL	*mass casualty
MASINT	*measurement and signature intelligence
MBA	***main battle area**
MCA	*military civic action
MCT	*movement control team
MCS	maneuver control system
MCOO	*modified combined obstacle overlay
MCPP	Marine Corps Planning Process (Marine Corps)
MCRP	Marine Corps reference publication
MCWP	Marine Corps warfighting publication
M-day	*mobilization day, unnamed day on which mobilization of forces begins
MDMP	***military decisionmaking process**
MEB	**maneuver enhancement brigade**
MEDEVAC	medical evacuation
MEF	Marine expeditionary force
MET	**mission-essential task**
METL	*mission-essential task list
METT-T	mission, enemy, terrain and weather, troops and support available-time available
METT-TC	**mission, enemy, terrain and weather, troops and support available, time available, and civil considerations [mission variables] (Army)**
MGRS	military grid reference system
MHE	materials handling equipment
MHQ	***music headquarters**
MI	military intelligence
MIA	missing in action
MIDS	**Multifunction Information Distribution System**
MILCON	military construction
MILDEC	*military deception
MIL-STD	military standard
MIS	**military information support**

Bolded entries apply only to the Army. An asterisk (*) marks terms shown in chapter 1.

MISG	military information support group
MISO	*military information support operations
MISOC	**military information support operations command**
MLC	military load classification
MLRS	multiple launch rocket system
MMT	military mail terminal
MNL	*multinational logistics
MOA	memorandum of agreement
MOB	*mobilization
MOE	*measure of effectiveness
MOP	*measure of performance
MOPP	*mission-oriented protective posture
MOS	military occupational specialty
MOU	memorandum of understanding
MP	military police
MPA	**Military Personnel, Army**
MPD	***music performance detachment**
MPT	***music performance team**
MPU	***music performance unit**
MRE	meal, ready to eat
MRL	**multiple rocket launcher**
MSC	*Military Sealift Command
MSE	mobile subscriber equipment
MSF	*mobile security force
MSL	**mean sea level**
MSR	*main supply route
MST	**maintenance support team**
MTF	medical treatment facility
MTOE	modified table of organization and equipment
MTON	*measurement ton
MTS	Movement Tracking System
MTT	*mobile training team
MTTP	multi-Service tactics, techniques, and procedures
MUM-T	*manned unmanned teaming
MWR	morale, welfare, and recreation

—N—

NAI	*named area of interest
NATO	North Atlantic Treaty Organization
NAR	*nonconventional assisted recovery
NARP	Nuclear Weapon Accident Response Procedures
NATO	North Atlantic Treaty Organization
NBC	nuclear, biological, and chemical

Bolded entries apply only to the Army. An asterisk (*) marks terms shown in chapter 1.

NCO	noncommissioned officer
NCOIC	noncommissioned officer in charge
N-day	day an active duty unit is notified for deployment or redeployment
NDS	*national defense strategy
NDT	***net division tonnage**
NEO	*noncombatant evacuation operations
NETOPS	*network operations
NFA	no-fire area
NGA	National Geospatial-Intelligence Agency
NGIC	National Ground Intelligence Center
NGO	*nongovernmental organization
NIMS	*National Incident Management System
NIPRNET	Nonsecure Internet Protocol Router Network
NIST	**national intelligence support team**
NLT	not later than
NMS	*national military strategy
NORAD	North American Aerospace Defense Command
NRF	National Response Framework
NSA	National Security Agency
NSC	National Security Council
NSFS	*naval surface fire support
NSL	*no-strike list
NSN	national stock number
NSS	*national security strategy
NSSE	*national special security event
NTC	National Training Center
NTL	***net trainload**
NTTP	Navy tactics, techniques, and procedures
NVD	*night vision device
NVG	*night vision goggle(s)
NWDC	Navy Warfare Development Command
NWP	Navy warfare publication

—O—

O&M	operation and maintenance
OA	*operational area
OAKOC	**observation and fields of fire, avenues of approach, key terrain, obstacles, and cover and concealment [military aspects of terrain]**
OB	order of battle
OCA	*offensive counterair
OCO	*offensive cyberspace operations
OCONUS	outside the continental United States
OCS	*operational contract support

Bolded entries apply only to the Army. An asterisk (*) marks terms shown in chapter 1.

OE	*operational environment
OGA	**other government agency**
OHDACA	Overseas Humanitarian, Disaster, and Civic Aid
OIC	officer in charge
OMA	**Operations and Maintenance, Army**
OP	**observation post**
OPCON	*operational control
OPDS	*offshore petroleum discharge system
OPIR	*overhead persistent infrared
OPLAN	*operation plan
OPNAVINST	Chief of Naval Operations instruction
OPORD	*operation order
OPSEC	*operations security
OPTEMPO	operating tempo
ORSA	operations research and systems analysis
OSD	Office of the Secretary of Defense
OSHA	**Occupational Safety and Health Administration**
OSINT	*open-source intelligence
OSUT	**one station unit training**

—P—

PA	*public affairs **personnel accountability**
PAA	*position area for artillery
PAO	public affairs officer
PB	*peace building
PBO	**property book officer**
PCC	**precombat check**
PCI	**precombat inspection**
PCS	permanent change of station
PDSS	predeployment site survey
PE	*preparation of the environment
PFC	**private first class**
PGM	*precision-guided munition
PID	positive identification
PIR	priority intelligence requirement
PKI	*public key infrastructure
PL	phase line
PLD	***probable line of deployment**
PLT	platoon
PM	**project manager**
PMCS	**preventive maintenance checks and services**
PMESII-PT	**political, military, economic, social, information, infrastructure, physical environment, and time [operational variables]**

Bolded entries apply only to the Army. An asterisk (*) marks terms shown in chapter 1.

PN	*partner nation
PO	**psychological objective,** *peace operations
POC	point of contact
POD	*port of debarkation
POE	*port of embarkation
POL	petroleum, oils, and lubricants
PPD	**purified protein derivative**
PPE	*personal protective equipment
PPO	**project purchasing officer**
PR	personnel recovery
PRT	provincial reconstruction team
PSA	*port support activity
PVNTMED	preventive medicine
PWS	*performance work statement
PZ	pickup zone

—Q—

QA	quality assurance
QC	quality control
QRF	quick reaction force

—R—

R	**reinforcing**
R&S	reconnaissance and surveillance
R-day	redeployment day
RADC	*regional air defense commander
RC	Reserve Component
RCA	*riot control agent
RCT	regimental combat team
RDD	*required delivery date
RDSP	**rapid decisionmaking and synchronization process**
RECON	*reconnaissance
RED HORSE	*Rapid Engineer Deployable Heavy Operational Repair Squadron Engineer
RF	radio frequency
RFA	request for assistance, *restrictive fire area
RF CM	*radio frequency countermeasures
RFF	request for forces
RFI	*request for information
RFL	restrictive fire line
RLD	*ready-to-load date
RM	risk management
ROE	rules of engagement

Bolded entries apply only to the Army. An asterisk (*) marks terms shown in chapter 1.	
ROZ	restricted operations zone
RP	***rally point,** *release point, **red phosphorus**
RPG	rocket propelled grenade
RR	*reattack recommendation, ***rolling resistance**
RS	religious support
RSO	reception, staging, and onward movement
RSOI	reception, staging, onward movement, and integration
RSTA	reconnaissance, surveillance, and target acquisition
RTL	*restricted target list
RUF	rules for the use of force

—S—

S-1	**battalion or brigade personnel staff officer**
S-2	battalion or brigade intelligence staff officer
S-2X	**battalion or brigade counterintelligence and human intelligence staff officer**
S-3	battalion or brigade operations staff officer
S-4	battalion or brigade logistics staff officer
S-5	**battalion or brigade plans staff officer**
S-6	**battalion or brigade signal staff officer**
S-9	**battalion or brigade civil affairs operations staff officer**
SA	*security assistance, situational awareness, *staging area
SAA	**satellite access authorization,** *senior airfield authority
SACC	*supporting arms coordination center
SADC	*sector air defense commander
SALT	**size, activity, location, time**
SALUTE	size, activity, location, unit, time, and equipment
SAM	surface-to-air missile
SAMS-E	**standard Army maintenance system-enhanced**
SAR	search and rescue, synthetic aperture radar
SARSS	**Standard Army Retail Supply System**
SATCOM	satellite communications
S&TI	*scientific and technical intelligence
SB (SO) (A)	**sustainment brigade (special operations) (airborne)**
SBCT	Stryker brigade combat team
SC	*security cooperation, *strategic communication
SCA	*space coordinating authority
SCAR	*strike coordination and reconnaissance
SCI	sensitive compartmented information
SCO	*security cooperation organization
S-day	*day the President authorizes selective reserve call-up
SDDC	Surface Deployment and Distribution Command
SE	*site exploitation

Bolded entries apply only to the Army. An asterisk (*) marks terms shown in chapter 1.

SEAD	*suppression of enemy air defenses
SecDef	Secretary of Defense
SERE	survival, evasion, resistance, and escape
SF	*special forces
SFA	security force assistance
SGT	**sergeant**
SGM	**sergeant major**
SHF	super-high frequency
SIGACT	**significant activity**
SIGINT	*signals intelligence
SINCGARS	single-channel ground and airborne radio system
SIPRNET	SECRET Internet Protocol Router Network
SIR	specific information requirement
SITREP	situation report
SJA	staff judge advocate
SME	subject matter expert
SMO	***Spectrum Management Operations**
SO	*special operations
SOCCE	*special operations command and control element
SOF	*special operations forces
SOFA	*status-of-forces agreement
SOI	signal operating instructions
SOLE	*special operations liaison element
SO-peculiar	*special operations-peculiar
SOP	*standard operating procedure
SOR	statement of requirement
SOTF-A	*special operations task force A
SP	***start point**
SPINS	special instructions
SPM	*single port manager
SPO	**support operations**
SPOD	seaport of debarkation
SPOE	seaport of embarkation
SPOTREP	*spot report
SR	*special reconnaissance
SROE	standing rules of engagement
SRP	**soldier readiness processing**
SRUF	standing rules for the use of force
SSA	**supply support activity**
SSR	*security sector reform
STANAG	standardization agreement
STB	**special troops battalion**

Bolded entries apply only to the Army. An asterisk (*) marks terms shown in chapter 1.

STE	secure telephone equipment, ***starting tractive effort**
STT	*special tactics team
SWEAT-MSO	**sewage, water, electricity, academics, trash, medical, safety, other considerations**
SWO	staff weather officer

—T—

TA	*target acquisition, target audience, *threat assessment
TAA	*tactical assembly area
TAC	**tactical command post (graphics), tactical (graphics)**
TAC(A)	*tactical air coordinator (airborne)
TACC	*tactical air command center (Marine), *tactical air control center (Navy)
TACLAN	tactical local area network
TACON	*tactical control
TACP	*tactical air control party
TACS	theater air control system
TAI	*target area of interest
TAMD	theater air and missile defense
TB MED	**technical bulletin (medical)**
TBM	theater ballistic missile
TC	**training circular**
TC-AIMS II	Transportation Coordinator's Automated Information for Movement System II
TCC	*transportation component command
TCF	*tactical combat force
TDA	Table of Distribution and Allowance
T-day	*effective day coincident with Presidential declaration of a National Emergency and authorization of partial.
TD	***train density**
TDMA	time division multiple access
TDY	temporary duty
TECHINT	*technical intelligence
TEMPER	tent extendible modular personnel
TES	*theater event system
TF	task force
TG	**technical guide**
TGO	*terminal guidance operations
THAAD	terminal high-altitude area defense
TIB	*toxic industrial biological
TIC	*toxic industrial chemical
TIM	*toxic industrial material
TIP	target intelligence package

Bolded entries apply only to the Army. An asterisk (*) marks terms shown in chapter 1.

TJAG	the judge advocate general
TLE	target location error
TLP	**troop leading procedures**
TM	technical manual
TMD	theater missile defense
TO	*theater of operations
TOC	tactical operations center
TO&E	table of organization and equipment
TOF	time of flight
TOW	tube launched, optically tracked, wire guided
TPFDD	*time-phased force and deployment data
TPFDL	*time-phased force and deployment list
TQ	*tactical questioning
TRADOC	United States Army Training and Doctrine Command
TRP	target reference point
TSC	theater sustainment command
TSCP	theater security cooperation plan
TSOC	*theater special operations command
TST	*time-sensitive target
TTP	tactics, techniques, and procedures

—U—

UA	*unmanned aircraft
UAR	*unconventional assisted recovery
UARCC	*unconventional assisted recovery coordination cell
UAS	*unmanned aircraft system
UCMJ	Uniform Code of Military Justice
UCP	*Unified Command Plan
UGR	**unitized group ration**
UH	**utility helicopter**
UIC	unit identification code
ULN	*unit line number
UMD	*unit movement data
UMT	unit ministry team
U.S.	**United States**
USA	United States Army
USACE	United States Army Corps of Engineers
USAF	United States Air Force
USAID	United States Agency for International Development
USAJFKSWC	United States Army John F. Kennedy Special Warfare Center
USAMC	United States Army Materiel Command
USAMEDCOM	**United States Army Medical Command**

Bolded entries apply only to the Army. An asterisk (*) marks terms shown in chapter 1.

USAPHC	**United States Army Public Health Command**
USAR	United States Army Reserve
USASOC	United States Army Special Operations Command
USC	United States Code
USCENTCOM	United States Central Command
USCG	United States Coast Guard
USEUCOM	United States European Command
USG	United States Government
USMC	United States Marine Corps
USMTF	United States message text format
USN	United States Navy
USNORTHCOM	United States Northern Command
USPACOM	United States Pacific Command
USSOCOM	United States Special Operations Command
USSOUTHCOM	United States Southern Command
USSTRATCOM	United States Strategic Command
USTRANSCOM	United States Transportation Command
UTM	universal transverse Mercator, **unit training management**
UTP	**unit training plan**
UW	*unconventional warfare
UXO	*unexploded explosive ordnance

—V—

VI	*visual information
VISA	*Voluntary Intermodal Sealift Agreement
VoIP	**voice over internet protocol**
VBIED	*vehicle-borne improvised explosive device
VRC	**vehicle radio communication**
VSAT	**very small aperture terminal**
VT	**variable time**
VTC	video teleconference

—W—

WAN	wide-area network
WARM	*wartime reserve modes
WARNORD	*warning order
W-day	*declared by the President, W-day is associated with an adversary decision to prepare for war
WGS	World Geodetic System
WEZ	*weapon engagement zone
WMD	*weapons of mass destruction
WP	white phosphorous

Bolded entries apply only to the Army. An asterisk (*) marks terms shown in chapter 1.

WPS	Worldwide Port System
WTBD	**warrior tasks and battle drills**
WTI	*weapons technical intelligence

—X—

XO	executive officer

—Z—

ZF	*zone of fire

SECTION II — GEOGRAPHICAL ENTITY CODES

2-2. On 19 February 2004, NATO Standardization Agreement (STANAG) 1059 (Edition 8), *Letter Codes for Geographical Entities*, was promulgated. The aim of this agreement is to provide unique three-letter codes for use within NATO to distinguish geographical entities. However, due to implementation difficulties, a 28 January 2005 memorandum from the NATO Standardization Agency declared that until STANAG 1059 could be fully implemented, the two-letter code would be used for the technical automated information system domain, while in all other (nontechnical) administrative areas, the three-letter code would be used. The list on pages 2-26 through 2-32 provides both the two-letter and three-letter codes.

Geographical entity	Two-letter code	Three-letter code
	—A—	
Afghanistan	AF	AFG
Albania	AL	ALB
Algeria	DZ	DZA
American Samoa	AS	ASM
Andorra	AD	AND
Angola	AO	AGO
Anguilla	AI	AIA
Antarctica	AQ	ATA
Antigua and Barbuda	AG	ATG
Argentina	AR	ARG
Armenia	AM	ARM
Aruba	AW	ABW
Australia	AU	AUS
Austria	AT	AUT
Azerbaijan	AZ	AZE
	—B—	
Bahamas	BS	BHS
Bahrain	BH	BHR
Bangladesh	BD	BGD
Barbados	BB	BRB
Belarus	BY	BLR
Belize	BZ	BLZ
Belgium	BE	BEL
Benin	BJ	BEN
Bermuda	BM	BMU
Bhutan	BT	BTN
Bolivia	BO	BOL
Bosnia and Herzegovina	BA	BIH
Botswana	BW	BWA
Bouvet Island	BV	BVT
Brazil	BR	BRA
British Indian Ocean Territory	IO	IOT

Geographical entity	Two-letter code	Three-letter code
Brunei Darussalam	BN	BRN
Bulgaria	BG	BGR
Burkina Faso	BF	BFA
Burundi	BI	BDI
—C—		
Cambodia	KH	KHM
Cameroon	CM	CMR
Canada	CA	CAN
Cape Verde	CV	CPV
Cayman Islands	KY	CYM
Central African Republic	CF	CAF
Chad	TD	TCD
Chile	CL	CHL
China	CN	CHN
Christmas Island	CX	CXR
Cocos (Keeling) Islands	CC	CCK
Colombia	CO	COL
Comoros	KM	COM
Congo	CG	COG
Congo, The Democratic Republic of the	CD	COD
Cook Islands	CK	COK
Costa Rica	CR	CRI
Cote d'Ivoire (Ivory Coast)	CI	CIV
Croatia (Hrvatska)	HR	HRV
Cuba	CU	CUB
Cypress	CY	CYP
Czech Republic	CZ	CZE
—D—		
Denmark	DK	DNK
Djibouti	DJ	DJI
Dominica	DM	DMA
Dominican Republic	DO	DOM
—E—		
Ecuador	EC	ECU
El Salvador	SV	SLV
Egypt	EG	EGY
Equatorial Guinea	GQ	GNQ
Eritrea	ER	ERI
Estonia	EE	EST
Ethiopia	ET	ETH

Geographical entity	*Two-letter code*	*Three-letter code*
—F—		
Falkland Islands (Malvinas)	FK	FLK
Faroe Islands	FO	FRO
Fiji	FJ	FJI
Finland	FI	FIN
France	FR	FRA
French Guiana	GF	GUF
French Polynesia	PF	PYF
French Southern Territories	TF	ATF
—G—		
Gabon	GA	GAB
Gambia	GM	GMB
Georgia	GE	GEO
Germany	DE	DEU
Ghana	GH	GHA
Gibraltar	GI	GIB
Greece	GR	GRC
Greenland	GL	GRL
Grenada	GD	GRD
Guadaloupe	GP	GLP
Guam	GU	GUM
Guatamala	GT	GTM
Guinea	GN	GIN
Guinea–Bissau	GW	GNB
Guyana	GY	GUY
—H—		
Haiti	HT	HTI
Heard Island and McDonald Islands	HM	HMD
Holy See (Vatican City State)	VA	VAT
Honduras	HN	HND
Hong Kong	HK	HKG
Hungary	HU	HUN
—I—		
Iceland	IS	ISL
India	IN	IND
Indonesia	ID	IDN
Iran, Islamic Republic of	IR	IRN
Iraq	IQ	IRQ
Ireland	IE	IRL

Geographical entity	*Two-letter code*	*Three-letter code*
Israel	IL	ISL
Italy	IT	ITA
—J—		
Jamaica	JM	JAM
Japan	JP	JPN
Jordan	JO	JOR
—K—		
Kazakhstan	KZ	KAZ
Kenya	KE	KEN
Kiribati	KI	KIR
Korea, Democratic People's Republic of	KP	PRK
Korea, Republic of	KR	KOR
Kuwait	KW	KWT
Kyrgyzstan	KG	KGZ
—L—		
Lao People's Democratic Republic	LA	LAO
Latvia	LV	LVA
Lebanon	LB	LBN
Lesotho	LS	LSO
Liberia	LR	LBR
Libyan	LY	LBY
Liechtenstein	LI	LIE
Lithuania	LT	LTU
Luxembourg	LU	LUX
—M—		
Macao	MO	MAC
Madagascar	MD	MDG
Malawi	MW	MWI
Malaysia	MY	MYS
Maldives	MV	MDV
Mali	ML	MLI
Malta	MT	MLT
Martinique	MQ	MTQ
Mauritania	MR	MRT
Mauritius	MU	MUS
Mexico	MX	MEX
Micronesia, Federated States of	FM	FSM
Moldova, Republic of	MD	MDA
Monoco	MC	MCO

Geographical entity	*Two-letter code*	*Three-letter code*
Mongolia	MN	MNG
Montserrat	MS	MSR
Morocco	MA	MAR
Mozambique	MZ	MOZ
Myanmar	MM	MMR
	—N—	
Namibia	NA	NAM
Nauru	NR	NRU
Nepal	NP	NPL
Netherlands	NL	NLD
Netherlands Antilles	AN	ANT
New Caledonia	NC	NCL
New Zealand	NZ	NZL
Nicaragua	NI	NIC
Niger	NE	NER
Nigeria	NG	NGA
Niue	NU	NIU
Norfolk Island	NF	NFK
Northern Mariana Islands	MP	MNP
Norway	NO	NOR
	—O—	
Oman	OM	OMN
	—P—	
Pakistan	PK	PAK
Palau	PW	PLW
Panama	PA	PAN
Papua New Guinea	PG	PNG
Paraguay	PY	PRY
Peru	PE	PER
Philippines	PH	PHL
Pitcairn	PN	PCN
Poland	PL	POL
Portugal	PT	PRT
Puerto Rico	PR	PRI
	—Q—	
Qatar	QA	QAT
	—R—	
Reunion	RE	REU

Geographical entity	*Two-letter code*	*Three-letter code*
Romania	RO	ROU
Russian Federation	RU	RUS
Rwanda	RW	RWA
—S—		
Saint Helena	SH	SHL
Saint Kitts and Nevis	KN	KNA
Saint Lucia	LC	LCA
Saint Pierre and Miquelone	PM	SPM
Saint Vincent and the Grenadines	VC	VCT
Samoa	WS	WSM
San Marino	SM	SMR
Sao Tome and Principe	ST	STP
Saudi Arabia	SA	SAU
Senegal	SN	SEN
Serbia and Montenegro	CS	SCG
Seychelles	SC	SYC
Sierra Leone	SL	SLE
Singapore	SG	SGP
Slovakia	SK	SVK
Slovenia	SI	SVN
Solomon Islands	SB	SLB
Somalia	SO	SOM
South Africa	ZA	ZAF
South Georgia and South Sandwich Islands	GS	SGS
Spain	ES	ESP
Sri Lanka	LK	LKA
Sudan	SD	SDN
Suriname	SR	SUR
Svalbard and Jan Mayen Islands	SJ	SJM
Swaziland	SZ	SWZ
Sweden	SE	SWE
Switzerland	CH	CHE
Syrian Arab Republic	SY	SYR
—T—		
Taiwan, Province of China	TW	TWN
Tajikistan	TJ	TJK
Tanzania, United Republic of	TZ	TZA
Thailand	TH	THA
Timor–Leste	TL	TLS
Togo	TG	TGO
Tokelau	TK	TKL

Geographical entity	*Two-letter code*	*Three-letter code*
Tonga	TO	TON
Trinidad and Tobago	TT	TTO
Tunisia	TN	TUN
Turkey	TR	TUR
Turkmenistan	TM	TKM
Turks and Caicos Islands	TC	TCA
Tuvalu	TV	TUV
—U—		
Uganda	UG	UGA
Ukraine	UA	UKR
United Arab Emirates	AE	AER
United Kingdom	GB	GBR
United States	US	USA
United States Minor Outlying Islands	UM	UMI
Uruguay	UY	URY
Uzbekistan	UZ	UZB
—V—		
Vanuatu	VU	VUT
Venezuala	VE	VEN
Viet Nam	VN	VNM
Virgin Islands (British)	VG	VGB
Virgin Islands (US)	VI	VIR
—W—		
Wallis and Futuna Islands	WF	WLF
Western Sahara	EH	ESH
—Y—		
Yemen	YE	YEM
Yugoslavia, Federal Republic of	YU	YUG
—Z—		
Zambia	ZM	ZMB
Zimbabwe	ZW	ZWE

Chapter 3
Military Symbology Basics

This chapter discusses framed symbols, locations of amplifiers, the bounding octagon, and the locations of icons and modifiers. It also discusses the building process for framed symbols and unframed symbols.

FRAMED SYMBOLS

3-1. A military symbol is a graphic representation of a unit, equipment, installation, activity, control measure, or tactical task relevant to military operations that is used for planning or to represent the common operational picture on a map, display, or overlay. Military symbols are governed by the rules in Military Standard (MIL-STD) 2525D. Military symbols fall into two categories: framed, which includes unit, equipment, installation, and activity symbols; and unframed, which includes control measure and tactical symbols.

3-2. A framed symbol is composed of a frame, color (fill), icon, modifiers, and amplifiers. Paragraphs 3-3 through 3-13 discuss framed symbols.

FRAME

3-3. The frame is the border of a symbol. It does not include associated information inside or outside of the border. The frame serves as the base to which other symbol components are added. The frame indicates the standard identity, physical domain, and status of the object being represented.

Standard Identity

3-4. Standard identity reflects the relationship between the viewer and the operational object being monitored. The standard identity categories are unknown, pending, assumed friend, friend, neutral, suspect, and hostile. In the realm of surface operation symbols, a circle or rectangle frame denotes friend or assumed friend standard identity, a diamond frame denotes hostile or suspect standard identity, a square frame denotes neutral standard identity, and a quatrefoil frame denotes unknown and pending standard identity. Table 3-1 (on page 3-2) shows frame shapes for standard identities for land symbols.

Physical Domain

3-5. The physical domain defines the primary mission area for the object within the operational environment. An object can have a mission area above the earth's surface (in the air domain or space domain), on the earth's surface, or below the earth's surface (that is, in the land domain or maritime domain). The land domain includes those mission areas on the land surface or close to the surface (such as caves, mines, and underground shelters). Maritime surface units are depicted in the sea surface dimension. Aircraft, regardless of Service ownership, are depicted in the air dimension while air facilities are depicted as land installations. Land equipment is depicted in the land dimension. Likewise, a landing craft whose primary mission is ferrying personnel or equipment to and from shore are represented in the sea surface dimension. However, a landing craft whose primary mission is to fight on land is a ground asset and is represented in the land dimension.

Table 3-1. Frame shapes for standard identities

Standard identity	*Friendly*	*Hostile*	*Neutral*	*Unknown*
	Assumed friend	*Suspect*		*Pending*
Unit				
Equipment				
Installation				
Activity				

Status

3-6. Status indicates whether an operational object exists at the location identified (status is "present" or "confirmed"), will in the future reside at that location (status is "planned" or "anticipated"), or is thought to reside at that location (suspected). The symbol frame is a solid line when indicating a present status and a dashed line when indicating anticipated, planned, or suspected status. When the standard identity of the frame is uncertain, as is the case for assumed friend, suspect, or pending, the status cannot be displayed. Additionally, the status cannot be shown when the symbol is unframed (equipment only) or is displayed as a dot. Table 3-2 shows examples of status.

Table 3-2. Examples of status

Friendly	*Present*	*Planned*
Hostile	*Present*	*Suspect*

Color (Fill)

3-7. In framed symbols, color provides a redundant clue with regard to standard identity. The fill is the interior area within a symbol. If color is not used, the fill is transparent. In unframed symbols (equipment), color is the sole indicator of standard identity, excluding text amplifiers. Blue for friendly or assumed friend, red for hostile or suspect, green for neutral, and yellow for unknown or pending are the default colors used to designate standard identity. Affiliation color without the fill may also be used for the frame, main icon, and modifiers.

Icons for Framed Symbols

3-8. The icon is the innermost part of a symbol. The icon provides an abstract pictorial or alphanumeric representation of units, equipment, installations, or activities. This publication distinguishes between icons that must be framed and icons for which framing is optional.

Modifiers for Framed Symbols

3-9. A modifier provides an abstract pictorial or alphanumeric representation, displayed in conjunction with an icon. The modifier provides additional information about the icon (unit, equipment, installation, or activity) being displayed. Modifiers conform to the bounding octagon and are placed either above or below the icon. This publication defines various types of modifiers and indicates where each is to be placed in relation to the icon within the symbol.

Amplifiers for Framed Symbols

3-10. An amplifier provides additional information about the symbol being portrayed and is displayed outside the frame. Figure 3-1, on page 3-4, shows the essential amplifier fields around a friendly land unit symbol frame. To avoid cluttering the display, only essential amplifiers should be used. Arabic numerals are normally used to show the unique designation of units. However, Roman numerals are used to show corps echelon units.

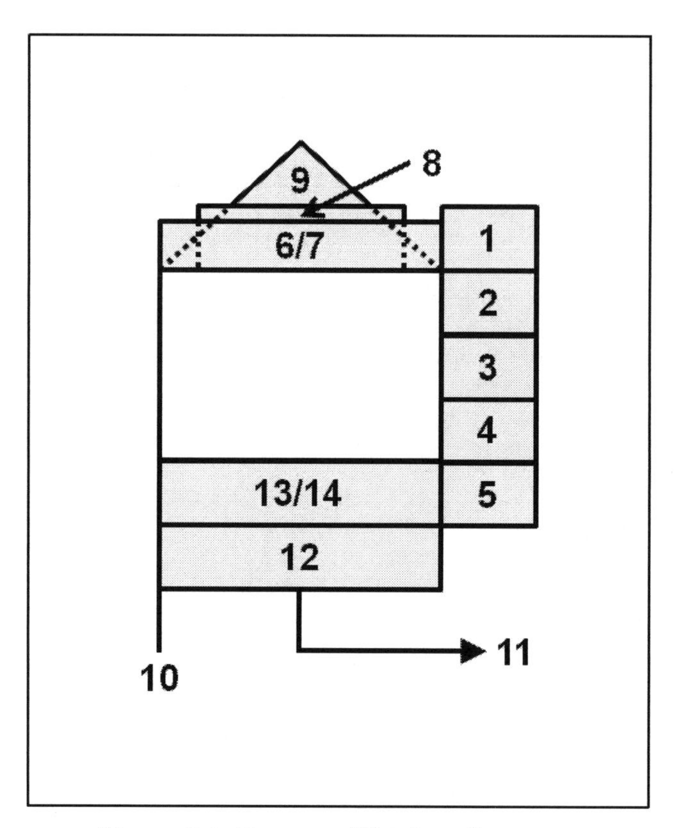

Figure 3-1. New amplifier locations

3-11. The amplifier locations in figure 3-1 are changed from previous editions of this manual. The amplifier locations also vary from MIL-STD-2525D. The old amplifier locations are shown in figure 3-2. The new amplifier locations are designed to reduce the amount of space used by a framed symbol and provide only critical information around the symbol.

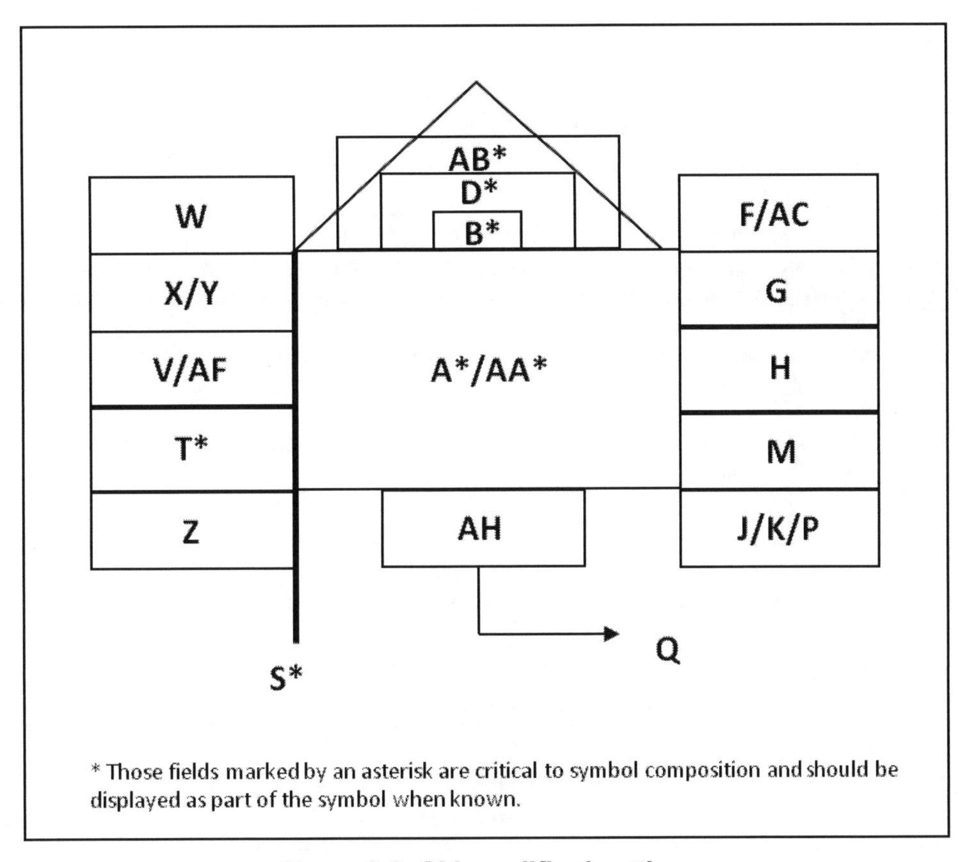

Figure 3-2. Old amplifier locations

LOCATION OF AMPLIFIERS FOR FRAMED SYMBOLS

3-12. The purpose of amplifier placement is to standardize the location of information. Figure 3-1 also illustrates the placement of amplifiers around a frame. The placement of amplifier information is the same regardless of frame shape. Table 3-3 (on page 3-6) provides a description of amplifiers for framed symbols.

Table 3-3. Description of amplifier fields

Field	*Description*
1	**Attached and detached** indicates that one or more sub-elements of a similar function have been attached or detached to a headquarters.(see table 4-9 on page 4-36)
2	**Country indicator** is an accepted code that indicates country of origin of the organization (see chapter 2, section II).
3	A **unique alphanumeric designation** that identifies the unit being displayed. ***Note:*** When showing unique alphanumeric designations for combat arms regimental units (air defense artillery, armor, aviation, cavalry, field artillery, infantry, and special forces) the following rules apply: **No regimental headquarters**: A dash (-) will be used between the battalion and the regimental designation where there is no regimental headquarters. (Example: A/6-37 for A Battery, 6th Battalion, 37th Field Artillery) **Regimental headquarters**: A slash (/) will be used between the battalion and the regimental designation where there is a regimental headquarters of an active operational unit to show continuity of the units. (Example: F/2/11 for F Troop, 2d Squadron/11th Armored Cavalry Regiment)
4	Number or title of the next **higher formation** of the unit being displayed.
5	Free text **staff comments** for information required by the commander. Can also be used for unit location if required.
6	**Echelon** indicator of the symbol. (See table 4-7 on page 4-33.)
7	**Quantity** that identifies the number of items present.
8	**Task force** amplifier placed over the echelon. (See table 4-8 on page 4-35.)
9	**Feint or dummy indicator** shows that the element is being used for deception purposes. ***Note***: The dummy indicator appears as shown in figure 3-1 on page 3-4 and can be used for all framed symbol sets. For control measures, it is a control measure symbol used in conjunction with other control measures. (See table 8-6 on page 8-71 for feint or dummy symbols.)
10	**Headquarters staff offset locator indicator** identifies symbol as a headquarters. (See figure 4-7 on page 4-40.)
	Offset location indicator is used to denote precise location of headquarters or to declutter multiple unit locations and headquarters. (See figure 4-7 on page 4-40)
11	The **direction of movement arrow** indicates the direction the symbol is moving or will move.
	The **offset location indicator** without the arrow is used to denote precise location of units or to declutter multiple unit locations, except for headquarters. (See figure 4-6 on page 4-39)
12	**Combat effectiveness** of unit or equipment displayed. (See table 4-11 on page 4-39.)
13	**Mobility indicator** of the equipment being displayed. (See figure 5-1 and table 5-3 on page 5-15)
14	**Command post and command group** indicates a unit headquarters where the commander and staff perform their activities. (See table 4-10 on page 4-37)

LETTERING FOR ALL SYMBOLS

3-13. The lettering for all military symbols will always be upper case. The lettering for all point, line, and area symbols will be oriented to the top of the display (north). In some cases the lettering may be tilted slightly to follow the contour of a line, but it must never be tilted so much that readers must tilt their heads to read it. The lettering for the bounding octagon will be the same as the orientation of the octagon. The lettering for the horizontal bounding octagon will be horizontal from left to right and the lettering for the vertical octagon will be vertical from top to bottom.

THE BOUNDING OCTAGON AND THE LOCATION OF ICONS AND MODIFIERS FOR FRAMED SYMBOLS

3-13. The bounding octagon serves as the spatial reference for placement of icons and modifiers within the frame of a symbol. It is divided into three sectors. The three sectors specify where icons and modifiers are positioned and how much space is available for sizing of icons and modifiers. The vertical bounding octagon allows for effective use of the space when dealing with vertical icons. Table 3-4 (on page 3-8) provides examples showing the horizontal and vertical bounding octagons and all examples for all frame shapes.

Table 3-4. Examples of horizontal and vertical bounding octagons

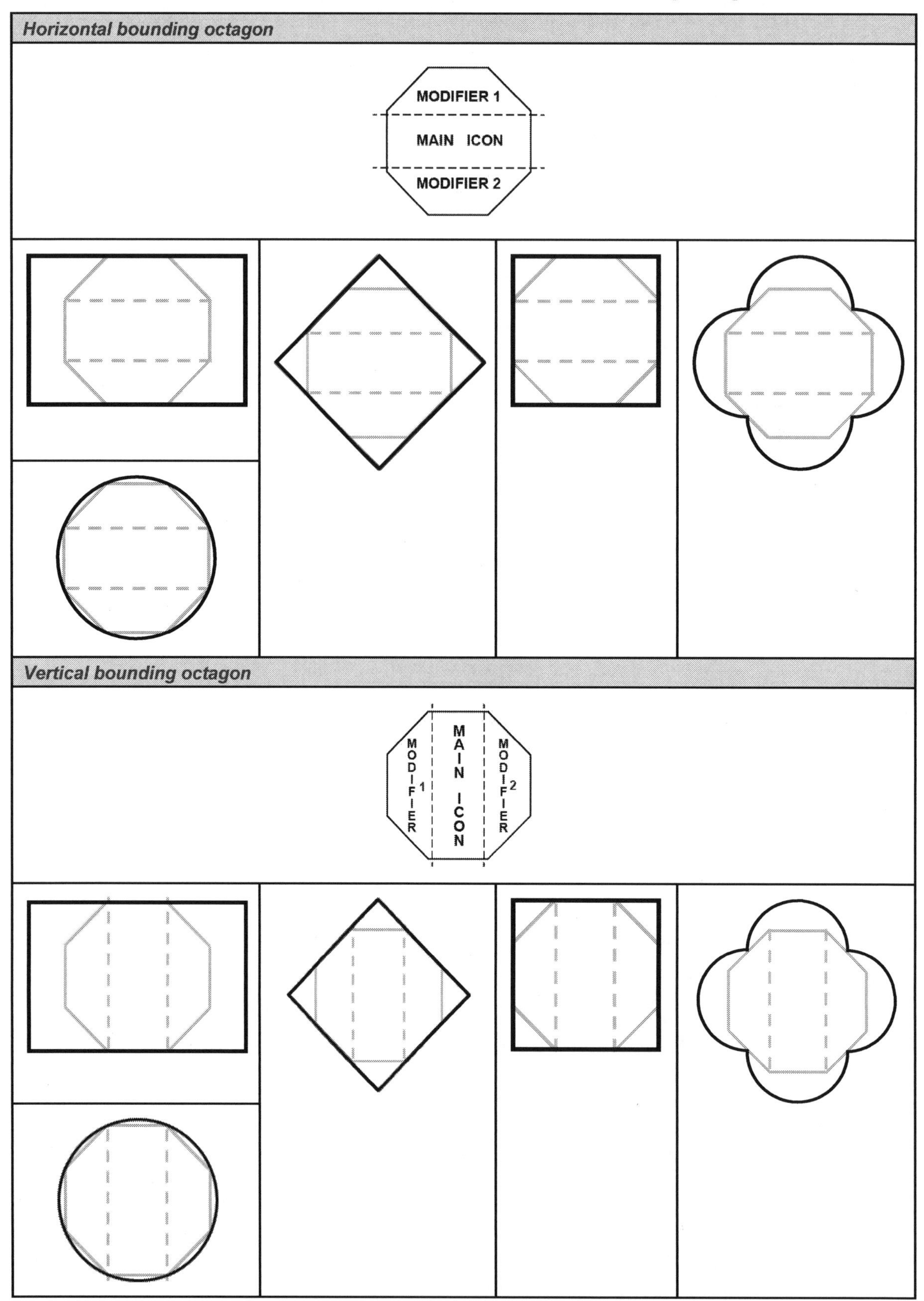

3-14. In general, icons should not be so large as to exceed the dimensions of the main sector of the bounding octagon or touch the interior border of the frame. However, there are exceptions to this size rule. In those cases the icons will occupy the entire frame and must, therefore, exceed the dimensions of the main sector of the bounding octagon and touch the interior border of the frame. These are called full-frame icons and occur only in land domain symbols. Figure 3-3 shows an example of a full-frame icon for all frame shapes.

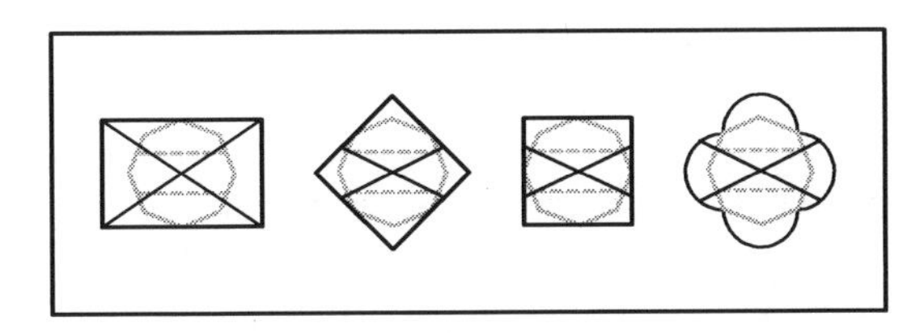

Figure 3-3. Example of full-frame icon

THE BUILDING PROCESS FOR FRAMED SYMBOLS

3-15. Chapters 4 through 7 provide an extensive number of icons and modifiers for building a wide variety of framed symbols. No attempt has been made to depict all possible combinations. Instead, a standard method for constructing symbols is presented. Once the user is familiar with the prescribed system, any desired symbol can be developed using the logical sequence provided in this chapter. The icons shown in this publication are adequate for depicting all standard identities for framed symbols. When representing unorthodox framed symbols, users select the most appropriate icon or modifier contained herein. Soldiers should avoid using any icon or modifiers or combinations and modifications that differ from those in this publication. If, after searching doctrinal icons and modifiers, it is necessary to create a new symbol, users should explain the symbol in an accompanying legend. Computer-generated systems will have difficulty in passing nonstandard symbols. Table 3-5 (on page 3-10) shows the steps in the building process for framed symbols.

Table 3-5. Building process for framed symbols

Step	Description	Example
1	Choose appropriate frame shape from table 3-1 on page 3-2.	
2	Choose appropriate main icon from chapters 2 through 5 and combine it with frame.	
3	Choose appropriate sector 1 modifier from chapters 2 through 5.	
4	Choose appropriate sector 2 modifier from chapters 2 through 5.	
5	Choose minimum essential amplifiers from those listed in table 3-3 on page 3-6.	US 6-37

UNFRAMED SYMBOLS

3-16. Paragraphs 3-17 through 3-21 discuss unframed symbols. Control measure symbols and mission task verb symbols are unframed symbols. They conform to special rules for their own elements.

CONTROL MEASURE SYMBOLS

3-17. A control measure is a means of regulating forces or warfighting functions. Control measures may be boundaries, special area designations, or other unique markings related to an operational environment's geometry and necessary for planning and managing operations. Control measure symbols represent control measures that can be graphically portrayed and provide operational information. They can be displayed as points, lines, and areas. Control measure symbols can be combined with other military symbols, icons, and amplifiers to display operational information. Control measure symbols follow the same basic building rules as framed symbols, but they are built in accordance with their template. Control measure symbols can be black or white, depending on display background. Display backgrounds can be blue (for friendly), red (for hostile), green (for obstacles), or yellow (for a chemical, biological, radiological, and nuclear [CBRN] contaminated area fill). Description, placement, and further details of control measure symbols are addressed in chapter 8. Table 3-6 (on page 3-12) shows descriptions of control measure amplifier fields. Table 3-7 (on page 3-13) shows the steps in the building process for control measure symbols. (See chapter 8 for more information about control measure symbols.)

Icons for Control Measures

3-18. The icon provides an abstract pictorial representation of the control measure. Icons can be depicted as points, lines, or areas.

Modifiers for Control Measures

3-19. Only chemical, biological, radiological, and nuclear events and contaminated areas have modifiers. The remaining control measures use amplifiers.

Amplifiers for Control Measures

3-20. As with the framed symbols, the amplifier in a control measure provides additional information about the icon being displayed. However, the location of the amplifiers for control measures varies and is dependent on the control measure symbol being displayed. Because the location of amplifiers varies, there is no standardized amplifier placement location for all types of control measures. Chapter 8 provides numerous figures and tables that identify the location of amplifiers for each of the different types of control measures. For multiple entries of the same type or similar information, the field number will be followed by a dash and a number designating the second or more uses. For example, a from-to date-time group may use 16 for the start time and 16-1 for the end time, or an airspace coordination area may use 3 to name an operational name or designation for the airspace coordination area and use 3-1 for the unique alphanumeric designation that identifies the unit establishing the airspace control area. Arabic numerals are normally used when showing the unique designation of units. However, corps echelon units are identified with Roman numerals.

Table 3-6. Description of control measure symbol amplifier fields

Field	*Description*
2	An accepted code that shows the **country indicator**.
3/3-1	A **unique alphanumeric designation** that identifies the unit being displayed. *Note:* When showing unique alphanumeric designations for combat arms regimental units (air defense artillery, armor, aviation, cavalry, field artillery, infantry, and special forces) the following rules apply: **No regimental headquarters**: A dash (-) will be used between the battalion and the regimental designation where there is no regimental headquarters. (Example: A/6-37 for A Battery, 6th Battalion, 37th Field Artillery) **Regimental headquarters**: A slash (/) will be used between the battalion and the regimental designation where there is a regimental headquarters of an active operational unit to show continuity of the units. (Example: F/2/11 for F Troop, 2d Squadron/11th Armored Cavalry Regiment)
	An **operational name/designation** given to a control measure to clearly identify it.
	For targets, this is a **target number** as described in appendix H of ATP 3-60. The target number is comprised of six alphanumeric characters of two letters followed by four numbers (for example, AB1234).
5	Free text **staff comments** for information required by the commander.
6	**Echelon** indicator of the symbol.
7	**Quantity** that identifies the number of items present. For a nuclear event, identifies the actual or estimated **size of the nuclear weapon** used in kilotons (KT) or megatons (MT).
11	The **direction of movement arrow** indicates the direction the symbol is moving or will move. For chemical, biological, radiological or nuclear events, the direction of movement arrow indicates **downwind direction**.
	The **offset location indicator** without the arrow is used to denote precise location of units or to declutter multiple unit locations, except for headquarters.
15	Denotes **enemy** symbol. The letters "ENY" are used when the color red is not an option.
16	An alphanumeric designator for displaying a **date-time group** (for example, DDHHMMSSZMONYYYY) or "O/O" for on order.
16-1	Used with 16 for displaying a **date-time group** for a from-to specified time period.
17	Identifies unique designation for **type of equipment**.
18	Denotes the **location** in latitude and longitude or grid coordinates.
19	Denotes the **altitude**
20/20-1	Denotes the **range**
21/21-1	Denotes the **azimuth**
22	Denotes the width

Table 3-7. Building process for control measure symbols

Step	Description	Example
1	Choose the appropriate control measure symbol.	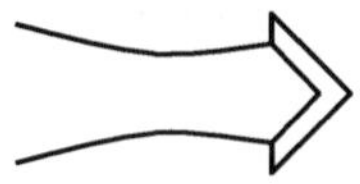
2	Choose the appropriate control measure template that will show the possible amplifiers.	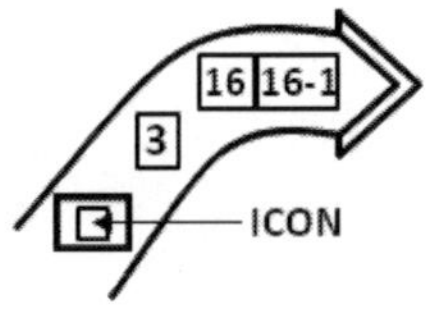
3	Choose the appropriate amplifier information by field.	**3** | **GOLD** 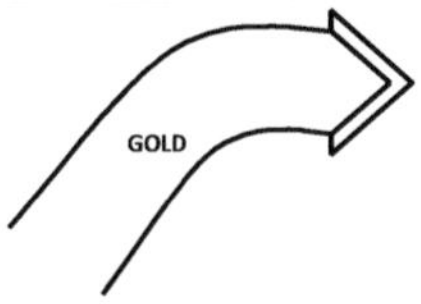
4	Choose the next appropriate amplifier information by field.	**16** **140600MAR2010** **16-1** Not required 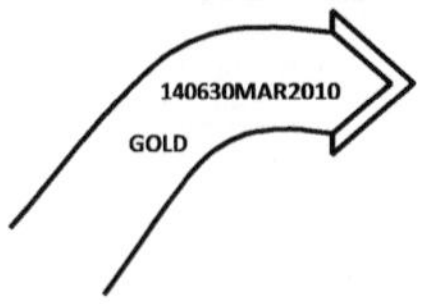
5	Choose the appropriate framed icon.	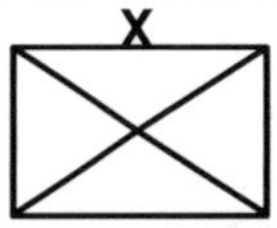 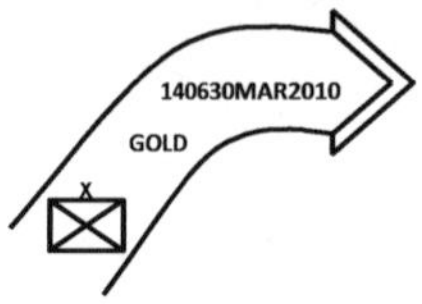

TACTICAL MISSION TASK SYMBOLS

3-21. The tactical mission task symbols are graphic representations of many of the tactical tasks. However, not all tactical tasks have an associated symbol. Tactical task symbols are for use in course of action sketches, synchronization matrixes, and maneuver sketches. They do not replace any part of the operation order. The tactical task symbols should be scaled to fit the map scale and the size of unit represented. Chapter 9 discusses tactical mission task symbols.

This page intentionally left blank.

Chapter 4

Units, Individuals, and Organizations

This chapter discusses symbols for units, individuals, and organizations.

MAIN ICONS FOR UNITS

4-1. A *unit* is a military element whose structure is prescribed by a competent authority, such as a table of organization and equipment; specifically, part of an organization (JP 1-02). Icons in the main sector of the bounding octagon reflect the main function of the symbol. (See table 3-4 on page 3-7). Table 4-1 (on pages 4-1 through 4-8) shows the main icons for units. There are exceptions to the main icon sector. These exceptions are full frame icons. Full frame icons are not limited to the main sector of the bounding octagon. Full-frame icons may reflect the main function of the symbol or may reflect modifying information. Full frame modifiers will be found in the modifier 1 and 2 tables. Table 4-2 (on pages 4-9 through 4-10) shows main icons for named units.

Table 4-1. Main icons for units

Function ***Note***: Unless otherwise noted (Marine Corps, Navy, Joint, or the North Atlantic Treaty Organization (NATO) these functions are applicable to United States Army units.	Icon Note: US Army icons were determined by table of organization and equipment and modified table of organization and equipment descriptions in the Force Management System of the United States Army Force Management Support Agency.	Orientation ***Note***: The bounding octagon is not part of the symbol. It is for spatial reference only.	Example ***Note***: Provides a basic unit symbol without modifiers or amplifiers.
Note: The main icons that are in the shaded rows are full frame icons. Full frame icons exceed the limits of the bounding octagon and are not confined to the area for main icons.			
Administrative ***Note***: No longer used as a United States Army table of organization and equipment or modified table of organization and equipment unit. North Atlantic Treaty Organization (NATO)	ADM	ADM	ADM
Air defense artillery *(radar dome)*			
Air and missile defense ***Note***: Change to MIL-STD-2525D. New symbol to reflect both functions.	MD	MD	MD

Table 4-1. Main icons for units (continued)

Function **Note**: Unless otherwise noted (as Marine Corps, Navy, Joint, or the North Atlantic Treaty Organization [NATO]) these functions are applicable to United States Army units.	*Icon* **Note**: US Army icons were determined by table of organization and equipment and modified table of organization and equipment descriptions in the Force Management System of the United States Army Force Management Support Agency.	*Orientation* **Note**: The bounding octagon is not part of the symbol. It is for spatial reference only.	*Example* **Note**: Provides a basic unit symbol without modifiers or amplifiers.
Note: The main icons that are in the shaded rows are full frame icons. Full frame icons exceed the limits of the bounding octagon and are not confined to the area for main icons.			
Air-naval gunfire liaison company (ANGLICO) **Note**: As a main icon this is a Marine Corps unit and is not a United States Army table of organization and equipment or modified table of organization and equipment unit.			
Anti-armor (anti-tank) *(upside down V)*			
Armored (armor) *(tank track)* **Note**: Armored protection and mobility.			
Armored cavalry **Note**: This main icon is the combination of two other main icons.			
Mechanized (armored) infantry **Note**: This main icon is the combination of two other main icons.			
Army aviation or rotary wing aviation **Note**: This icon is used for army aviation and rotary wing aviation.			
Fixed wing aviation			
Music performance			
Cavalry (reconnaissance) (cavalry bandoleer)			

Table 4-1. Main icons for units (continued)

Function *Note*: Unless otherwise noted (as Marine Corps, Navy, Joint, or the North Atlantic Treaty Organization [NATO]) these functions are applicable to United States Army units.	*Icon* *Note*: US Army icons were determined by table of organization and equipment and modified table of organization and equipment descriptions in the Force Management System of the United States Army Force Management Support Agency.	*Orientation* *Note*: The bounding octagon is not part of the symbol. It is for spatial reference only.	*Example* *Note*: Provides a basic unit symbol without modifiers or amplifiers.
Note: The main icons that are in the shaded rows are full frame icons. Full frame icons exceed the limits of the bounding octagon and are not confined to the area for main icons.			
Chemical (chemical, biological, radiological, and nuclear) *(crossed retorts)*			
Chemical, biological, radiological, nuclear, and high-yield explosives *Note*: This is a change to MIL-STD-2525D. New symbol.	E	E	E
Civil affairs *(abbreviation)*	CA	CA	CA
Civil-military cooperation North Atlantic Treaty Organization (NATO)			
Chaplain (religious support)	REL	REL	REL
Combined arms *(modified cross straps and tank track)*			
Engineer *(bridge)*			
Field artillery *(cannon ball)*			

Table 4-1. Main icons for units (continued)

Function **Note**: Unless otherwise noted (as Marine Corps, Navy, Joint, or the North Atlantic Treaty Organization [NATO]) these functions are applicable to United States Army units.	*Icon* **Note**: US Army icons were determined by table of organization and equipment and modified table of organization and equipment descriptions in the Force Management System of the United States Army Force Management Support Agency.	*Orientation* **Note**: The bounding octagon is not part of the symbol. It is for spatial reference only.	*Example* **Note**: Provides a basic unit symbol without modifiers or amplifiers.
Note: The main icons that are in the shaded rows are full frame icons. Full frame icons exceed the limits of the bounding octagon and are not confined to the area for main icons.			
Finance *(strong box)*			
Hospital (medical treatment facility)			
Infantry *(crossed straps)*			
Information operations (*abbreviation)*	IO	IO	IO
Interpreter or translator **Note**: This is a change to MIL-STD-2525D, new symbol.			
Judge advocate general (*abbreviation)*	JAG	JAG	JAG
Liaison North Atlantic Treaty Organization (NATO)	LO	LO	LO
Maintenance *(double end wrench)*			
Maneuver enhancement			

Table 4-1. Main icons for units (continued)

Function ***Note***: Unless otherwise noted (as Marine Corps, Navy, Joint, or the North Atlantic Treaty Organization [NATO]) these functions are applicable to United States Army units.	***Icon*** ***Note***: US Army icons were determined by table of organization and equipment and modified table of organization and equipment descriptions in the Force Management System of the United States Army Force Management Support Agency.	***Orientation*** ***Note***: The bounding octagon is not part of the symbol. It is for spatial reference only.	***Example*** ***Note***: Provides a basic unit symbol without modifiers or amplifiers.
Note: The main icons that are in the shaded rows are full frame icons. Full frame icons exceed the limits of the bounding octagon and are not confined to the area for main icons.			
Medical *(Geneva cross)*			
Military history *(abbreviation)* ***Note***: This is a change to MIL-STD-2525D. New symbol.	MH	MH	MH
Military intelligence *(abbreviation)*	MI	MI	MI
Military police *(abbreviation)*	MP	MP	MP
Missile ***Note***: This is not a United States Army table of organization and equipment or modified table of organization and equipment unit. North Atlantic Treaty Organization (NATO)			
Missile defense ***Note***: This is a change to MIL-STD-2525D and reflects current air domain icon for missile defense.	MD	MD	MD
Mortar ***Note***: This is not a United States Army table of organization and equipment or modified table of organization and equipment unit. North Atlantic Treaty Organization (NATO)			

Table 4-1. Main icons for units (continued)

Function ***Note***: Unless otherwise noted (as Marine Corps, Navy, Joint, or the North Atlantic Treaty Organization [NATO]) these functions are applicable to United States Army units.	*Icon* ***Note***: US Army icons were determined by table of organization and equipment and modified table of organization and equipment descriptions in the Force Management System of the United States Army Force Management Support Agency.	*Orientation* ***Note***: The bounding octagon is not part of the symbol. It is for spatial reference only.	*Example* ***Note***: Provides a basic unit symbol without modifiers or amplifiers.
Note: The main icons that are in the shaded rows are full frame icons. Full frame icons exceed the limits of the bounding octagon and are not confined to the area for main icons.			
Naval ***Note***: As a main icon this is a Navy unit and is not a United States Army table of organization and equipment or modified table of organization and equipment unit.			
Ordnance *(bursting bomb)*			
Personnel (personnel services or human resources)	PS	PS	PS
Psychological operations ***Note***: Psychological operations have been renamed to military information support operations (MISO) in United States doctrine.			
Public affairs	PA	PA	PA
Quartermaster (key to the stores)			
Ranger ***Note***: This is a change to MIL-STD-2525D. Main icon instead of modifier 1.	RGR	RGR	RGR

Table 4-1. Main icons for units (continued)

Function **Note**: Unless otherwise noted (as Marine Corps, Navy, Joint, or the North Atlantic Treaty Organization [NATO]) these functions are applicable to United States Army units.	***Icon*** **Note**: US Army icons were determined by table of organization and equipment and modified table of organization and equipment descriptions in the Force Management System of the United States Army Force Management Support Agency.	***Orientation*** **Note**: The bounding octagon is not part of the symbol. It is for spatial reference only.	***Example*** **Note**: Provides a basic unit symbol without modifiers or amplifiers.
Note: The main icons that are in the shaded rows are full frame icons. Full frame icons exceed the limits of the bounding octagon and are not confined to the area for main icons.			
Sea, air, land (SEAL) Navy	SEAL	SEAL	SEAL
Security (internal security forces) North Atlantic Treaty Organization (NATO)	SEC	SEC	SEC
Security police	SP	SP	SP
Signal (lightning flash)			
Space *(star)* **Note**: This is a change to MIL-STD-2525D. New symbol.			
Special forces	SF	SF	SF
Special operations forces joint	SOF	SOF	SOF

Table 4-1. Main icons for units (continued)

Function *Note*: Unless otherwise noted (as Marine Corps, Navy, Joint, or the North Atlantic Treaty Organization [NATO]) these functions are applicable to United States Army units.	*Icon* *Note*: US Army icons were determined by table of organization and equipment and modified table of organization and equipment descriptions in the Force Management System of the United States Army Force Management Support Agency.	*Orientation* *Note*: The bounding octagon is not part of the symbol. It is for spatial reference only.	*Example* *Note*: Provides a basic unit symbol without modifiers or amplifiers.
Note: The main icons that are in the shaded rows are full frame icons. Full frame icons exceed the limits of the bounding octagon and are not confined to the area for main icons.			
Special troops *Note*: This is a change to MIL-STD-2525D. New symbol.	ST	ST	ST
Support *Note*: This is a change to MIL-STD 2525D. Also used as a modifier 2.	SPT	SPT	SPT
Surveillance (battlefield surveillance)	▲	▲	▲
Sustainment	SUST	SUST	SUST
Transportation *(wheel)*	⊛	⊛	⊛

Table 4-2. Main icons for named units

Named Unit	*Icon*	*Orientation*	*Example*
Combatant Commands			
United States Africa Command	AFRICOM	AFRICOM	AFRICOM
			XXXXXX AFRICOM United States Africa Command
United States Central Command	CENTCOM	CENTCOM	CENTCOM
United States European Command	EUCOM	EUCOM	EUCOM
United States Northern Command	NORTHCOM	NORTHCOM	NORTHCOM
United States Pacific Command	PACOM	PACOM	PACOM
United States Southern Command	SOUTHCOM	SOUTHCOM	SOUTHCOM
Army elements for combatant commands			
United States Army, Africa Command	USARAF	USARAF	USARAF

Table 4-2. Main icons for named units (continued)

Named Unit	*Icon*	*Orientation*	*Example*
United States Army, Central Command	USARCENT	USARCENT	USARCENT
			XXXX USARCENT United States Army Central Command
United States Army, Europe Command	USAREUR	USAREUR	USAREUR
United States Army, North	USARNORTH	USARNORTH	USARNORTH
United States Army, Pacific Command	USARPAC	USARPAC	USARPAC
United States Army, Southern Command	USARSO	USARSO	USARSO
North Atlantic Treaty Organizations			
Allied Command Operations	ACO	ACO	ACO

Sector 1 Modifiers for Units

4-2. Table 4-3 (on pages 4-11 through 4-24) shows sector 1 modifiers for unit capabilities. These modifiers show the specific functions that the unit is organized and equipped to perform.

Table 4-3. Sector 1 modifiers for units

Function	*Icon* *Note*: The icon has been enlarged for better visibility and is not proportional to the orientation or example.	*Orientation*	*Example*
Aviation			
Assault	ASLT	ASLT	ASLT Assault aviation unit
Attack	A	A	A Attack helicopter unit
Search and rescue	SAR	SAR	SAR Aviation search and rescue unit
Unmanned aerial system			Aviation unmanned aerial system unit
Utility	U	U	U Utility helicopter unit

Table 4-3. Sector 1 modifiers for units (continued)

Function	*Icon* *Note*: The icon has been enlarged for better visibility and is not proportional to the orientation or example.	*Orientation*	*Example*
Chemical (chemical, biological, radiological, and nuclear)			
Biological	B	B	B
Chemical	C	C	C
Decontamination	D	D	D
Nuclear	N	N	N
Radiological	R	R	R
Smoke (obscuration)	S	S	S
Engineer			
Bridging			Bridge unit
Combat	CBT	CBT	CBT Combat engineer unit

Table 4-3. Sector 1 modifiers for units (continued)

Function	*Icon* ***Note***: The icon has been enlarged for better visibility and is not proportional to the orientation or example.	*Orientation*	*Example*
Construction North Atlantic Treaty Organization (NATO)	CON	CON	CON Engineer construction unit
Diving			Engineer diving unit
Drilling			Drilling unit
Naval construction (Seabee) ***Note***: Not a US Army unit, but is the modifier 1 to reflect a United States Navy engineer unit.			
General ***Note***: Replaces construction for US Army and is a change from MIL-STD-2525D.	GEN	GEN	GEN General engineer unit
Topographic			Engineer topographic unit

Table 4-3. Sector 1 modifiers for units (continued)

Function	***Icon*** *Note*: The icon has been enlarged for better visibility and is not proportional to the orientation or example.	***Orientation***	***Example***
Field Artillery			
Fire direction center	FDC	FDC	FDC Fire direction center
Meteorological	MET	MET	MET Meteorological unit
Multiple rocket launcher			Multiple rocket launcher unit
Single rocket launcher			Single rocket launcher unit
Sound ranging *Note*: The United States Army no longer has sound ranging units.	SDR	SDR	SDR Sound ranging unit
Survey			Survey unit
Target acquisition	TA	TA	TA Target acquisition unit

Table 4-3. Sector 1 modifiers for units (continued)

Function	*Icon* **Note**: The icon has been enlarged for better visibility and is not proportional to the orientation or example.	*Orientation*	*Example*
Infantry			
Mortar			
Sniper			
Weapons **Note**: This is a change to MIL-STD-2525D.	W	W	W
Maintenance			
Electro-optical	EO	EO	EO Electro-optical Maintenance Unit
Medical **Note**: Modifiers for medical units are offset to the right to avoid overlapping with the main icon.			
North Atlantic Treaty Organization (NATO) medical role 1	1	1	1
North Atlantic Treaty Organization (NATO) medical role 2	2	2	2

Table 4-3. Sector 1 modifiers for units (continued)

Function	*Icon* *Note*: The icon has been enlarged for better visibility and is not proportional to the orientation or example.	*Orientation*	*Example*
North Atlantic Treaty Organization (NATO) medical role 3	3	3	3
North Atlantic Treaty Organization (NATO) medical role 4	4	4	4
Military Intelligence			
Counterintelligence	CI	CI	CI MI
Electronic warfare	EW	EW	EW MI
Sensor	✦	✦	✦ MI
Signals intelligence			MI
Tactical exploitation	TE	TE	TE MI
Military Police			
Criminal investigation division *Note*: This is a change to MIL-STD-2525D.	CID	CID	CID MP
Detention	DET	DET	DET MP

Table 4-3. Sector 1 modifiers for units (continued)

Function	*Icon* *Note*: The icon has been enlarged for better visibility and is not proportional to the orientation or example.	*Orientation*	*Example*
Dog (military working dog)	DOG	DOG	DOG MP
Special weapons and tactics	SWAT	SWAT	SWAT MP
Ordnance			
Explosive ordnance disposal	EOD	EOD	EOD
Quartermaster			
Mortuary affairs *Note*: Change to MIL-STD-2525D. Moved from a main icon to a modifier 1.			
Pipeline			
Water			
Personnel (Personnel Services and Human Resources)			
Postal			PS

Table 4-3. Sector 1 modifiers for units (continued)

Function	*Icon* ***Note***: The icon has been enlarged for better visibility and is not proportional to the orientation or example.	*Orientation*	*Example*
Security			
Border	BOR	BOR	BOR
			BOR SEC Border security unit
Signal			
Digital ***Note***: This is a change to MIL-STD-2525D.	DIG	DIG	DIG
Enhanced	ENH	ENH	ENH
Mobile subscriber equipment	MSE	MSE	MSE
Network or network operations ***Note***: This is a change to MIL-STD-2525D.	NET	NET	NET
Tactical satellite			Signal tactical satellite unit

Table 4-3. Sector 1 modifiers for units (continued)

Function	*Icon* *Note*: The icon has been enlarged for better visibility and is not proportional to the orientation or example.	*Orientation*	*Example*
Video imagery (combat camera)			Signal combat camera unit
Cyberspace			
Command Post Node	CPN	CPN	CPN
Joint Node Network	JNN	JNN	JNN
Retransmission	RTNS	RTNS	RTNS
Transportation			
Airfield, aerial port of debarkation, or aerial port of embarkation *Note*: This is a change to MIL-STD-2525D.			Airfield unit
Movement control center	MCC	MCC	MCC
Railway or railhead *Note*: This is a change to MIL-STD-2525D.			Railway Unit
Seaport, seaport of debarkation, or seaport of embarkation *Note*: Uses same symbol as Naval. This is a change to MIL-STD-2525D.			Seaport Unit
Watercraft *Note*: This is a change to MIL-STD-2525D.			

Table 4-3. Sector 1 modifiers for units (continued)

Function	*Icon* *Note*: The icon has been enlarged for better visibility and is not proportional to the orientation or example.	*Orientation*	*Example*
Not specific to one branch or function			
Area	AREA	AREA	AREA SPT Area support unit
Armored (protection)			 Stryker unit (infantry in armor protected wheeled vehicles)
Close protection North Atlantic Treaty Organization (NATO)	CLP	CLP	CLP
Command and control	C2	C2	C2 C2
Cross cultural communication NATO	CCC	CCC	CCC
Crowd and riot control NATO	CRC	CRC	CRC

Table 4-3. Sector 1 modifiers for units (continued)

Function	*Icon* *Note:* The icon has been enlarged for better visibility and is not proportional to the orientation or example.	*Orientation*	*Example*
Direct communications North Atlantic Treaty Organization (NATO)			
Echelon of Support *Note*: This is a change to MIL-STD-2525D. Eliminates the current modifiers that are outside the bounding octagon.			
Theater *Note*: This is a change to MIL-STD-2525D. Moved from a main icon to a modifier 2.	XXXXX	XXXXX	XXXXX Theater aviation unit
Army or Theater Army *Note*: This is a change to MIL-STD-2525D. Moved from a main icon to a modifier 2.	XXXX	XXXX	XXXX
			XXXX MI
Corps *Note*: This is a change to MIL-STD-2525D. Moved from a main icon to a modifier 2.	XXX	XXX	XXX Corps echelon support unit
			XXX ST Corps special troops unit

Table 4-3. Sector 1 modifiers for units (continued)

Function	*Icon* ***Note***: The icon has been enlarged for better visibility and is not proportional to the orientation or example.	*Orientation*	*Example*
Division ***Note***: This is a change to MIL-STD-2525D. Marine Corps	XX	XX	XX
			XX Division reconnaissance unit
Brigade ***Note***: This is a change to MIL-STD-2525D.	X	X	X SPT Brigade support unit
			II X SPT Brigade support battalion
Force Marine Corps	F	F	F
			F
Forward	FWD	FWD	FWD SPT Forward support unit

Table 4-3. Sector 1 modifiers for units (continued)

Function	*Icon* *Note*: The icon has been enlarged for better visibility and is not proportional to the orientation or example.	*Orientation*	*Example*
Headquarters or headquarters element Moved from a main icon to a modifier 1.			XXX X
Maintenance			Aviation maintenance unit
Medical evacuation			Aviation medical evacuation unit
Mobile advisor and support North Atlantic Treaty Organization (NATO)			
Mobility support	MS	MS	MS

Table 4-3. Sector 1 modifiers for units (continued)

Function	*Icon* *Note*: The icon has been enlarged for better visibility and is not proportional to the orientation or example.	*Orientation*	*Example*
Multinational	MN	MN	MN
Multinational specialized unit North Atlantic treaty Organization (NATO)	MSU	MSU	MSU
Operations	OPS	OPS	OPS
Petroleum, oil, and lubricants (POL)			
Radar			
			Field artillery radar unit

Sector 2 Modifiers for Units

4-3. Tables 4-4 through 4-6 (on pages 4-25 through 4-32) show sector 2 icons. Sector 2 modifiers reflect the mobility (from table 4-3); size, range, or altitude of unit equipment; or additional capability of units.

Table 4-4. Sector 2 modifiers for unit mobility

Function	*Icon*	*Orientation*	*Example*
Air assault ***Note***: This is a change to MIL-STD-2525D. It was moved from sector 1 to sector 2 for consistency with other mobility modifiers.			 Air assault field artillery unit
Air assault with organic lift			
Airborne			Airborne unit Infantry airborne unit
Amphibious ***Note***: This is a change to MIL-STD-2525D. It was moved from main sector to sector 2 for consistency for mobility modifiers.			LS Amphibious landing support unit
Arctic (sled)			
Bicycle-equipped			

Table 4-4. Sector 2 modifiers for unit mobility (continued)

Function	*Icon*	*Orientation*	*Example*
Mountain			Mountain unit
			Mountain infantry unit
Motorized ***Note***: Not used for Army units.			
Over-snow (prime mover)			
Pack animal			
Railroad			
Riverine or floating			
Ski			
Towed			Towed unit
			Field artillery towed unit

Table 4-4. Sector 2 modifiers for unit mobility (continued)

Function	Icon	Orientation	Example
Tracked or self-propelled			
Vertical take-off and landing	VTOL	VTOL	VTOL
Wheeled	OOO	OOO	OOO Wheeled field artillery multiple launch rocket system unit

Table 4-5. Sector 2 modifiers for unit equipment size, range, or altitude

Function	*Icon*	*Orientation*	*Example*
Close range	CR	CR	CR CR
Heavy	H	H	H H

Table 4-5. Sector 2 modifiers for unit equipment size, range, or altitude (continued)

Function	*Icon*	*Orientation*	*Example*
High altitude	HA	HA	HA
			MD HA
Light	L	L	L
			L
Long range	LR	LR	LR
			LR
Low altitude	LA	LA	LA
			LA
Low to medium altitude	LMA	LMA	LMA
			LMA

Table 4-5. Sector 2 modifiers for unit equipment size, range, or altitude (continued)

Function	*Icon*	*Orientation*	*Example*
Medium	M	M	M
			MR
Medium altitude	MA	MA	MA
			MA
Medium range	MR	MR	MR
			MR
Medium to high altitude	MHA	MHA	MHA
			MHA
Short range	SR	SR	SR
			SR
Very heavy ***Note***: Applies to field artillery only. This is a change to MIL-STD-2525D.	VH	VH	VH
			VH

Table 4-6. Sector 2 modifiers for additional capability

Function	*Icon*	*Orientation*	*Example*
Aviation			
Launcher (unmanned aerial system)			
Recovery (unmanned aerial system)			
Chemical (including chemical, biological, radiological, and nuclear)			
Decontamination	D	D	D
			C D Chemical decontamination unit
Laboratory	LAB	LAB	LAB
Medical			
Casualty staging North Atlantic Treaty Organization (NATO)	CS	CS	CS
Dental	D	D	D
Psychological	P	P	P Psychological medical unit
Veterinary	V	V	V

Table 4-6. Sector 2 modifiers for additional capability (continued)

Function	*Icon*	*Orientation*	*Example*
		Military Intelligence	
Analysis			EW MI
Direction finding			EW MI
Electronic ranging			EW MI
Intercept			EW MI
Jamming			EW MI
Search			EW MI
		Quartermaster	
Supply ***Notes***: This is a change to MIL-STD-2525D. Moved from a main icon to a modifier 2. Classes of United States and North Atlantic Treaty Organization supply are found in the sustainment function in the control measure symbols.			Supply unit
		Transportation	
Intermodal			

Table 4-6. Sector 2 modifiers for additional capability (continued)

Function	Icon	Orientation	Example
Not specific to one branch or function			
Recovery *Note*: Uses the same icon as maintenance, but serves as recovery in the modifier 2 position.			
Strategic	STR	STR	STR
Support	SPT	SPT	SPT
Tactical	TAC	TAC	TAC

ECHELON AMPLIFIERS (FIELD 6)

4-4. An *echelon* is a separate level of command (JP 1-02). In addition, there is also a separate echelon known as a command. A command is a unit or units, an organization, or an area under the command of one individual. It does not correspond to any of the other echelons. Figure 4-1 shows the template for an echelon amplifier. The height of the echelon amplifier is one-fourth of the size of the height of the frame. Table 4-7 (on pages 4-33 through 4-34) shows the field 6 amplifiers for Army echelons and commands.

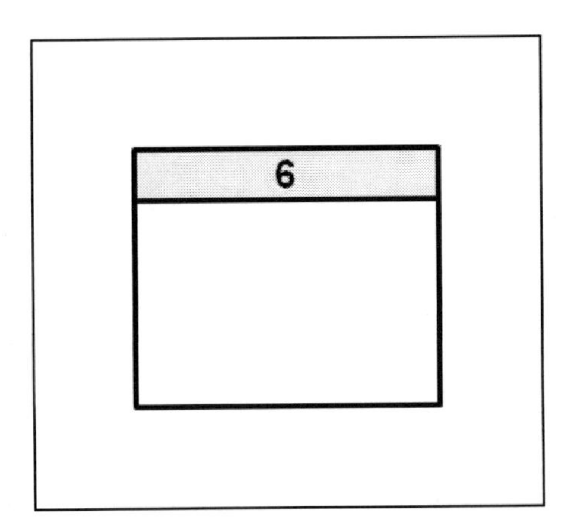

Figure 4-1.Template for an echelon amplifier

Table 4-7. Echelon amplifiers

Echelon	***Amplifier***	***Example of amplifier with friendly unit frame***
Team or crew ***Note***: This is the smallest echelon and should not be confused with company team and brigade combat team in the next paragraph.	Ø	Ø
Squad	●	●
Section	●●	●●
Platoon or detachment	●●●	●●●
Company, battery, or troop	I	I
Battalion or squadron	II	II
Regiment or group	III	III
Brigade	X	X
Division	XX	XX
Corps	XXX	XXX

Table 4-7. Echelon amplifiers (continued)

Echelon	*Amplifier*	*Example of amplifier with friendly unit frame*
Army	XXXX	XXXX
Army group	XXXXX	XXXXX
Theater	XXXXXX	XXXXXX
Nonechelon	***Amplifier***	***Example of amplifier with friendly unit frame***
Command	++	++

TASK FORCE AND TEAM AMPLIFIERS (FIELD 8)

4-5. This amplifier is used with a task force, company team, or brigade combat team. A task force is a temporary grouping of units under one commander formed to carry out a specific operation or mission, or a semipermanent organization of units under one commander formed to carry out a continuing specified task. Definitions for company team, and brigade combat team can be found in Chapter 1. Figure 4-2 shows the template for a task force or team amplifier. Table 4-8 shows the task force and team amplifier.

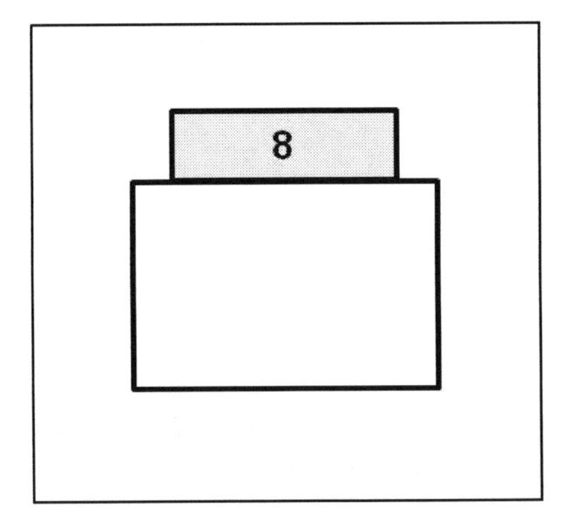

Figure 4-2.Template for task force or team amplifier

Table 4-8. Task force and team amplifier

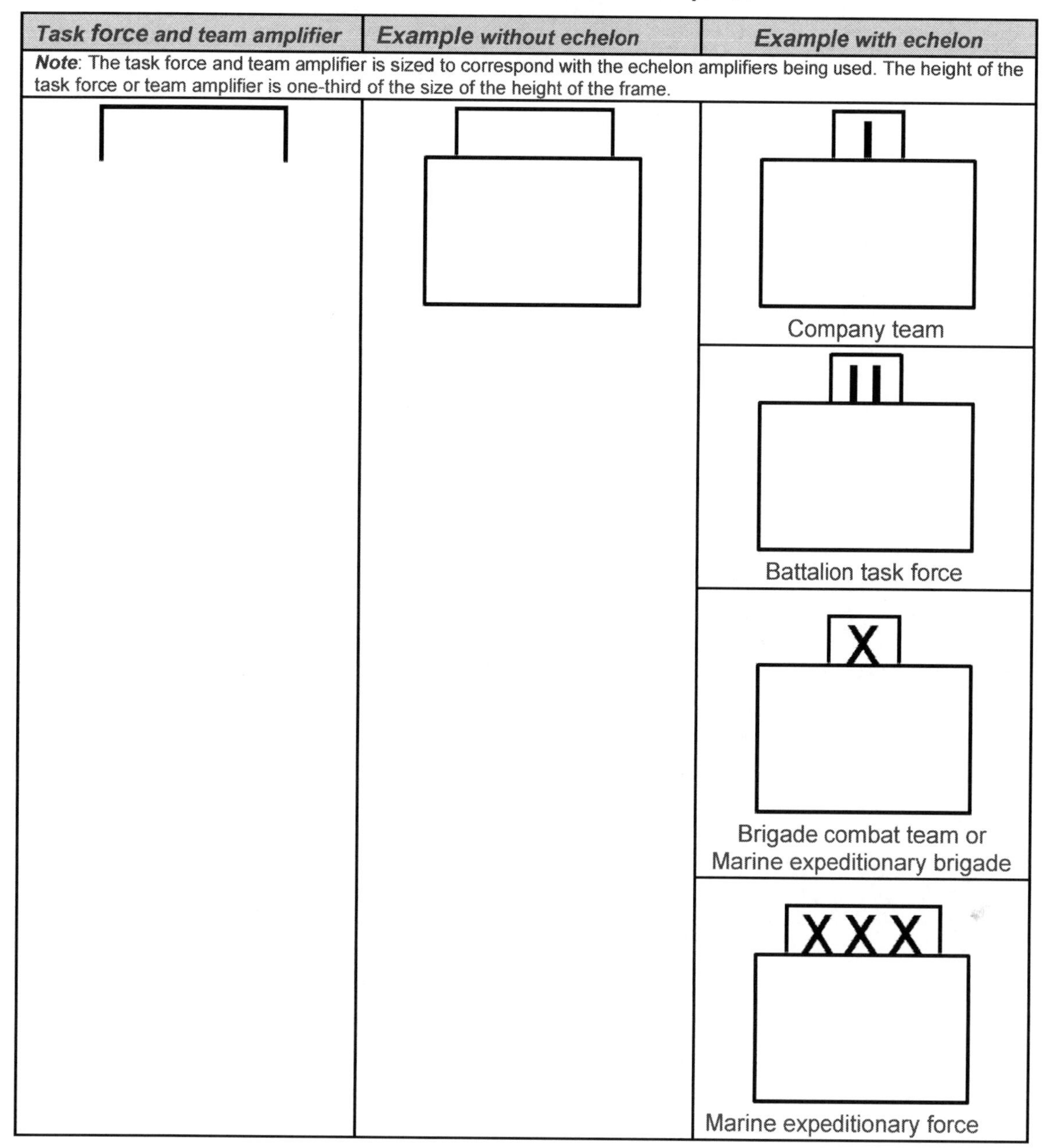

Task force and team amplifier	Example without echelon	Example with echelon
Note: The task force and team amplifier is sized to correspond with the echelon amplifiers being used. The height of the task force or team amplifier is one-third of the size of the height of the frame.		
		Company team
		Battalion task force
		Brigade combat team or Marine expeditionary brigade
		Marine expeditionary force

Attached and Detached Amplifiers (Field 1)

4-6. This amplifier is used at brigade echelon and below. Use a plus + symbol when attaching one or more subelements of a similar function to a headquarters. Use a minus symbol – when deleting one or more subelements of a similar function to a headquarters. Figure 4-3 shows a template for attached and detached amplifiers. Table 4-9 provides an explanation of attached and detached modifiers. (See FM 6-0 for additional information on attachment and detachment.)

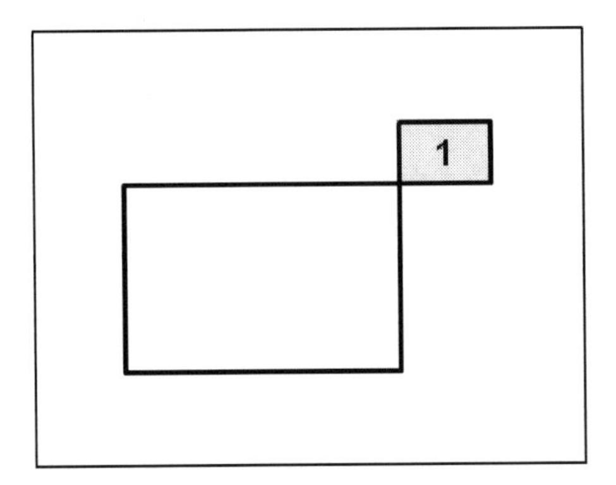

Figure 4-3.Template for attached and detached amplifiers

Table 4-9. Attached and detached amplifiers

Description	*Amplifier*	*Example of amplifier with friendly unit frame*
Attached	+	+
Detached	–	–

Command Posts and Command Group Amplifiers (Field 14)

4-7. A command post is a unit headquarters where the commander and staff perform their activities. A command group is the commander and selected staff members who accompany commanders and enable them to exercise mission command away from a command post. The headquarters staff indicator (field 10) is always used in conjunction with the command post and command group amplifiers. Figure 4-4 shows the template for the command post and command group amplifier. Table 4-10 (on pages 4-37 through 4-38) shows the command post and command group amplifiers (field 14).

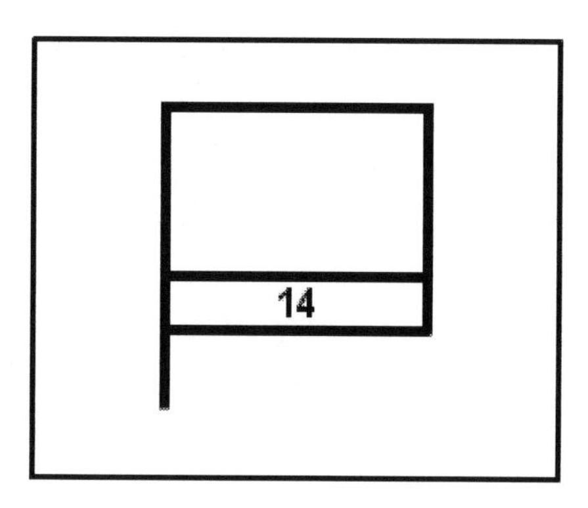

Figure 4-4.Template for command post and command group amplifier

Table 4-10. Command post and command group amplifiers

Description	*Amplifier*	*Example of amplifier with friendly unit frame*
Combat trains command post	CTCP	CTCP
Command group	CMD	CMD
Early entry command post	EECP	EECP
Emergency operations center	EOC	EOC

Table 4-10. Command post and command group amplifiers (continued)

Description	*Amplifier*	*Example of amplifier with friendly unit frame*
Field trains command post	**FTCP**	FTCP
Main command post	**MAIN**	MAIN
Tactical command post	**TAC**	TAC

Combat Effectiveness Amplifiers (Field 12)

4-8. Combat effectiveness is the ability of a unit to perform its mission. Factors such as ammunition, personnel, fuel status, and weapon systems are evaluated and rated. The ratings are—

- Fully operational - green (85 percent or greater).
- Substantially operational – amber (70 to 84 percent).
- Marginally operational – red (50 to 69 percent).
- Not operational – black (less than 50 percent).

Field 12 is used to display the level of combat effectiveness of the unit or equipment symbol. Figure 4-5 shows the template for the combat effectiveness amplifier. Table 4-11 shows the combat effectiveness amplifiers (field 12). (See chapter 10 for a discussion of combat effectiveness icons used with task organization composition symbols.)

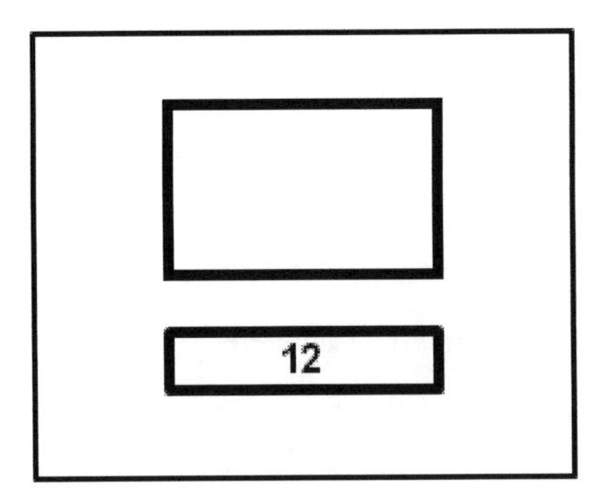

Figure 4-5.Template for combat effectiveness amplifier

Table 4-11. Combat effectiveness amplifiers

Description	*Amplifier*	*Example of amplifier with friendly unit frame*
Fully operational		
Substantially operational		
Marginally operational		
Not operational		

Offset Locator Indicator Amplifier (Field 11) and Headquarters Staff Offset Locator Indicator Amplifier (Field 10)

4-9. The center of mass of the unit symbol indicates the general vicinity of the center of mass of the unit. To indicate precise location or reduce clutter in an area with multiple units, a line (without an arrow) extends from the center of the bottom of the frame to the unit location displayed as field 11. The line may be extended or bent as needed. If a group of units (or installations) other than a headquarters is at one location, the grouping of the symbols may be enclosed with a bracket and the exact location indicated by a line from the centre of the bracket. Figure 4-6 shows examples of how to use the offset locator indicator.

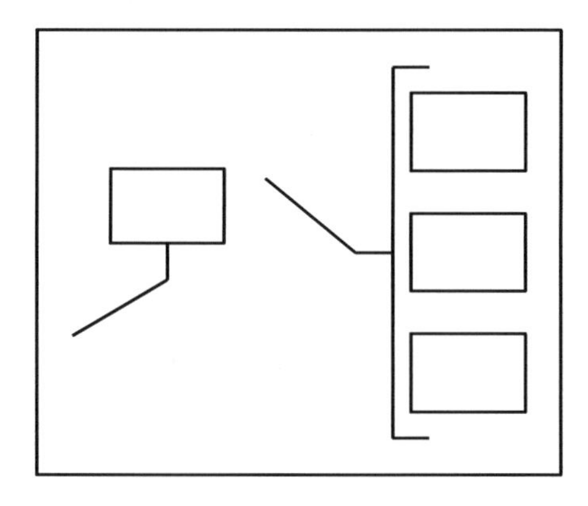

Figure 4-6. Offset locator indicators

4-10. To indicate a precise location or reduce the clutter of headquarters unit symbols, a staff extends from the bottom left hand corner to the headquarters location displayed as field 10. This staff may be bent or extended as needed. If several headquarters are at one location, more than one headquarters can be on a single staff. The highest echelon headquarters is placed on top, followed by the next echelons in descending order. Figure 4-7 shows examples of how the headquarters offset locator indicator is used. Table 4-12 gives examples of unit symbols.

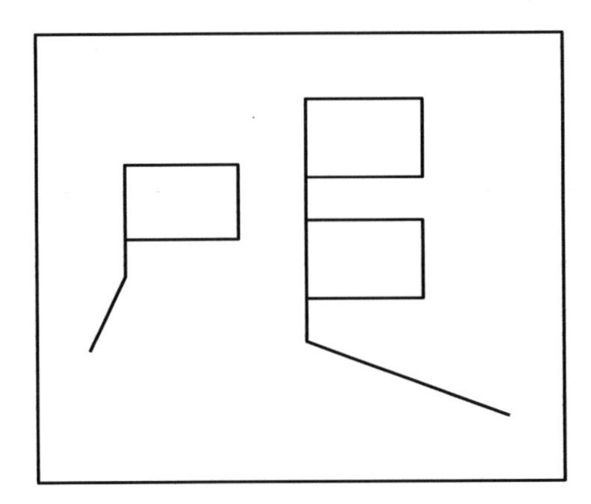

Figure 4-7. Headquarters offset locator indicators

Table 4-12. Examples of unit symbols

Description of Unit Unique Designation	Symbol
Special Troops Battalion, 2d Brigade Combat Team, 10th Mountain Division, III Corps	II X ST 2BCT/10 III
299 Support Battalion (Brigade), 2d Brigade Combat Team, 1st Infantry Division	II X SPT 299/2BCT 1ID
4th Battalion, 23d Infantry Regiment (Stryker), 2d Brigade, 7th Infantry Division	II 4-23 2/7ID
4th Platoon (Mobile Gun System), A Company, 4thBattalion, 23d Infantry Regiment (Stryker), 2d Brigade, 7th Infantry Division	●●● 4/A/4-23 2/7ID MGS
Division Artillery, 101st Airborne Division (Air Assault)	X XX 101AAD
6th Battalion, 37th Field Artillery Regiment, 2d Infantry Division	II 6-37 2ID
181 Transportation Battalion, 3d Expeditionary Sustainment Command	II 181 3ESC
67th Forward Support Company, 2d Battalion, 20th Field Artillery	I FWD SPT 67 2-20FA
C Battery, 3d Battalion, 319 Field Artillery (Air Assault)	I 3-319
III Marine Expeditionary Force	XXX III

MAIN ICONS FOR INDIVIDUALS AND ORGANIZATIONS (CIVILIAN)

4-11. These symbols are a recognition of the larger role of military forces beyond war fighting. Military forces are also engaged in stability activities and support to civil authorities around the world. Symbols for individuals and organizations represent civilians and normally do not have prescribed structures. Organization symbols can reflect civic, ethnic, religious, social, or other groupings. Icons in the main sector reflect the main function of the icon. Table 4-13 (on pages 4-42 through 4-44) shows the main icons for civilian individuals and organizations.

Table 4-13. Main icons for civilian individuals and organizations

Function	*Icon*	*Example*
Horizontal bounding octagon		
Vertical bounding octagon		
Border patrol		
Bureau of Alcohol, Tobacco, Firearms, and Explosives (ATF)	ATF	ATF
Coast Guard		
Customs service		
Department of Justice		
Drug Enforcement Agency (DEA)	DEA	DEA
Emergency operations (management)		

Table 4-13. Main icons for civilian individuals and organizations (continued)

Function	Icon	Example
Environmental protection		
Federal Bureau of Investigation (FBI)	FBI	FBI
Fire department		
Governmental	GO	GO
Internal security force	ISF	ISF
Law enforcement		
Nongovernmental	NGO	NGO
Pirates		
Police department		
Prisons		
Spy	SPY	SPY
Transportation Security Agency (TSA)	TSA	TSA

Table 4-13. Main icons for civilian individuals and organizations (continued)

Function	*Icon*	*Example*
United States Marshall Service		
United States Secret Service	USSS	USSS
Persons and organizations or groups		
Unspecified individual Change: Unspecified individual icon uses the vertical bounding octagon.		
Unspecified organization		
Criminal activities victim Change: Criminal activities victim icon uses the vertical bounding octagon.		
Criminal activities victims		
Attempted criminal activities victim Change: Attempted criminal activities victim icon uses the vertical bounding octagon.		
Attempted criminal activities victims		

SECTOR 1 MODIFIERS FOR INDIVIDUALS AND ORGANIZATIONS

4-12. Table 4-14 (on pages 4-45 through 4-48) shows sector 1 modifiers. Sector 1 modifiers reflect the function of civilian individuals or organizations.

Table 4-14. Sector 1 modifiers for civilian individuals and organizations

General groupings	*Characteristic*	*Modifier*	*Example of modifier with friendly unit frame:* *Note*: This does not imply that individuals and organizations are friendly, but only servers as a single frame reference for the symbol. / *Example of most common usage:*
Horizontal bounding octagon			
Vertical bounding octagon			
Types of killing victims	Assassinated	AS	AS
			A S Assassination victim
	Executed	EX	EX
			E X Execution victim
	Murdered	MU	MU
			M U Murder Victim

Table 4-14. Sector 1 modifiers for civilian individuals and organizations (continued)

General groupings	*Characteristic*	*Modifier*	*Example of modifier with friendly unit frame:* *Note*: This does not imply that individuals and organizations are friendly, but only servers as a single frame reference for the symbol. / *Example of most common usage:*
Types of criminal activities victims	Hijacked	H	H
			H
	Kidnapped	K	K
			K
	Piracy	PI	PI
			PI Piracy victims
	Rape	RA	RA
			RA Rape victim

Table 4-14. Sector 1 modifiers for civilian individuals and organizations (continued)

General groupings	Characteristic	Modifier	Example of modifier with friendly unit frame: *Note*: This does not imply that individuals and organizations are friendly, but only servers as a single frame reference for the symbol. / Example of most common usage:
Types of civilian individuals and organizations	Displaced persons, refugees, and evacuees	DPRE	DPRE
			DPRE
	Foreign fighters	FF	FF
			FF
	Gang	GANG	GANG
			GANG
	Leader	LDR	LDR
			L D R Leader
	Religious	REL	REL
			REL
	Speaker	S P K	S P K
			S P K
	Targeted	TGT	TGT

Table 4-14. Sector 1 modifiers for civilian individuals and organizations (continued)

General groupings	*Characteristic*	*Modifier*	*Example of modifier with friendly unit frame:* **Note**: This does not imply that individuals and organizations are friendly, but only servers as a single frame reference for the symbol. / *Example of most common usage:*
Types of civilian individuals and organizations	Targeted	**TGT**	TGT
	Terrorist	**TER**	TER
			TER
Types of recruitment	Unwilling or coerced	**CR**	CR
			CR
	Willing	**WR**	WR
			WR

Sector 2 Modifiers for Individuals and Organizations

4-13. Table 4-15 shows sector 2 modifiers for individuals and organizations. Sector 2 modifiers reflect the nature of the relationship of civilian individuals or organizations. Table 4-16 (on page 4-50) shows examples of symbols for civilian individuals and organizations.

Table 4-15. Sector 2 modifiers for civilian individuals and organizations

Characteristic	***Modifier***	***Example of modifier with friendly unit frame:*** ***Note:*** This does not imply that individuals and organizations are friendly, but only servers as a single frame reference for the symbol. / ***Example of most common usage***
Horizontal bounding octagon		
Vertical bounding octagon		
Types of recruitment		
Coerced	**CR**	**CR**
		FF **CR**
Willing	**WR**	**WR**
		FF **WR**
Leader	L D R	L D R
		T E R L D R

Table 4-16. Examples of symbols for civilian individuals and organizations

Description	*Symbol*
Terrorist	TER
Murdered criminal activities victim	MU
Attempted rape criminal activities victim	RA

Chapter 5
Equipment

This chapter discusses main icons, sector 1 modifiers, and mobility indicator amplifiers for equipment.

MAIN ICONS FOR EQUIPMENT

5-1. Equipment is all nonexpendable items needed to outfit or equip an individual or organization. Equipment symbols can be used with or without frames. When frames are not used, then standard identity colors must be used. Icons in the main sector reflect the main function of the symbol. Equipment can use either the horizontal or vertical bounding octagon depending on the icon. Table 5-1 (on pages 5-1 through 5-10) shows the main icons for equipment.

Table 5-1. Main icons for equipment

Description	*Icon or symbol without frame*	*Orientation within the bounding octagon*	*Example of modifier with friendly equipment frame*
Note: Systems that use these indicators are shown in the second line of the entry in order of light, low altitude, or short-range; as medium, medium altitude, or medium-range, and a heavy, high-altitude, or long-range.			
Weapon systems			
Note. Weapon systems, missile launchers, and nonlethal weapons use the vertical bounding octagon and a unique system for indicating size, altitude, or range. Weapons size is indicated by a horizontal line(s) perpendicular to the weapon icon. If an equipment symbol has no lines, it is a basic equipment symbol. Adding one line designates it as light, low altitude, or short-range. Adding two lines designates it as medium, medium altitude, or medium-range. Finally, adding three lines designates it as heavy, high altitude, or long-range. If a weapon system is designated as greater than heavy, high altitude, or long-range, then a heavy, high-altitude, or long-range indicator is used.			
Unspecified weapon			
Flame thrower			
Grenade launcher			

Table 5-1. Main icons for equipment (continued)

Description	*Icon or symbol without frame*	*Orientation within the bounding octagon*	*Example of modifier with friendly equipment frame*
Guns			
Air defense gun ***Note***: The use of the air defense dome similar to the unit icon at the base of the shaft indicates that it is primarily an air defense weapon.			
Antitank gun ***Note***: The use of the inverted V similar to the unit icon at the base of the shaft indicates that it is primarily an antitank weapon.			
Direct fire gun			
Recoilless gun			

Table 5-1. Main icons for equipment (continued)

Description	*Icon or symbol without frame*	*Orientation within the bounding octagon*	*Example of modifier with friendly equipment frame*
Howitzer			
Howitzer ***Note***: The use of the circle similar to the unit icon for field artillery at the base of the shaft indicates that it is primarily a high trajectory.			
	120 millimeters or less	Greater than 120 millimeters but less than 160 millimeters	Greater than 160 millimeters but less than 210 millimeters
	Greater than 210 millimeters very heavy ***Note***: This icon is historical and is not available in computer-based military symbols software.		
Machine gun			
Machine gun			
Missile launchers			
Missile launcher ***Note***: The use of the dome covering most or the entire shaft similar to the unit icon indicates that it is a missile launcher.			

Table 5-1. Main icons for equipment (continued)

Description	*Icon or symbol without frame*	*Orientation within the bounding octagon*	*Example of modifier with friendly equipment frame*
Air defense missile launcher or surface-to-air missile launcher			
		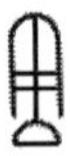	
Antitank missile launcher			
			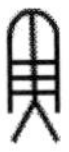
Surface-to-surface missile launcher			
			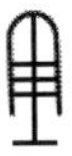
Mortars			
Mortar		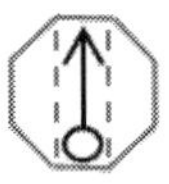	
	 60 millimeters or less	 Greater than 60 millimeters but less than 107 millimeters	 Greater than 107 millimeters

Table 5-1. Main icons for equipment (continued)

Description	*Icon or symbol without frame*	*Orientation within the bounding octagon*	*Example of modifier with friendly equipment frame*
Rifles			
Rifle			
	Single shot rifle	Semi-sutomatic rifle	Automatic rifle
Rockets			
Single rocket launcher ***Note***: The use of the double inverted V's similar to the multiple rocket launcher unit icon indicates that it is a rocket launcher.			
Multiple rocket launcher			
Antitank rocket launcher			

Table 5-1. Main icons for equipment (continued)

Description	*Icon or symbol without frame*	*Orientation within the bounding octagon*	*Example of modifier with friendly equipment frame*
Nonlethal weapons			
Nonlethal weapon			
Taser			
Water cannon			
Vehicles			
Note: Vehicle systems use a unique system for indicating size or range. Vehicle size is indicated by either horizontal or vertical line(s) within the icon depending on the orientation of the symbol. If an equipment symbol has no lines, it is a basic equipment symbol. Adding one line designates it as light or short-range. Adding two lines designates it as medium or medium-range. Finally, adding three lines designates it as heavy or long-range. Armored fighting vehicles, armored personnel carriers, earthmovers, and tanks use the vertical bounding octagon. All remaining equipment icons use the horizontal bounding octagon.			
Armored protected			
Armored fighting vehicle			
Armored personnel carrier			
Armored protected vehicle			
Tank			

Table 5-1. Main icons for equipment (continued)

Description	*Icon or symbol without frame*	*Orientation within the bounding octagon*	*Example of modifier with friendly equipment frame*
Aircraft ***Note***: These are aircraft on the ground. Aircraft in flight use the air domain frame. Change: This brings the land equipment icons in line with air domain icons.			
Helicopter (rotary wing aircraft)			
Fixed wing aircraft			
Unmanned aircraft			
Train Cars ***Note***: The addition of a mobility modifier to the icon is the key to identification of the symbol.			
Train locomotive			
Railcar or boxcar			
Railcar or flatcar			

Table 5-1. Main icons for equipment (continued)

Description	***Icon or symbol without frame***	***Orientation within the bounding octagon***	***Example of modifier with friendly equipment frame***
Wheeled ***Note:*** The addition of a mobility modifier to the icon is the key to identification of the symbol.			
Utility (personnel or cargo carrying) vehicles (without mobility modifier)			
Limited cross-country utility vehicle (truck)			
Cross-country utility vehicle (truck)			
Semi-trailer truck ***Note:*** Semi-trailer truck has a unique mobility modifier.			
Engineer equipment			
Bridge			
Fixed bridge			

Table 5-1. Main icons for equipment (continued)

Description	*Icon or symbol without frame*	*Orientation within the bounding octagon*	*Example of modifier with friendly equipment frame*
Folding girder bridge			
Hollow deck bridge			
Drill			
Earthmover			
Mine clearing			
Mine laying			
Other equipment			
Antenna			
Chemical, biological, radiological, or nuclear equipment			
Computer			
Generator	G	G	G
Laser			

Table 5-1. Main icons for equipment (continued)

Description	*Icon or symbol without frame*	*Orientation within the bounding octagon*	*Example of modifier with friendly equipment frame*
Military information support operations			
Radar			
Sensor			
Note: Systems that use these indicators are shown in the second line of the entry in order of light, low altitude, or short-range; as medium, medium altitude, or medium-range, and a heavy, high-altitude, or long-range.			

SECTOR 1 MODIFIERS FOR EQUIPMENT

5-2. Table 5-2 (on pages 5-10 through 5-13) shows sector 1 modifiers for equipment. This is a change to the previous system.

Table 5-2. Sector 1 modifiers for equipment

Description	*Modifier*	*Example of modifier with friendly equipment frame* / *Example of most common usage* / *Icon or symbol without frame*
Horizontal bounding octagon		
Vertical bounding octagon		
Attack	A	A
		A Attack helicopter
		A

Table 5-2. Sector 1 modifiers for equipment (continued)

Description	*Modifier*	*Example of modifier with friendly equipment frame* / *Example of most common usage* / *Icon or symbol without frame*
Bus Change: Bus is spelled out to avoid duplication with biological.	BUS	BUS BUS Bus BUS
Cargo	C	C C Cargo helicopter C
Command and control	C2	C2 C 2 Command and control armored personnel carrier C 2
Medical evacuation	✚	✚ ✚ Medical evacuation helicopter ✚

Table 5-2. Sector 1 modifiers for equipment (continued)

Description	*Modifier*	*Example of modifier with friendly equipment frame* / *Example of most common usage* / *Icon or symbol without frame*
Multifunctional	M F	M F M F Multifunctional earthmover M F
Petroleum, oils, and lubricants	∇	∇ ∇ Cross country petroleum, oils, and lubricants truck ∇
Utility	U	U U Utility helicopter U

Table 5-2. Sector 1 modifiers for equipment (continued)

Description	*Modifier*	*Example of modifier with friendly equipment frame* / *Example of most common usage* / *Icon or symbol without frame*
Water		Cross country water truck

SECTOR 2 MODIFIERS FOR EQUIPMENT

5-3. Table 5-3 (on pages 5-13 through 5-14) shows sector 2 modifiers for equipment.

Table 5-3. Sector 2 modifiers for equipment

Description	*Modifier*	*Example of modifier with friendly equipment frame* / *Example of most common usage* / *Icon or symbol without frame*
Horizontal bounding octagon		
Vertical bounding octagon		
Light	L	L L Light bridge L

Table 5-3. Sector 2 modifiers for equipment (continued)

Description	Modifier	Example of modifier with friendly equipment frame / Example of most common usage / Icon or symbol without frame
Medium	M	M C M Medium cargo helicopter C M
Heavy	H	H H Heavy semi-trailer greater than 40 tons H
Mine layer launcher	▬	
Recovery	]—[	]—[Tank recovery vehicle

MOBILITY INDICATOR AMPLIFIER (FIELD 13)

5-4. Mobility indicator amplifiers are used for both framed and unframed icons but are displayed in different locations for each. Figure 5-1 shows examples of how each is displayed. Table 5-4 (on pages 5-15 through 5-16) shows mobility indicator amplifiers.

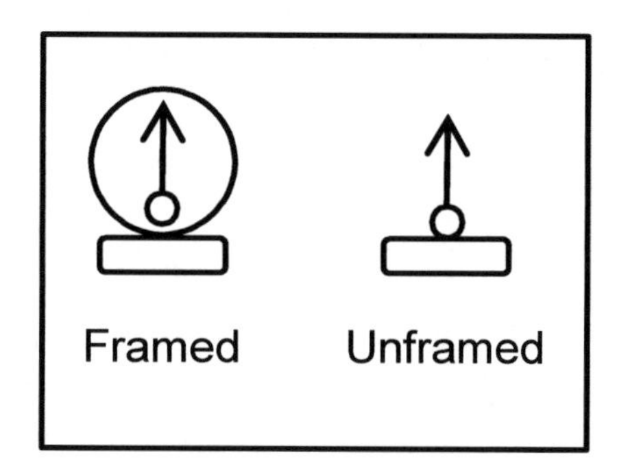

Figure 5-1. Examples of mobility indicator amplifiers for framed and unframed equipment symbols

Table 5-4. Mobility indicator amplifiers for equipment

Description	*Amplifier*	*Amplifier with friendly equipment frame*	*Example of amplifier use without frame*
Amphibious			Amphibious armored personnel carrier
Barge			Floating bridge
Over snow (prime mover)			Utility snow prime mover
Pack animal			Pack howitzer
Railway			Boxcar
Sled			Utility sled

Table 5-4. Mobility indicator amplifiers for equipment (continued)

Description	*Amplifier*	*Amplifier with friendly equipment frame*	*Example of amplifier use without frame*
Towed		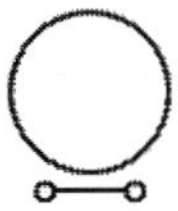	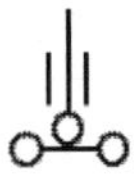 Towed howitzer
Tracked			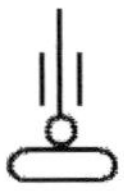 Tracked howitzer
Tractor trailer ***Note***: This mobility modifier only used with tractor trailer.		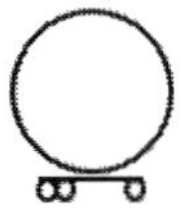	 Tractor trailer
Wheeled (cross-country)		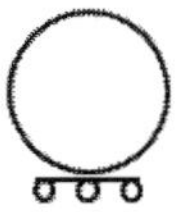	 Utility vehicle
Wheeled (limited mobility)		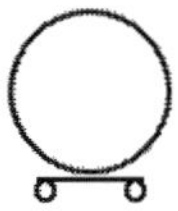	 Utility vehicle
Wheeled and tracked		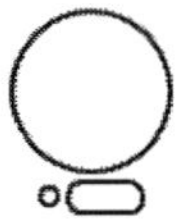	 Utility vehicle

Chapter 6
Installations

This chapter discusses main icons, sector 1 modifiers, and sector 2 modifiers for installations.

MAIN ICONS FOR INSTALLATIONS

6-1. Installations are sites that incorporate permanent, semipermanent, and temporary structures. Icons in the main sector reflect the main function of the symbol. Table 6-1 (on pages 6-1 through 6-3) shows the main icons for installations.

Table 6-1. Main icons for installations

Function	*Icon*	*Example*
Horizontal bounding octagon		
Vertical bounding octagon		
Aircraft ***Note***: This is a change from MIL-STD-2525D and STANAG 2019.		
Airport ***Note***: This is a change from MIL-STD -2525D and STANAG 2019.		
Ammunition		
Black list location	BLK	BLK
Broadcast transmitter antenna		
Chemical, biological, radiological, or nuclear		

Table 6-1. Main icons for installations (continued)

Function	Icon	Example
Economic	ECON	ECON
Electric power plant ***Note***: This is a change from MIL-STD-2525D and STANAG 2019.		
Food		
Gray list location	GRAY	GRAY
Mass grave		
Medical		
Medical treatment facility		
Mine		
Nuclear (non-chemical, biological, radiological, or nuclear defense)		
Printed media		
Prison or jail		

Table 6-1. Main icons for installations (continued)

Function	*Icon*	*Example*
Railhead or railroad station ***Note***: This is a change from MIL-STD-2525D and STANAG 2019.		
Safe house	SAFE	SAFE
Sea port ***Note***: This is a change from MIL-STD-2525D and STANAG 2019		
Telecommunications		
School or educational institution		
Water		
Tented camp		
Industrial building		
White list location	WHT	WHT

SECTOR 1 MODIFIERS FOR INSTALLATIONS

6-2. Table 6-2 (on pages 6-4 through 6-7) shows sector 1 modifiers. Sector 1 modifiers reflect the specific capability of the installation.

Table 6-2. Sector 1 modifiers for installations

Description	Modifier	Example
Horizontal bounding octagon		
Vertical bounding octagon		
Chemical, biological, radiological, and nuclear		
Biological	B	B
		B
Chemical	C	C
		C
Nuclear	N	N
		N
Radiological	R	R
		R

Table 6-2. Sector 1 modifiers for installations (continued)

Description	Modifier	Example
Energy sources		
Coal	CO	CO
		CO Coal electric power plant
Geothermal	GT	GT
		GT Geothermal electric power plant
Hydroelectric	HY	HY
		HY Hydroelectric power plant
Natural gas	NG	NG
		NG
Nuclear (non-chemical, biological, radiological, nuclear defense) ***Note***: This is a change from MIL-STD-2525D and STANAG 2019.		
		Nuclear electric power plant

Table 6-2. Sector 1 modifiers for installations (continued)

Description	Modifier	Example
Petroleum	∀	
		Petroleum electric power plant
Telecommunications		
Radio	RAD	RAD
		RAD Radio facility
Telephone	T	T
		T Telephone facility
Television	TV	TV
		TV Television facility

Table 6-2. Sector 1 modifiers for installations (continued)

Description	Modifier	Example
Other		
College or university	C O L	C O L
		C O L
Displaced persons, refugees, or evacuees *Note*: This is a change from MIL-STD-2525D and STANAG 2019.	DPRE	DPRE
		DPRE
Shipyard *Note*: This is a change from MIL-STD-2525D and STANAG 2019.	YRD	YRD
		YRD
Training *Note*: This is a change from MIL-STD-2525D and STANAG 2019.	TNG	TNG
		TNG
Water treatment (purification) *Note*: This is a change from MIL-STD-2525D and STANAG 2019.	PURE	PURE
		PURE

SECTOR 2 MODIFIERS FOR INSTALLATIONS

6-3. Table 6-3 shows sector 2 modifiers. Sector 2 modifiers reflect the specific type of installation.

Note: This is a change from MIL-STD-2525D and STANAG 2019.

Table 6-3. Sector 2 modifiers for installations

Description	Modifier	Example
Horizontal bounding octagon		
Vertical bounding octagon		
Production	PROD	PROD
		PROD Ammunition production
Repair	RPR	RPR
		YRD RPR Shipyard repair
Research	RSH	RSH
		RSH Medical research facility

Table 6-3. Sector 2 modifiers for installations (continued)

Description	*Modifier*	*Example*
Service	**SVC**	SVC
		SVC Airplane service facility
Storage ***Note***: This is a change from MIL-STD-2525D and STANAG 2019.	STOR	STOR
		STOR Storage facility or warehouse
Test	TEST	TEST
		TEST Airplane test facility

This page intentionally left blank.

Chapter 7
Activities

This chapter discusses main icons and sector 1 modifiers for activities.

MAIN ICONS FOR ACTIVITIES

7-1. Activities symbols are applicable across the range of military operations, but they normally focus on stability activities and defense support of civil authorities' activities. Activities can affect military operations. Activities represented by icons can include acts of terrorism, sabotage, organized crime, a disruption of the flow of vital resources, and the uncontrolled movement of large numbers of people. Many of these icons represent emergency first response activities used in the civilian community. Icons in the main sector reflect the main function of the symbol. Table 7-1 (on pages 7-1 through 7-5) shows the main icons for activities. Many of the icons in this chapter are also found in individuals and organizations (see chapter 4) and installations (see chapter 6). The icons in table 4-11 (on page 4-39) have been omitted from table 7-1; however, all the icons in table 4-11 can be used to build activities icons.

Table 7-1. Main icons for activities

Function	*Icon*	*Example*
Horizontal bounding octagon		
Vertical bounding octagon		
Arrest Change: The stick figure in the center of the arrest icon has been changed to reflect the individual icon.		
Attempted criminal activity against an individual		
Attempted criminal activity against multiple individuals or an organization		
Bombing	BOMB	BOMB
Booby trap		

Table 7-1. Main icons for activities (continued)

Function	*Icon*	*Example*
Demonstration	MASS	MASS
Drug related activity (Illegal)	DRUG	DRUG
Election, voting, or polling place	VOTE	VOTE
Emergency management operations This is a change from MIL-STD-2525D.		
Emergency medical operations This is a change from MIL-STD-2525D.	*	*
Exfiltration	EXFL	EXFL
Explosion		
Extortion	$ Dollars € Euros £ Pounds ¥ Yuan	$
Fire		
Fire hot spot		
Fire origin		

Table 7-1. Main icons for activities (continued)

Function	Icon	Example
Graffiti		
Home or house		
Criminal activities victim Change: Criminal activity victim icon uses the vertical bounding octagon.		
Criminal activity victims		
Improvised explosive device activity	IED	IED
Infiltration	INFL	INFL
Patrolling	P	P
Poisoning		
Riot	RIOT	RIOT
Black market This is a change from MIL-STD-2525D.	BM	BM
Jail break		
Burglary	BUR	BUR

Table 7-1. Main icons for activities (continued)

Function	*Icon*	*Example*
Robbery	ROB	ROB
Searching		
Shooting		
Sniping		
Spying	SPY	SPY
Theft	THF	THF
Rock throwing		
Smuggling	SMGL	SMGL
Sabotage	SAB	SAB
Meeting	MTG	MTG
Military information support operations		
School		

Table 7-1. Main icons for activities (continued)

Function	*Icon*	*Example*
Unexploded ordnance	UXO	UXO
Triage This is a change from MIL-STD-2525D.		
Warrant served	WNT	WNT

SECTOR 1 MODIFIERS FOR ACTIVITIES

7-2. Table 7-2 (on pages 7-5 through 7-9) shows sector 1 modifiers. Sector 1 modifiers reflect the specific type of activity.

Table 7-2. Sector 1 modifiers for activities

	Characteristic	*Modifier*	*Examples*
Horizontal bounding octagon			
Vertical bounding octagon			
Types of killings	Assassination	**AS**	A S
			A S
	Wrongful execution	**EX**	E X
			E X
	Murder	**MU**	MU
			MU Murdered group

Table 7-2. Sector 1 modifiers for activities (continued)

	Characteristic	Modifier	Examples
Criminal activities	Arson	A S N	A S N A S N Arson fire
	Hijacking	H	H H
	Laboratory	LAB	LAB LAB DRUG Illegal drug laboratory
	Kidnapping	K	K K Kidnapping
	Piracy	PI	PI PI Boat piracy
	Rape	R A	R A R A Rape

Table 7-2. Sector 1 modifiers for activities (continued)

	Characteristic	*Modifier*	*Examples*
Criminal activities	Trafficking	**TFK**	TFK
			TFK DRUG Drug trafficking
	Meeting	MTG	MTG
			MTG Organizational meeting
Explosions	Bomb	BOMB	BOMB
			BOMB Bomb explosion
	Grenade This is a change from MIL-STD-2525D.	**GR**	GR
			GR Grenade explosion
	Improvised explosive device	**IED**	IED
			IED Improvised explosive device explosion
	Incendiary device This is a change from MIL-STD-2525D.	**IN**	IN
			IN Incendiary device explosion

Table 7-2. Sector 1 modifiers for activities (continued)

	Characteristic	*Modifier*	*Examples*
	Mine This is a change from MIL-STD-2525D.		
	Mortar This is a change from MIL-STD-2525D.		
	Rocket This is a change from MIL-STD-2525D.		
	Emergency collection evacuation point This is a change from MIL-STD-2525D.	ECEP	ECEP ECEP
	Eviction	EV	EV EV Home eviction
	Foraging		Foraging for food
	Raid	RAID	RAID RAID Raid on a home

Table 7-2. Sector 1 modifiers for activities (continued)

	Characteristic	*Modifier*	*Examples*
	Suspicious activity or threat	?	?
			? BOMB Bomb threat

SECTOR 2 MODIFIERS FOR ACTIVITIES

7-3. While table 7-2 shows sector 2 modifiers, table 7-3 shows sector 1 modifiers that reflect the variations in modifier 1 specific types of activity.

Table 7-3. Sector 1 modifiers for activities

Characteristic	*Modifier*	*Examples*
Horizontal bounding octagon		
Vertical bounding octagon		
Drive by shooting		
		Drive by shooting
Premature	P	P
		IED P Premature improvised explosive device explosion

This page intentionally left blank.

Chapter 8
Control Measure Symbols

This chapter discusses basics, points, lines, boundary lines, areas, and abbreviations and acronyms for use with control measure symbols.

BASICS OF CONTROL MEASURE SYMBOLS

8-1. A control measure symbol is a graphic used on maps and displays to regulate forces and warfighting functions. Definitions of terms related to control measure symbols are provided in chapter 1. The control measure symbols in this chapter are organized by the six warfighting functions: mission command, movement and maneuver, fires, protection, sustainment, and intelligence. Also included are airspace control measures, which are a combination of movement and maneuver, fires, and protection. Control measure symbols generally fall into one of three categories: points, lines, or areas. The coloring and labeling of control measure symbols is almost identical to framed symbols.

COLOR OF CONTROL MEASURE SYMBOLS

8-2. Friendly graphic control measures are shown in black or blue. Hostile graphic control measures are shown in red. If red is not available, they are shown in black with the abbreviation "ENY" placed on the graphic in amplifier field 15. If a special requirement arises to show neutral or unknown graphic control measures, they are shown in black, and the abbreviations of "NEU" for neutral or "UNK" for unknown are used in amplifier field 15. All obstacles, regardless of standard identity, are shown in green. If green is not available, obstacles should be shown using black. Yellow is used for the cross-hatching of areas with chemical, biological, radiological, or nuclear (CBRN) contamination. The use of green and yellow for obstacles and CBRN contamination is in contradiction to the standard identities.

LETTERING FOR CONTROL MEASURE SYMBOLS

8-3. All lettering for control measure symbols must be in upper case (all capital letters). All lettering should be oriented horizontally, from left to right, so readers can see it easily without having to tilt their heads. All lettering should be sized as large as possible, so a reader can easily understand it. However, the lettering should not be so large that it interferes with other symbols or icons.

ACRONYMS AND ABBREVIATIONS FOR CONTROL MEASURE SYMBOLS

8-4. Acronyms and abbreviations for use with control measure symbols are shown in this chapter and must be used for Army control measure symbols. No abbreviations other than those provided in this publication may be used.

AMPLIFIER FIELDS FOR CONTROL MEASURE SYMBOLS

8-5. See table 3-6 (on page 3-12) for descriptions of all the amplifier fields for control measure symbols. For control measures, field 3 can represent either a unique alphanumeric designation that identifies the establishing unit, serviced unit, or a name, letter, or number. There is no requirement for all amplifier fields to be filled in for control measure symbols. Only required amplifier fields must be filled in.

POINT SYMBOLS

8-6. A point is a control measure symbol that has only one set of coordinates. Most Army point symbols follow a standard format. Figure 8-1 (on page 8-2) shows the composition and placement of an icon, its modifiers, and its amplifiers for a standard point and a supply point. The external amplifier field 3 is used to designate the unit being serviced or another unique designation, while the internal amplifier field 3 is used to designate the unit providing the service. Point symbols cannot be rotated; therefore, text must be written horizontally only (not on an angle or diagonal).

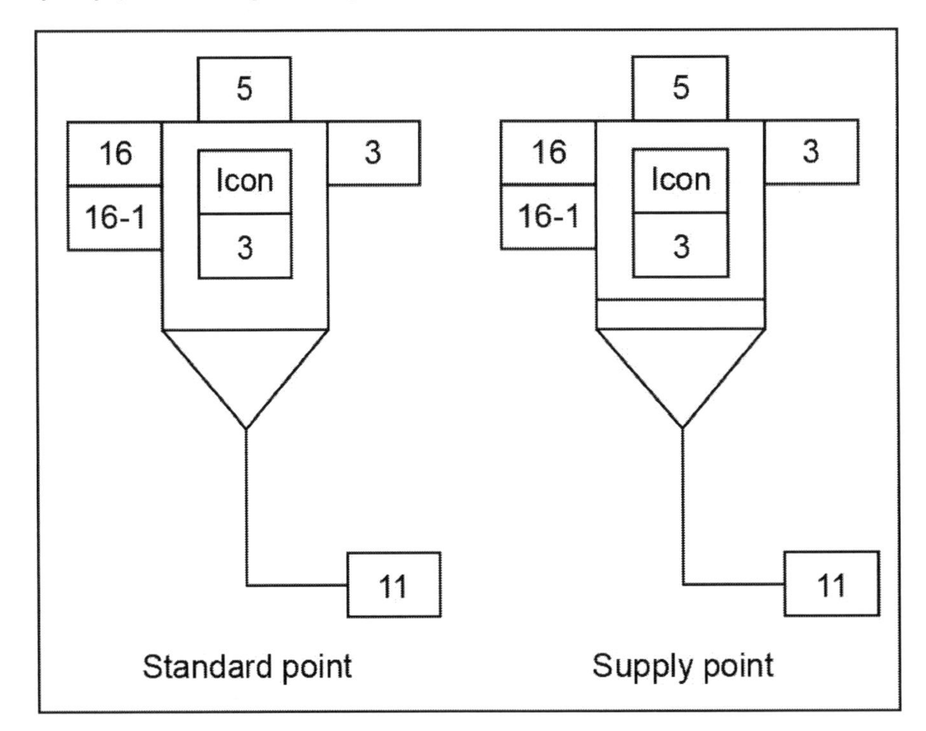

Figure 8-1. Standard point and supply point templates

LINE SYMBOLS

8-7. A line is a control measure symbol with multiple sets of coordinates. Figure 8-2 shows the composition and placement of an icon, its modifiers, and its amplifiers for a standard line. Most lines are also labelled as phase lines for easy reference in orders and during transmissions. A phase line is marked as PL, with the line's name in field 3. When lines representing other purposes are marked as phase lines, they should show their primary purpose in the icon field (such as NFL for no fire line). The purpose of the line is labelled on top of the line at both ends inside the lateral boundaries or as often as necessary for clarity. Field 3 is used for fire support coordination measures, to show the designation of the controlling headquarters. The use of phase lines to mark line control measure symbols is not mandatory.

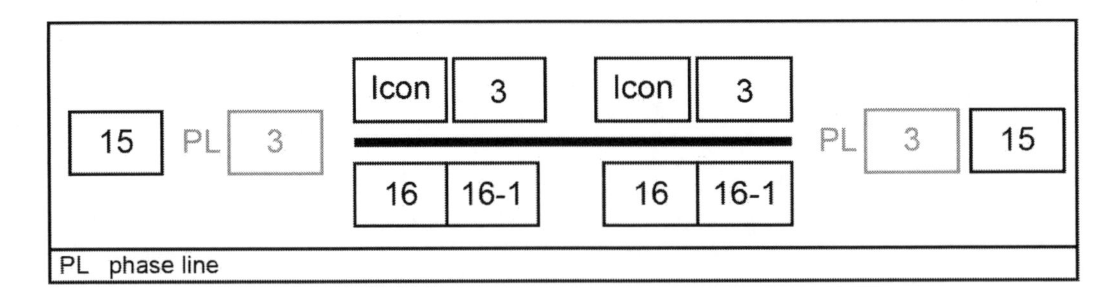

Figure 8-2. Standard line template

BOUNDARY LINE SYMBOLS

8-8. A *boundary* is a line that delineates surface areas for the purpose of facilitating coordination and deconfliction of operations between adjacent units, formations, or areas (JP 3-0). There are three types of boundary lines: lateral, rear, and forward. Amplifiers are displayed perpendicular to the boundary line. Figure 8-3 shows standard horizontal (east-west) and vertical (north-south) boundary lines and the orientation of their amplifiers. The graphic for the highest echelon (field 6) unit on lateral boundaries is used for the boundary line. The graphic for the lower echelon (field 6) unit on a rear or forward boundary is used for the boundary line. When units of the same echelon are adjacent to each other, the abbreviated echelon designator (field 3, such as CO, BN, or BDE) can be omitted from the alphanumeric designator. Table 8-9 (on page 8-86) provides a list of abbreviations and acronyms used in field 3. When the boundary is between units of different countries, the three-letter country code (field 2) is shown in parentheses behind or below the unit designation.

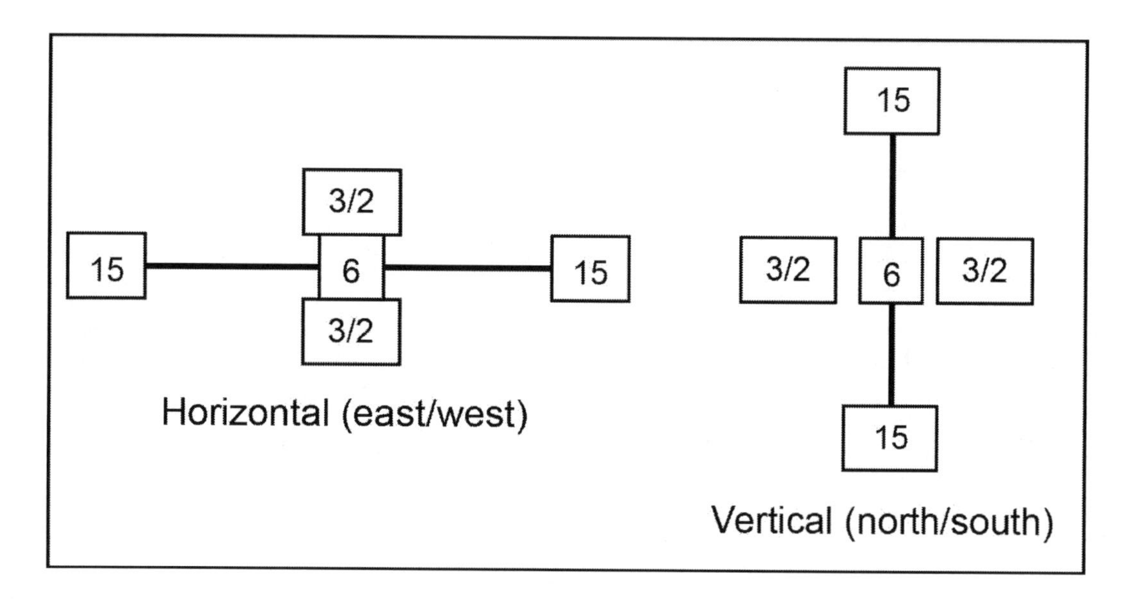

Figure 8-3. Horizontal and vertical boundary templates

AREA SYMBOLS

8-9. An area is a control measure symbol with multiple sets of coordinates that start and finish at the same point. Figure 8-4 (on page 8-4) shows the composition and placement of an icon, its modifiers, and its amplifiers for a standard area. Areas normally are marked with the abbreviation for the type of area in the icon field, followed by a name in field 3. This labeling should be in the center of the area unless the area is too small or the labeling would interfere with locating units. The type of area determines the number of fields being used. Not all fields are required for each area. Some areas may use only one field, while other areas will use several.

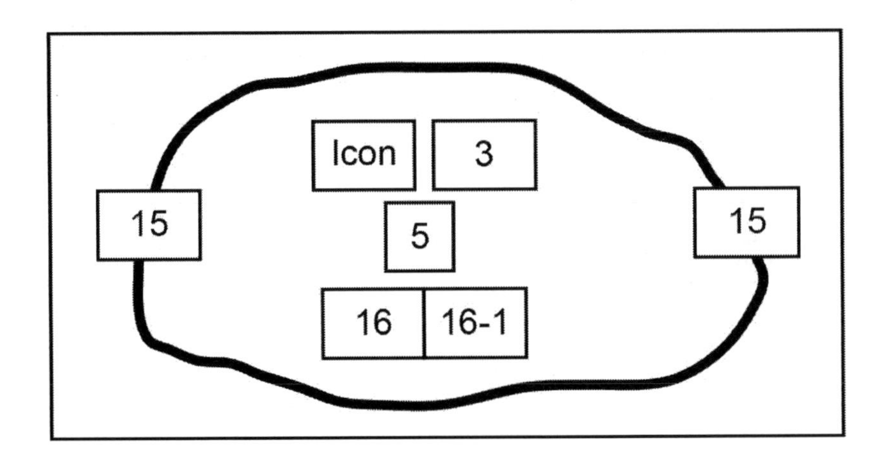

Figure 8-4. Standard area template

MISSION COMMAND (COMMAND AND CONTROL)

8-10. Table 8-1 (on pages 8-4 through 8-13) shows mission command (or command and control) control measure symbols. Table 8-8 on page 8-85 provides abbreviations and acronyms for use with boundaries.

Table 8-1. Mission command

Type *(Description of use publication)*	*Template and icons (found within the boxes)*	*Example*
	Boundaries	
Friendly present	3/2 6 3/2	2ID USA XX 52ID GBR
Friendly planned or on order	3/2 6 3/2	5 CAN X 2 FRA
Enemy known	3/2 15 6 15 3/2	12IN ENY II ENY 7IN

Table 8-1. Mission command (continued)

Type (Description of use publication)	*Template and icons (found within the boxes)*	*Example*
Enemy known	3/2 6 3/2	1AAB X 3ARBN
Enemy suspected	3/2 15 6 15 3/2	211AR ENY II ENY 12ARCOY
	3/2 6 3/2	3ABB X 8ABR
Examples		
Lateral ***Note***: Lateral boundaries use the echelon icon of the highest unit on boundary lines.	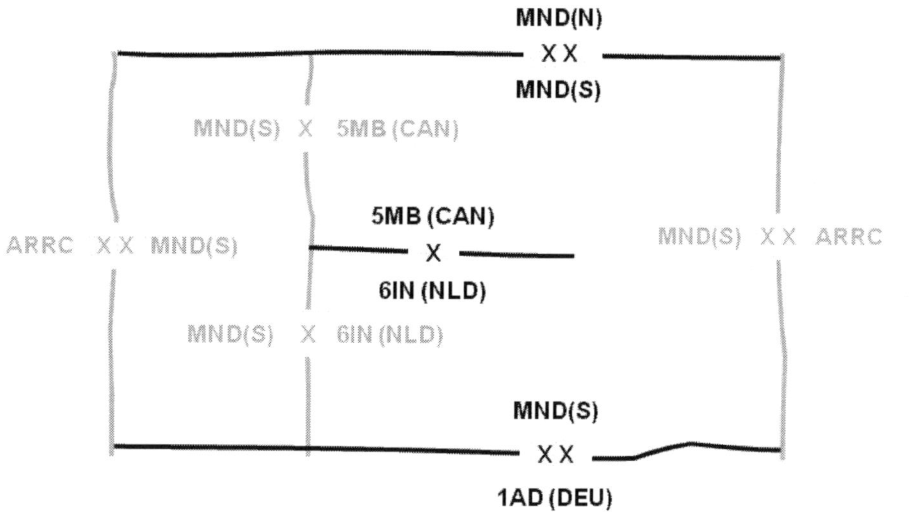 MND(N) XX MND(S) MND(S) X 5MB (CAN) ARRC XX MND(S) 5MB (CAN) X 6IN (NLD) MND(S) XX ARRC MND(S) X 6IN (NLD) MND(S) XX 1AD (DEU)	

Table 8-1. Mission command (continued)

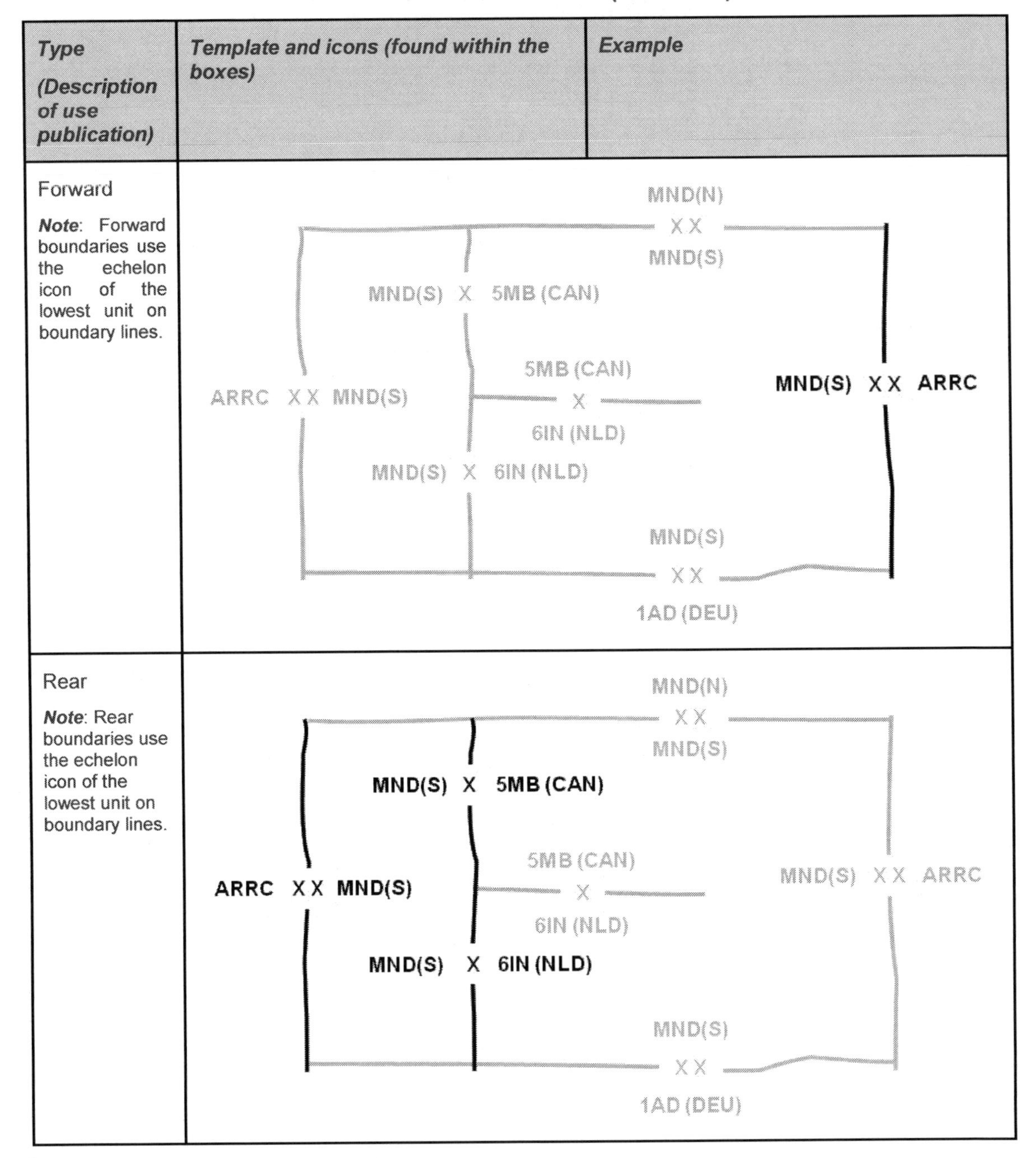

Type *(Description of use publication)*	*Template and icons (found within the boxes)*	*Example*
Forward ***Note***: Forward boundaries use the echelon icon of the lowest unit on boundary lines.	MND(N) XX MND(S) MND(S) X 5MB (CAN) ARRC XX MND(S) 5MB (CAN) X 6IN (NLD) MND(S) X 6IN (NLD) MND(S) XX 1AD (DEU)	**MND(S) XX ARRC**
Rear ***Note***: Rear boundaries use the echelon icon of the lowest unit on boundary lines.	MND(N) XX MND(S) **MND(S) X 5MB (CAN)** **ARRC XX MND(S)** 5MB (CAN) X 6IN (NLD) **MND(S) X 6IN (NLD)** MND(S) XX 1AD (DEU)	MND(S) XX ARRC

Table 8-1. Mission command (continued)

Type (Description of use publication)	*Template and icons (found within the boxes)*	*Example*
Note: Those symbols that are shown in gray are used to explain how the symbol might be displayed, but are not part of the control measure.		
Points		
Airfield	5 *Note*: Entries for amplifier are MILITARY, CIVILIAN, or JOINT. JOINT in this specific case refers to both civilian and military.	JOINT
Amnesty point	5 16 AMN 3 16-1 3-1	WEAPONS 080700ZMAY13-120700ZMAY13 AMN NZ UN
Center of main effort *Note*: This is a North Atlantic Treaty Organization symbol.		
Contact point	3	1
Coordination point		

Table 8-1. Mission command (continued)

Type (Description of use publication)	*Template and icons (found within the boxes)*	*Example*
Note: Those symbols that are shown in gray are used to explain how the symbol might be displayed, but are not part of the control measure.		
Decision point	3	1
Checkpoint ***Note***: This is a change from MIL-STD-2525D (from CKP to CP).	5 16 CP 3 16-1 3-1	MSR 5 140700ZMAR13 - 142200ZMAR13 CP 12 39 TPT
Distress call point	5 16 SOS 3 16-1 3-1	UH-60 161000ZJUN13 SOS 101AVN
Entry control point	5 16 EC 3 16-1 3-1	VEHICLES EC 1 ISAF

Table 8-1. Mission command (continued)

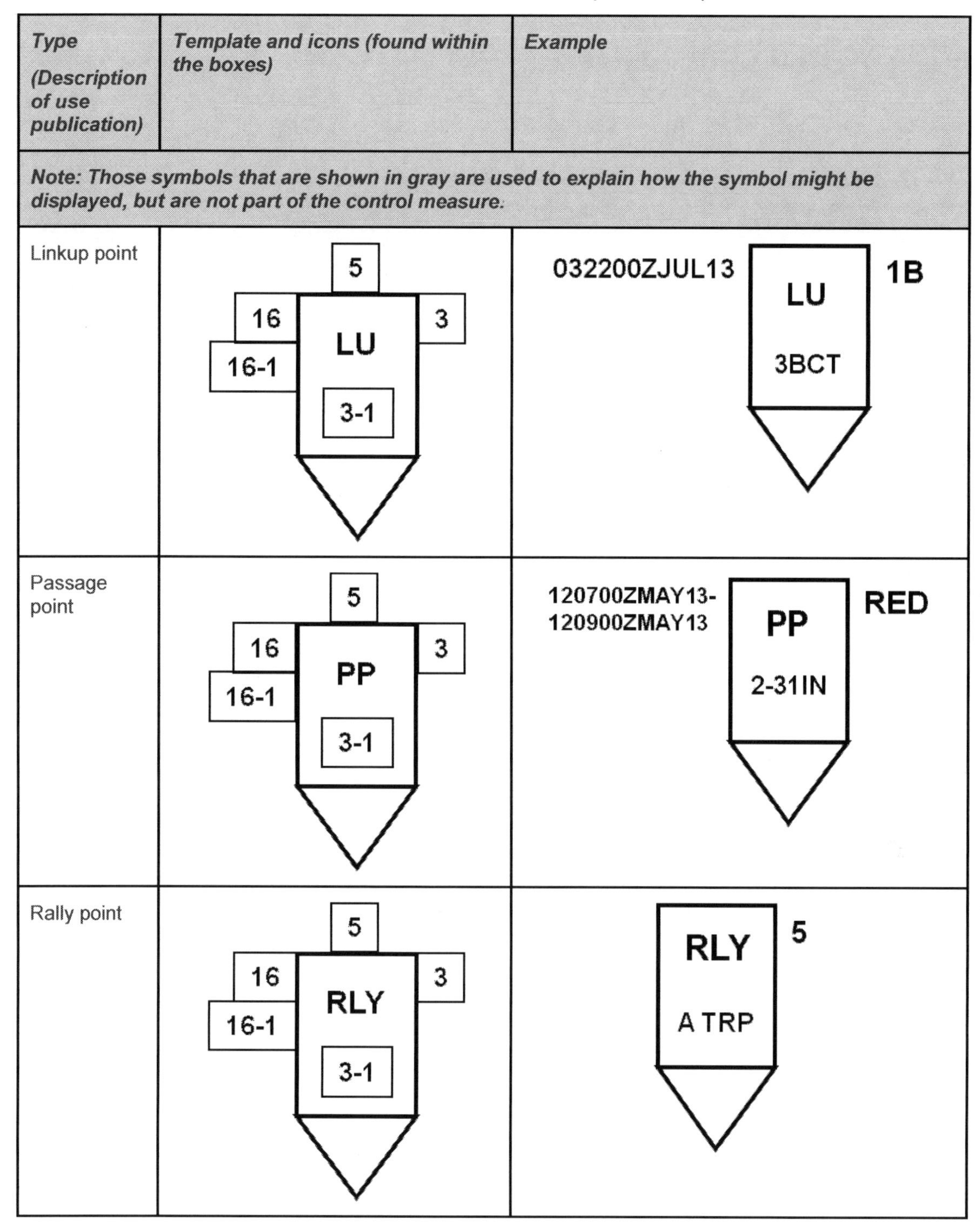

Type (Description of use publication)	*Template and icons (found within the boxes)*	*Example*
Note: Those symbols that are shown in gray are used to explain how the symbol might be displayed, but are not part of the control measure.		
Linkup point	5 16 16-1 LU 3 3-1	032200ZJUL13 LU 3BCT 1B
Passage point	5 16 16-1 PP 3 3-1	120700ZMAY13- 120900ZMAY13 PP 2-31IN RED
Rally point	5 16 16-1 RLY 3 3-1	RLY A TRP 5

Table 8-1. Mission command (continued)

Type (Description of use publication)	*Template and icons (found within the boxes)*	*Example*
Note: Those symbols that are shown in gray are used to explain how the symbol might be displayed, but are not part of the control measure.		
Release point	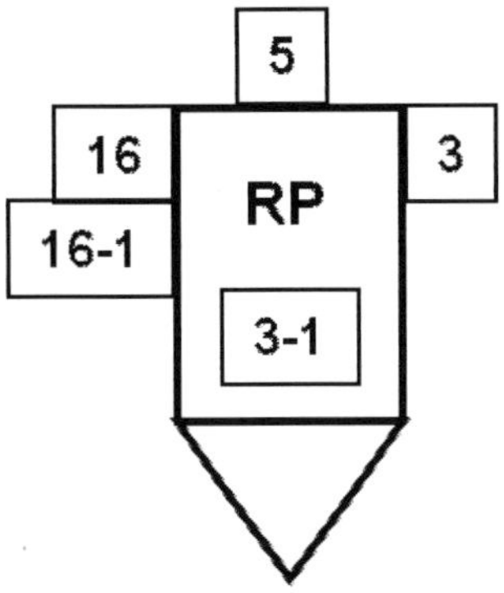 5 16 RP 3 16-1 3-1	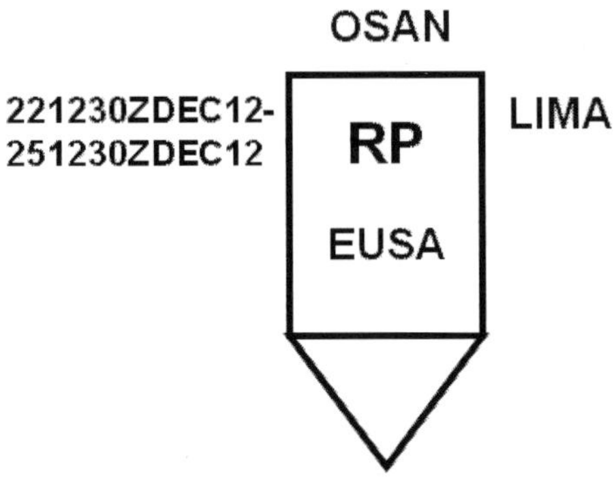 OSAN 221230ZDEC12- 251230ZDEC12 RP LIMA EUSA
Special point	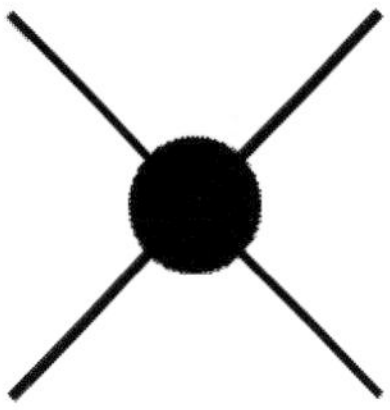	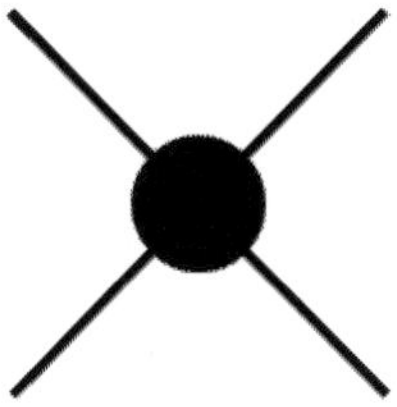
Start point	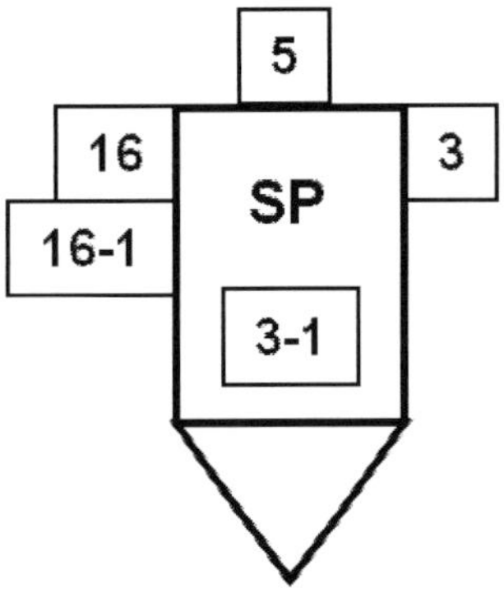 5 16 SP 3 16-1 3-1	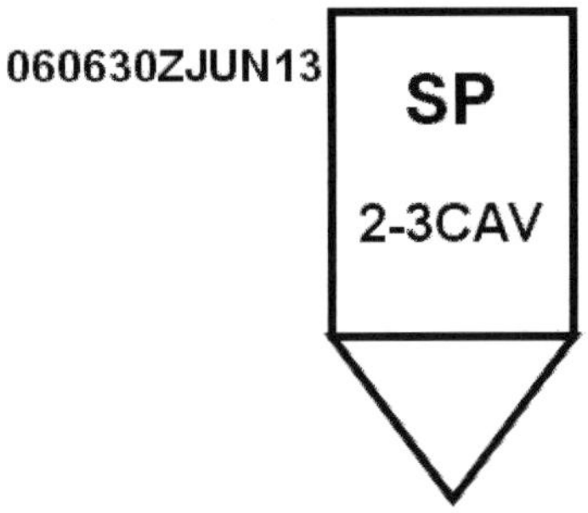 060630ZJUN13 SP 2-3CAV
Waypoint	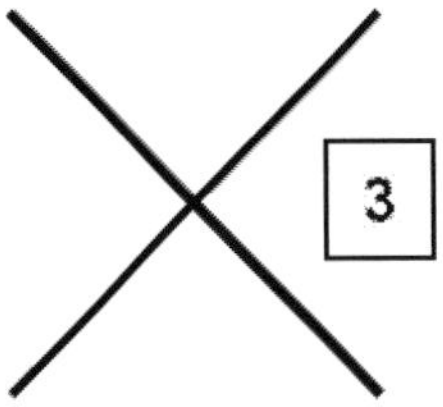 3	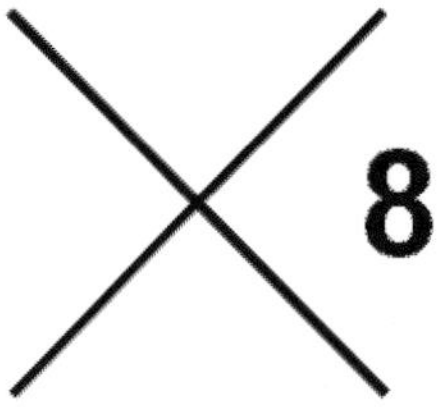 8

Table 8-1. Mission command (continued)

Type (Description of use publication)	*Template and icons (found within the boxes)*	*Example*
Note: Those symbols that are shown in gray are used to explain how the symbol might be displayed, but are not part of the control measure.		
Lines		
`	EWL 3/2 EWL 3/2	EWL 326EN BN (USA) EWL 127EN BN (USA)
Light line	PL 3 LL LL PL 3	PL CRAB LL LL PL CRAB
Areas		
Airfield zone	5	750M
Area of operations	AO 3	AO KANSAS
Area of operations as a boundary change	AO 3 6	AO LEAVENWORTH XXX

Table 8-1. Mission command (continued)

Type *(Description of use publication)*	*Template and icons (found within the boxes)*	*Example*
Note: Those symbols that are shown in gray are used to explain how the symbol might be displayed, but are not part of the control measure.		
Base camp	BC 3	BC CADD
Guerrilla base	GB 3	GB BOOGEYMEN
Joint operations area	JOA 3	JOA LIBERTY
Joint special operations area	JSOA 3	JSOA BLACK

Table 8-1. Mission command (continued)

Type *(Description of use publication)*	*Template and icons (found within the boxes)*	*Example*
Note: Those symbols that are shown in gray are used to explain how the symbol might be displayed, but are not part of the control measure.		
Named area of interest	NAI 3	NAI 2
Targeted area of interest	TAI 3	TAI YUKON

MOVEMENT AND MANEUVER

8-11. Table 8-2 (on pages 8-14 to 8-29) shows movement and maneuver control measure symbols. These symbols are further subdivided into general, defensive, and offensive symbols.

Table 8-2. Movement and maneuver

Type	*Template and icons*	*Example*
Note: Those symbols that are shown in gray are used to explain how the symbol might be displayed, but are not part of the control measure symbol.		
	General	
Points		
Point of interest	3	8
	Lines	
Forward edge of the battle area		
Actual forward edge of the battle area	FEBA FEBA	FEBA PL MOCA FEBA PL MOCA
Proposed or on order forward edge of the battle area	FEBA FEBA	FEBA PL INK FEBA PL INK
Forward line of own troops		
Friendly forward line of own troops		X X
Friendly planned or on order forward line of own troops		X X

Table 8-2. Movement and maneuver (continued)

Type	*Template and icons*	*Example*
Enemy forward line of own troops - for systems that display multiple colors, including red		
Enemy forward line of own troops - for systems that only display two colors	15 15	ENY ENY
Enemy suspected forward line of own troops - for systems that display multiple colors, including red		
Enemy suspected forward line of own troops - for systems that only display two colors	15 15	ENY ENY

Table 8-2. Movement and maneuver (continued)

Type	*Template and icons*	*Example*
Other lines		
Handover line	 PL 3 HL HL PL 3	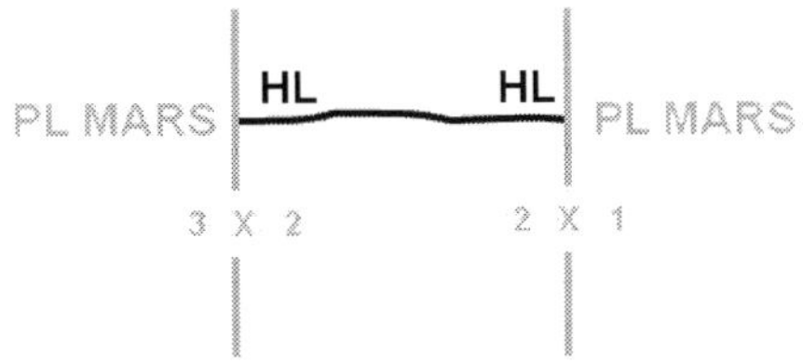 PL MARS HL HL PL MARS 3 X 2 2 X 1
Line of contact	 LC LC	 LC ENY LC ENY
Phase line	 PL 3 PL 3	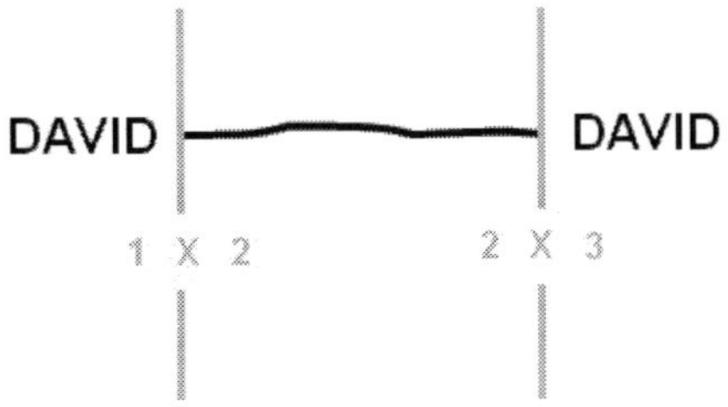 DAVID DAVID 1 X 2 2 X 3

Table 8-2. Movement and maneuver (continued)

Type (Description of use publication)	*Template and icons (found within the boxes)*	*Example*
Note: Those symbols that are shown in gray are used to explain how the symbol might be displayed, but are not part of the control measure.		
Areas		
Friendly area		
Friendly planned or on order area		
Enemy known or confirmed area ***Note***: If red is available, it is used for enemy control measures.	15 15	ENY ENY
Enemy suspected area	15 15	ENY ENY
Assembly area	AA 3	AA 2

Table 8-2. Movement and maneuver (continued)

Type (Description of use publication)	*Template and icons (found within the boxes)*	*Example*
Note: Those symbols that are shown in gray are used to explain how the symbol might be displayed, but are not part of the control measure.		
Occupied assembly area	AA 3 UNIT SYMBOL ***Note***: One or more unit symbols can be displayed in an assembly area or if required the symbols can be offset. If unit symbols are present, it indicates an occupied assembly area.	AA BLUE 1-3 AA PINE 4-7 3-66 1-23
Planned or on order assembly area	AA 3	AA DELTA
Planned or on order assembly area with unit to occupy	AA 3 UNIT SYMBOL	AA COAL 2-3
Forward assembly area	FAA 3	FAA NORTH
Drop zone	DZ 3	DZ HAWK

Table 8-2. Movement and maneuver (continued)

Type (Description of use publication)	*Template and icons (found within the boxes)*	*Example*
Note: Those symbols that are shown in gray are used to explain how the symbol might be displayed, but are not part of the control measure.		
Extraction zone	EZ T	EZ ROCK
Landing zone	LZ 3	LZ SILVER
Pickup zone	PZ T	PZ WOLF
Engagement area	EA 3	EA ROCK
Search area or reconnaissance area	X UNIT SYMBOL X	X X

Table 8-2. Movement and maneuver (continued)

Type (Description of use publication)	*Template and icons (found within the boxes)*	*Example*
Note: Those symbols that are shown in gray are used to explain how the symbol might be displayed, but are not part of the control measure.		
Points		
Combat outpost	3	3/C/2-15IN
Observation post	3	2/2/A/1-31IN
Observation post by unit type or equipment type	ICON 3	1/3/A/23
Lines		
Final protective line	PL 3 FPL FPL PL 3	PL EARL FPL FPL PL EARL 3 X 2 2 X 1

Table 8-2. Movement and maneuver (continued)

Type (Description of use publication)	*Template and icons (found within the boxes)*	*Example*
Note: Those symbols that are shown in gray are used to explain how the symbol might be displayed, but are not part of the control measure.		
Areas		
Occupied battle position	3 6 ***Note:*** The side opposite (field 6) always faces toward the hostile force.	XRAY II Battalion battle position
Prepared but not occupied battle position	(P) 3 6	(P) ZULU II
Planned battle position	3 6	YANKEE II Planned battalion battle position
Fortified area		

Table 8-2. Movement and maneuver (continued)

Type (Description of use publication)	*Template and icons (found within the boxes)*	*Example*
Note: Those symbols that are shown in gray are used to explain how the symbol might be displayed, but are not part of the control measure.		
Friendly strong point	3 6	TWO I
Known enemy strong point	15 3 15 6	ENY 3 ENY I
Principal direction of fire	Equipment Symbol	

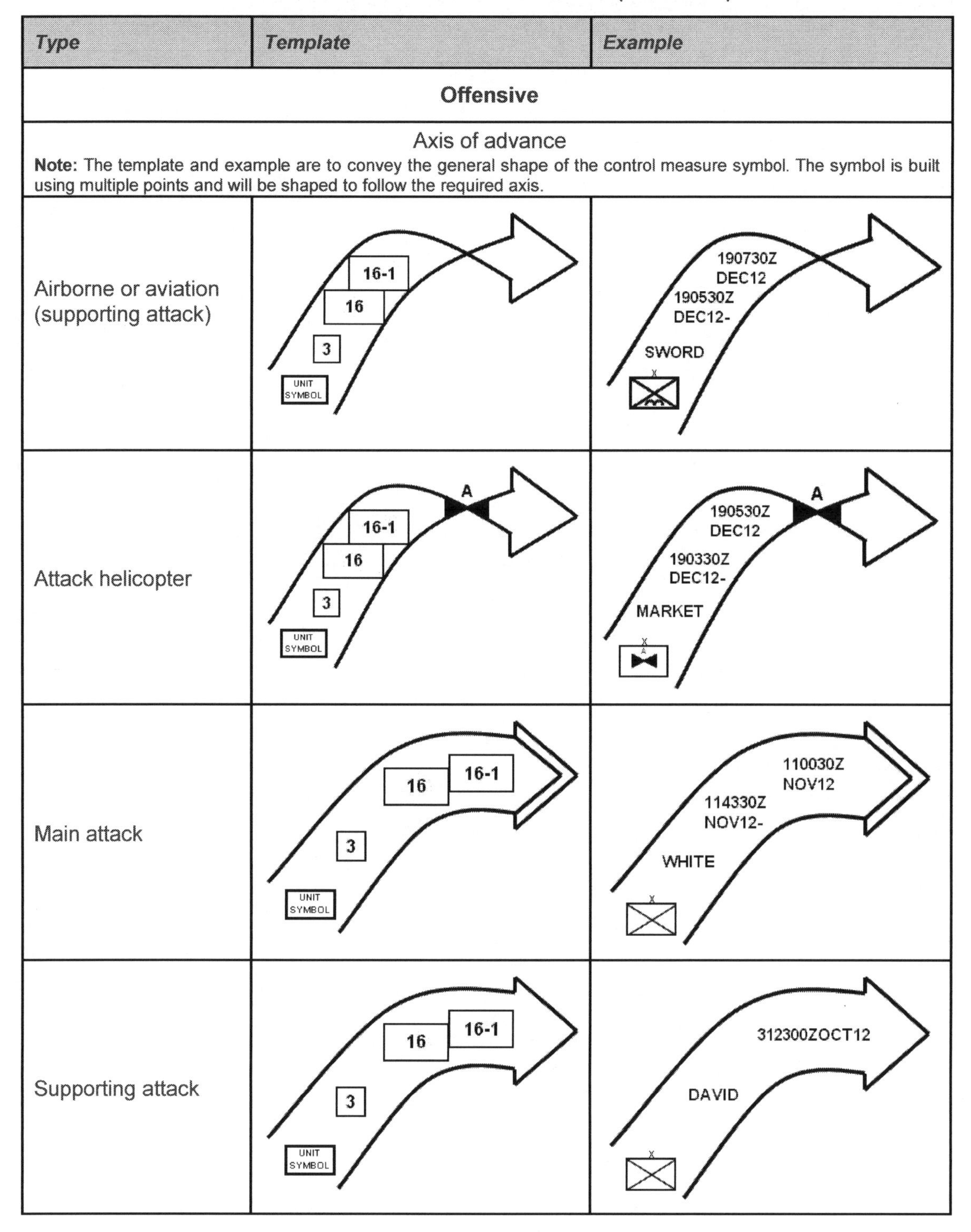

Table 8-2. Movement and maneuver (continued)

Type	Template	Example
Offensive		
Axis of advance **Note:** The template and example are to convey the general shape of the control measure symbol. The symbol is built using multiple points and will be shaped to follow the required axis.		
Airborne or aviation (supporting attack)	16-1 16 3 UNIT SYMBOL	190730Z DEC12 190530Z DEC12- SWORD
Attack helicopter	A 16-1 16 3 UNIT SYMBOL	A 190530Z DEC12 190330Z DEC12- MARKET
Main attack	16 16-1 3 UNIT SYMBOL	110030Z NOV12 114330Z NOV12- WHITE
Supporting attack	16 16-1 3 UNIT SYMBOL	312300ZOCT12 DAVID

Table 8-2. Movement and maneuver (continued)

Type	*Template*	*Example*
Planned or on order with date-time group (if known) effective	16 16-1 EFF 3 UNIT SYMBOL	EFF 150730ZNOV13 GRANT X
Enemy known or confirmed	15 16 16-1 3 15 UNIT SYMBOL	221330ZDEC12-230530ZDEC12 ENY MARX ENY X
Enemy suspected or template	15 16 16-1 3 15 UNIT SYMBOL	291030ZFEB13-291530ZFEB13 ENY RED 1 ENY X
Direction of attack		
Aviation	UNIT SYMBOL 3 16 16-1	X A AVON 011000ZJAN13-011100ZJAN13
Main attack	UNIT SYMBOL 3 16 16-1	II SABER 251100ZJUN12-251130ZJUN12
Supporting attack	UNIT SYMBOL 3 16 16-1	II YALU 290730ZFEB13

Table 8-2. Movement and maneuver (continued)

Type	*Template*	*Example*
Planned or on order with date-time group (if known) effective	UNIT SYMBOL 3 EFF 16 16-1	ORNE EFF260230ZAUG13
Enemy known or confirmed	UNIT SYMBOL 15 3 16 16-1	ENY 1 051200ZJAN13-051300ZJAN13
Enemy suspected or template	UNIT SYMBOL 15 3 16 16-1	ENY DELTA 071200ZJUN13-081200ZJUN13
Points		
Target reference point ***Notes: 1.*** Task force units and below use target reference points (TRPs). A TRP can delineate sectors of fire within an engagement area. TRPs are designated using the standard target symbol or numbers issued by the fire support officer. Once designated, TRPs can also constitute indirect fire targets. **2.** For indirect fire target identification, use the Point single target in table 8-3 (Fires).	3	201

Table 8-2. Movement and maneuver (continued)

Type	*Template*	*Example*
	Lines	
Final coordination line	PL 3 FCL FCL PL 3	PL OPAL FCL FCL PL OPAL 2 X 4 4 X 1
Infiltration lane	LANE 3	LANE APPLE
Limit of advance	PL 3 LOA LOA PL 3	PL RUBY LOA LOA PL RUBY 2 X 3 3 X 1
Line of departure	PL 3 LD LD PL 3	PL JADE LD LD PL JADE 2 X 3 3 X 1
Line of departure or line of contact	PL 3 LD/LC LD/LC PL 3	PL JADE LD/LC LD/LC PL JADE 2 X 3 3 X 1
Probable line of deployment	PL 3 PLD PLD PL 3 ***Note:*** Use the planned status (dashed) for the line.	PL LEAD PLD PLD PL LEAD 1 X 3 3 X 2
Release line	PL 3 RL RL PL 3	PL PINE RL RL PL PINE 3 X 2 2 X 1

Table 8-2. Movement and maneuver (continued)

Type	*Template*	*Example*
Areas		
Assault position	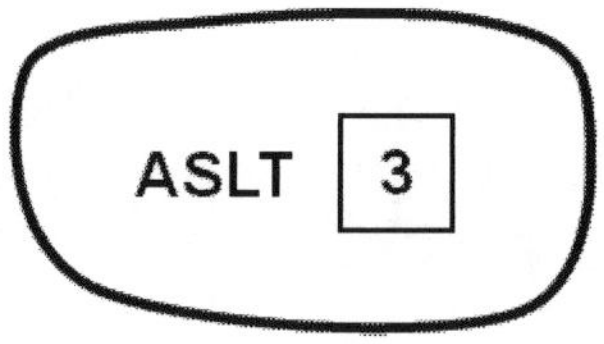 ASLT 3	 ASLT DANUBE
Attack position	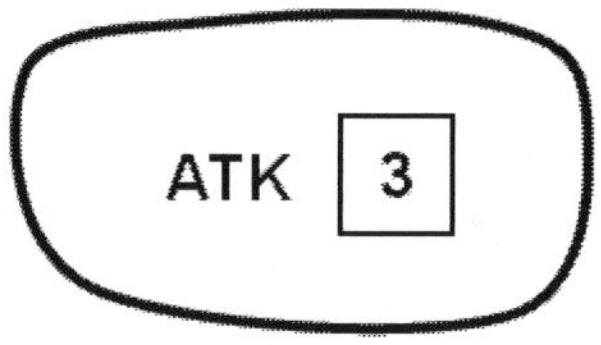 ATK 3	 ATK NILE
Friendly occupied attack position ***Note:*** Used only if a unit must stop in the attack position.	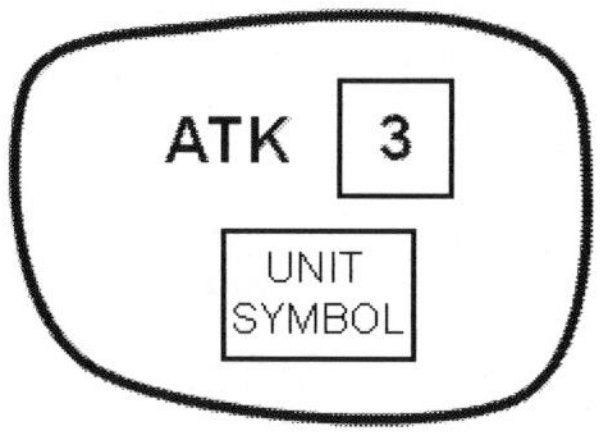 ATK 3 UNIT SYMBOL	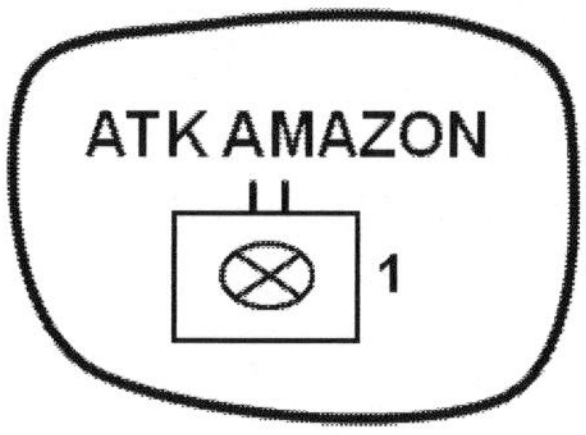 ATK AMAZON 1
Planned, proposed, or on order attack position	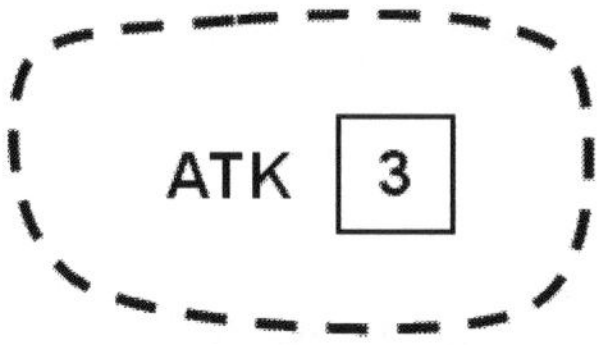 ATK 3	 ATK OHIO
Objective	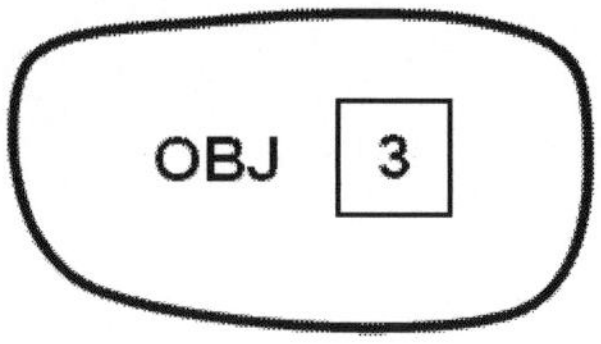 OBJ 3	 OBJ FIVE
Friendly forces encircled	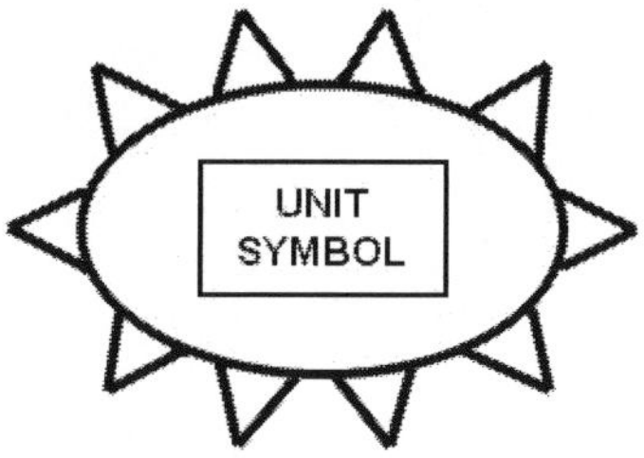 UNIT SYMBOL	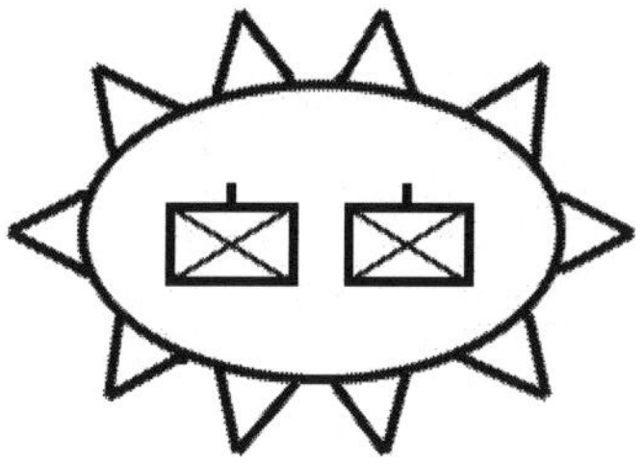

Table 8-2. Movement and maneuver (continued)

Type	*Template*	*Example*
Enemy forces encircled	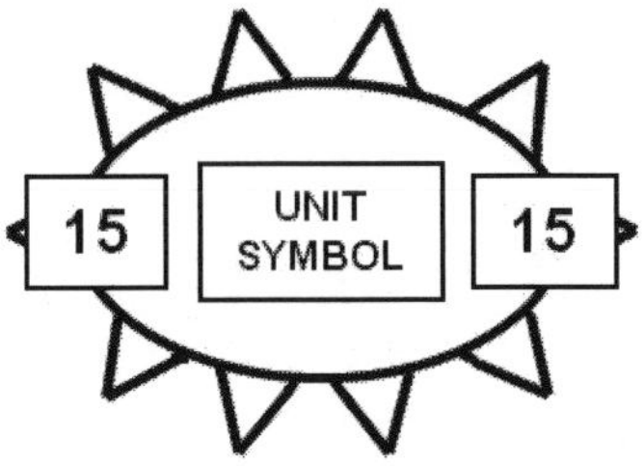 15 UNIT SYMBOL 15	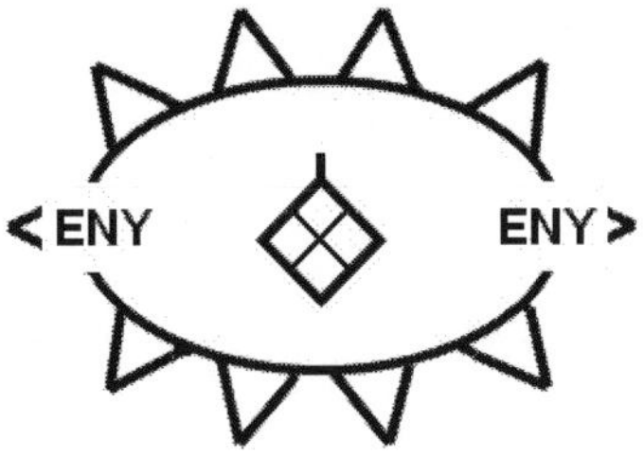 ENY ENY
Bridging operations		
Bridgehead line (bridgehead)	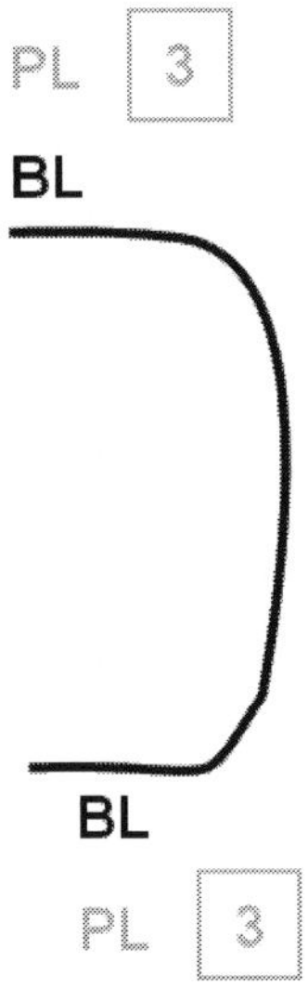 PL 3 BL BL PL 3	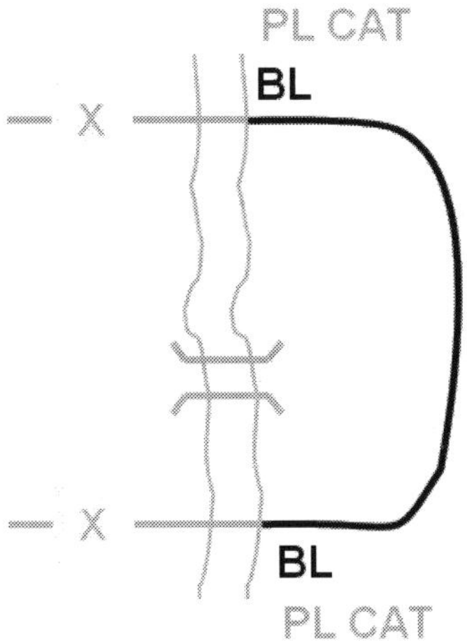 PL CAT BL X X BL PL CAT
Holding line	 PL 3 HL HL PL 3	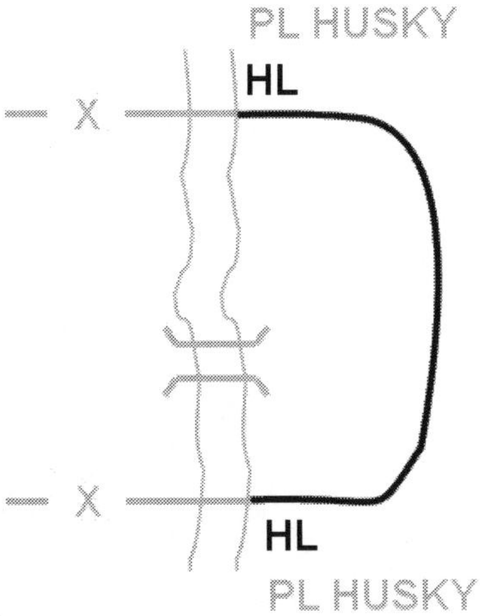 PL HUSKY HL X X HL PL HUSKY

Table 8-2. Movement and maneuver (continued)

Type	*Template*	*Example*
Release line	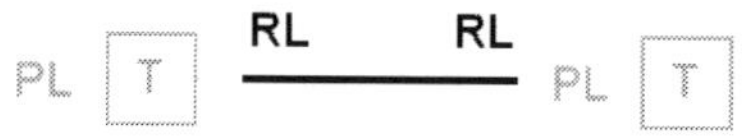 PL [T] RL RL PL [T]	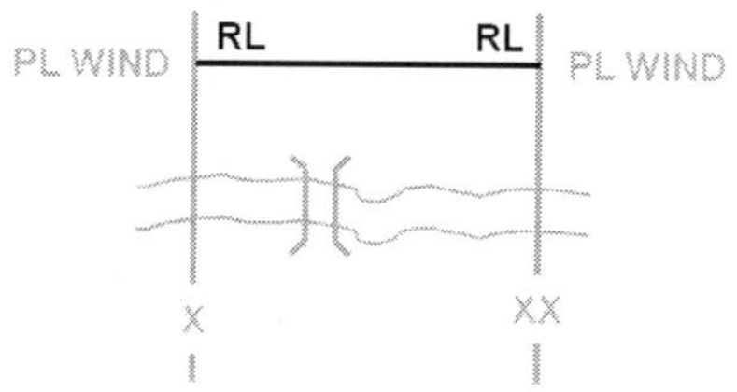 PL WIND RL RL PL WIND X XX
Special areas		
Airhead line	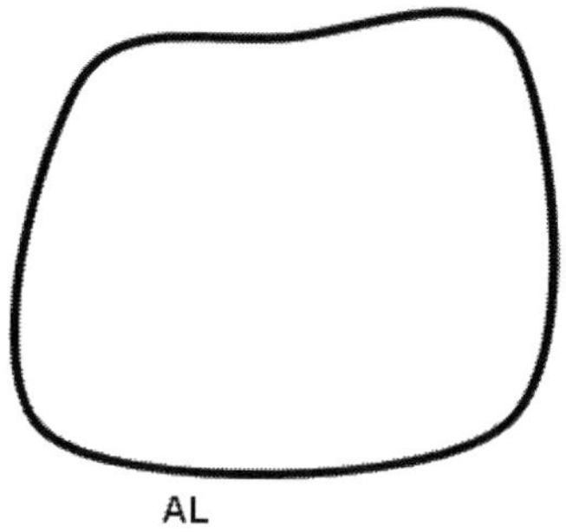 AL ***Note:*** An airhead line can be an area or a line.	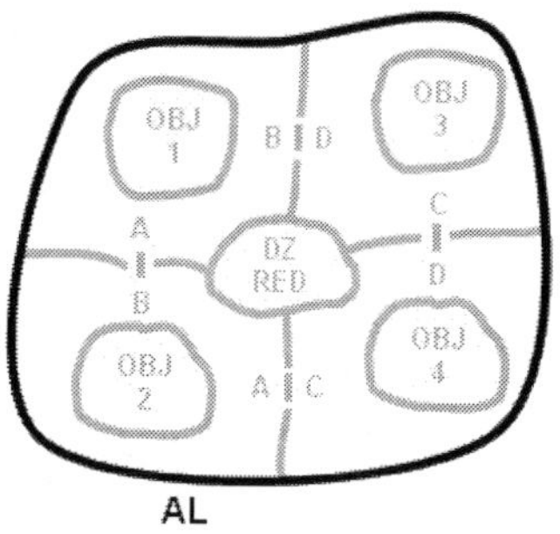 OBJ 1 B D OBJ 3 A B DZ RED C D OBJ 2 A C OBJ 4 AL

FIRES

8-12. Table 8-3 (on pages 8-30 through 8-40) shows fires control measure symbols.

Table 8-3. Fires

Type	*Icon*	*Example*
Fire support coordination measures		
Points		
Fire support station	FSS 3	FSS 5
Lines		
Coordinated fire line	CFL 3 CFL 3 16 16-1 16 16-1	CFL 2BCT 210800ZNOV12- 241200ZNOV12 CFL 2BCT 210800ZNOV12- 241200ZNOV12 PL MAPLE PL MAPLE X X
Fire support coordination line	FSCL 3 FSCL 3 16 16-1 16 16-1	FSCL 1AD 110800ZMAY98- 041200MAY13 FSCL 1AD 110800ZMAY98- 041200MAY13 PL FOX PL FOX XX XX
Restrictive fire line	RFL 3 RFL 3 16 16-1 16 16-1	X 3/1 RFL 23ID 110800ZMAY98- 041200MAY13 RFL 23ID 110800ZMAY98- 041200MAY13 PL RED PL RED XX X XX 2-23

Table 8-3. Fires (continued)

Type	*Icon*	*Example*
	Areas	
Airspace coordination area	ACA 3 3-1 MIN ALT MAX ALT 18 16 16-1	ACA ROVER 1 MND(N) MIN ALT 500 MAX ALT 3000 GRID FD1173, FD825, FD8211, FD1111 240000ZDEC12- 291100ZDEC12
Bomb	BOMB	BOMB
Free fire area	FFA 3 16 16-1	FFA 2ID 141230ZMAR13- 142330ZMAR13
Fire support area	FSA 3	FSA ZULU

Table 8-3. Fires (continued)

Type	Icon	Example
Kill box	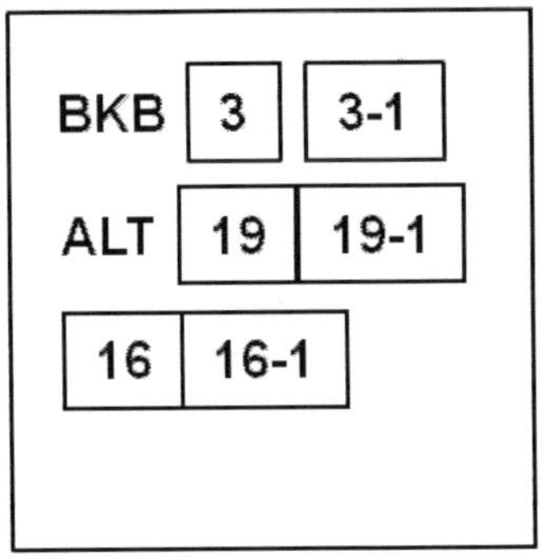 BKB 3 3-1 ALT 19 19-1 16 16-1 Blue kill box	 BKB044 JFLCC ALT 9000-30000FT 160600ZMAY13- 171200ZMAY13
	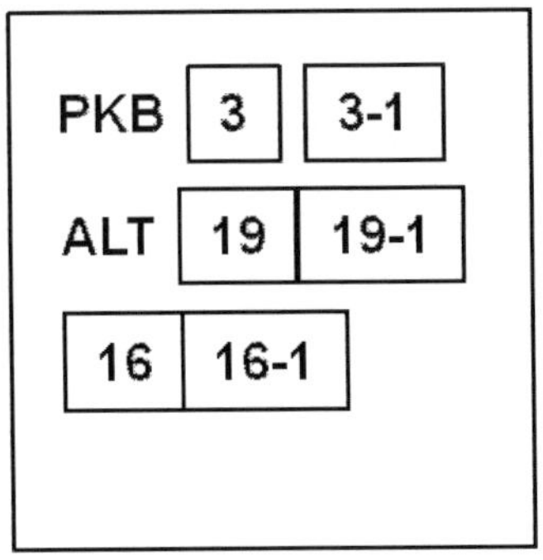 PKB 3 3-1 ALT 19 19-1 16 16-1 Purple kill box	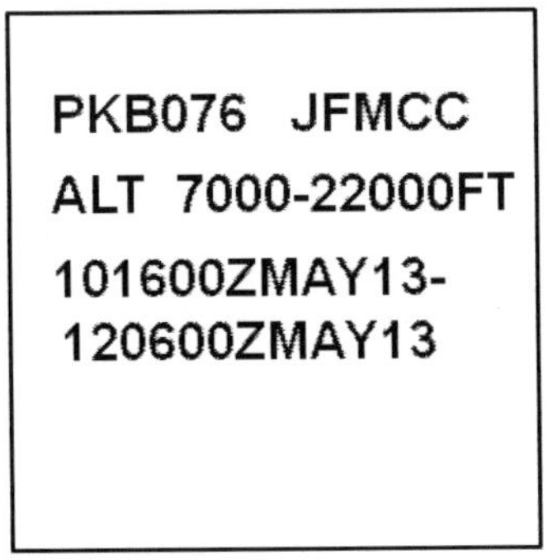 PKB076 JFMCC ALT 7000-22000FT 101600ZMAY13- 120600ZMAY13
No-fire area	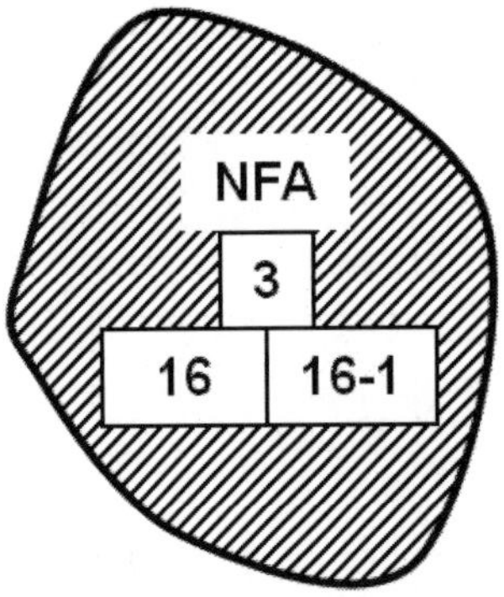 NFA 3 16 16-1 *Note*: No fire area has black hatching.	 NFA 52ID (GBR) 031230ZJUL13- 232330ZJUL13
Position area for artillery	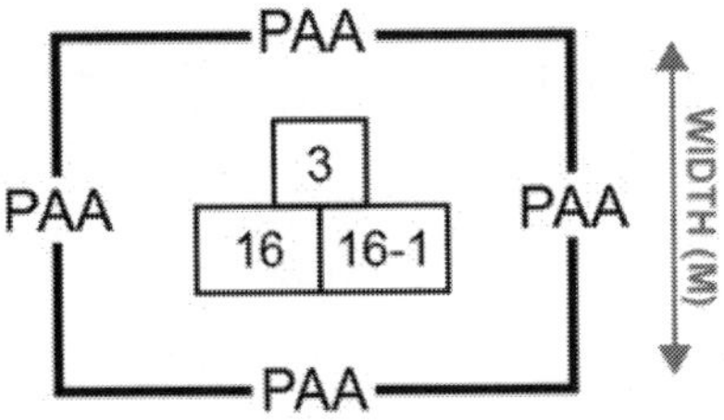 PAA PAA 3 PAA 16 16-1 PAA WIDTH (M)	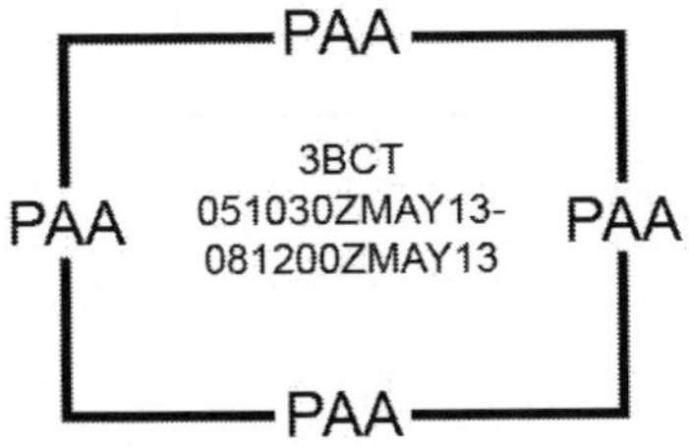 PAA PAA 3BCT PAA 051030ZMAY13- 081200ZMAY13 PAA

Table 8-3. Fires (continued)

Type	*Icon*	*Example*
Restrictive fire area	RFA 3 16 16-1	RFA 10MTND 230700ZMAY13- 232100ZMAY13
Targeting		
Points		
Nuclear target	3	DT0307
Point or single target	3 19 TARGET DESCRIPTION	ST1912 25 ROCKET LAUNCHER
Lines or linear targets		
Final protective fire (FPF)	3 FPF 3-1 17	DT0307 FPF 2-3CAV MORTAR
Linear target	3	MC2302

Table 8-3. Fires (continued)

Type	*Icon*	*Example*
Linear smoke target	3 SMOKE	CS1403 SMOKE
Areas ***Note***: All area target control measure symbols can be built using one of three ways. Area target control measure symbols can be circular rectangular, or irregular. Examples of how each area is determined are shown below. However, only the irregular area target control measure symbol will be used for the template and example.		
Circular area target	3 RADIUS ***Note***: A circular target consists of a center point and a radius.	ST2305
Rectangular area target	LENGTH 3 WIDTH ***Note***: A rectangular target consists of a corner point and a length and width.	BT2305
Irregular area target	3	MT2307

Table 8-3. Fires (continued)

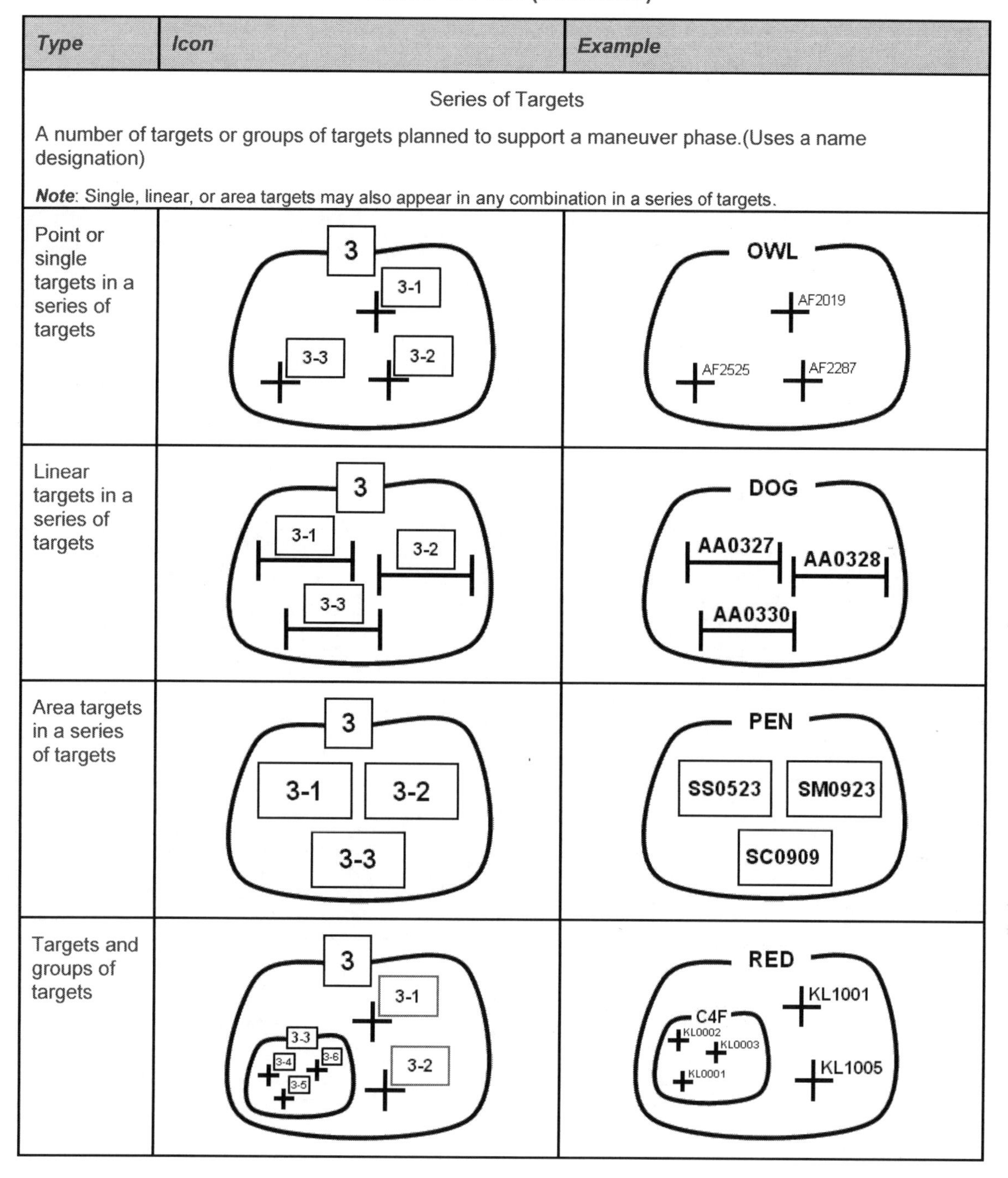

Type	*Icon*	*Example*
Series of Targets A number of targets or groups of targets planned to support a maneuver phase.(Uses a name designation) ***Note***: Single, linear, or area targets may also appear in any combination in a series of targets.		
Point or single targets in a series of targets		
Linear targets in a series of targets		
Area targets in a series of targets		
Targets and groups of targets		

Table 8-3. Fires (continued)

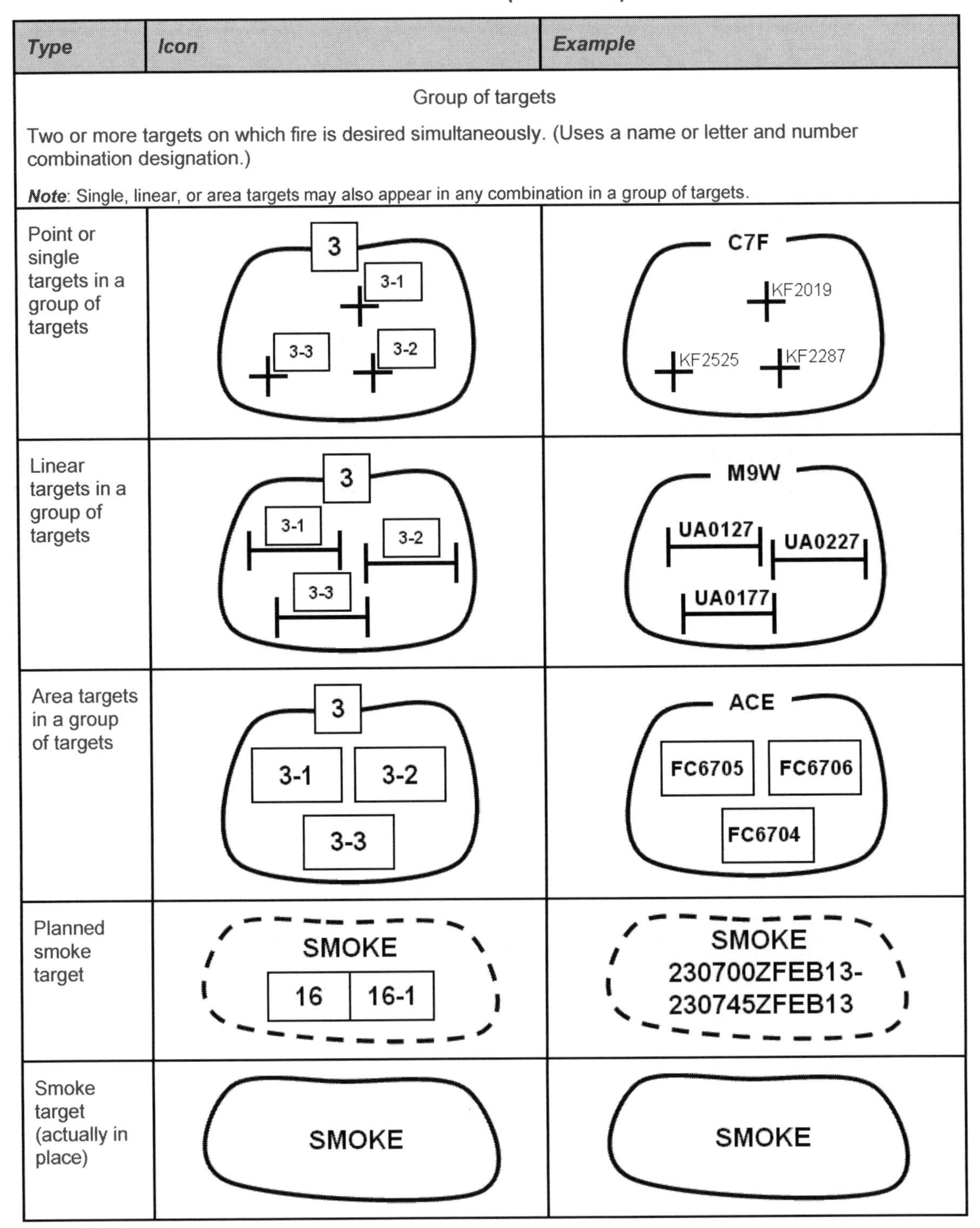

Type	*Icon*	*Example*
Group of targets Two or more targets on which fire is desired simultaneously. (Uses a name or letter and number combination designation.) ***Note***: Single, linear, or area targets may also appear in any combination in a group of targets.		
Point or single targets in a group of targets	3 3-1 3-3 3-2	C7F KF2019 KF2525 KF2287
Linear targets in a group of targets	3 3-1 3-2 3-3	M9W UA0127 UA0227 UA0177
Area targets in a group of targets	3 3-1 3-2 3-3	ACE FC6705 FC6706 FC6704
Planned smoke target	SMOKE 16 16-1	SMOKE 230700ZFEB13- 230745ZFEB13
Smoke target (actually in place)	SMOKE	SMOKE

Table 8-3. Fires (continued)

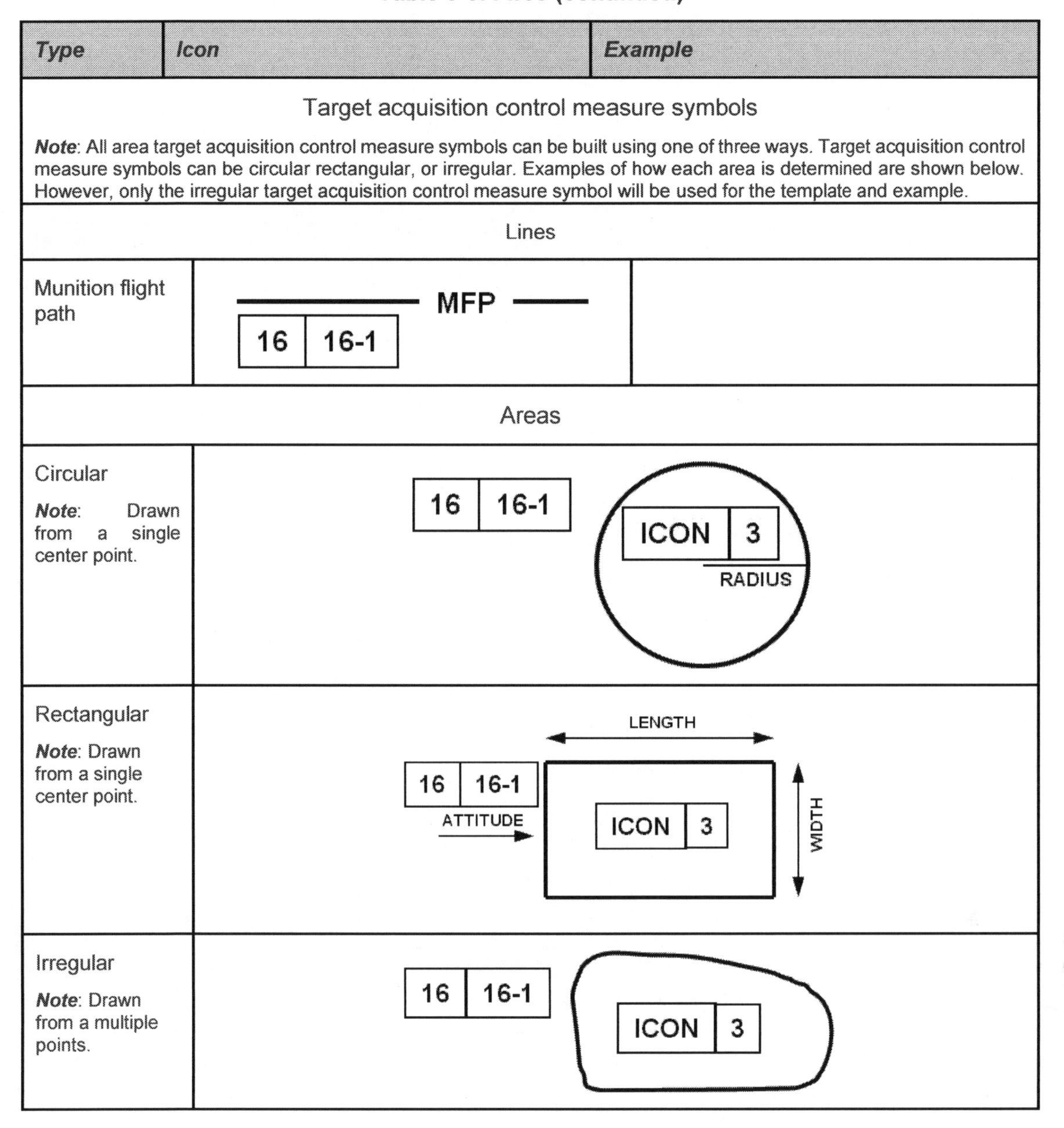

Type	*Icon*	*Example*
Target acquisition control measure symbols ***Note***: All area target acquisition control measure symbols can be built using one of three ways. Target acquisition control measure symbols can be circular rectangular, or irregular. Examples of how each area is determined are shown below. However, only the irregular target acquisition control measure symbol will be used for the template and example.		
Lines		
Munition flight path	MFP 16 16-1	
Areas		
Circular ***Note***: Drawn from a single center point.	16 16-1 ICON 3 RADIUS	
Rectangular ***Note***: Drawn from a single center point.	LENGTH 16 16-1 ATTITUDE ICON 3 WIDTH	
Irregular ***Note***: Drawn from a multiple points.	16 16-1 ICON 3	

Table 8-3. Fires (continued)

Type	*Icon*	*Example*
Target acquisition areas		
Artillery target intelligence zone	16 | 16-1 ATI 3	091200ZMAY12 ATI 1
Call for fire zone	16 | 16-1 CFF 3	120030ZFEB13-120030ZMAR13 CFF TANGO 2
Critical friendly zone	16 | 16-1 CF ZONE 3	190900ZDEC12-220900ZDEC12 CF ZONE RAIN
Dead space area	16 | 16-1 DA 3	140030ZAUG13-160300ZAUG13 DA RADAR 1
Sensor zone	16 | 16-1 SENSOR ZONE 3	SENSOR ZONE GREEN
Target build-up area	16 | 16-1 TBA 3	010001ZJAN13-032400ZJAN13 TBA CHIEFS

Table 8-3. Fires (continued)

Type	*Icon*	*Example*
Target value area	16 16-1 TVAR 3	070600ZJAN13- 071400ZJAN13 TVAR LAKERS
Zone of responsibility	16 16-1 ZOR 3	091400ZOCT12 ZOR 9
Terminally guided munitions footprint	TGMF	TGMF
Circular weapon or sensor range fan	UNIT OR EQUIPMENT SYMBOL MIN RG 20 ALT 19 MAX RG (1) 20-1 ALT 19-1 MAX RG (2) 20-2 ALT 19-2	MIN RG 500 ALT GL MAX RG (1) 9000 ALT 100 MAX RG (2) 16500 ALT 150

Table 8-3. Fires (continued)

Type	*Icon*	*Example*
Weapon or sensor range fan	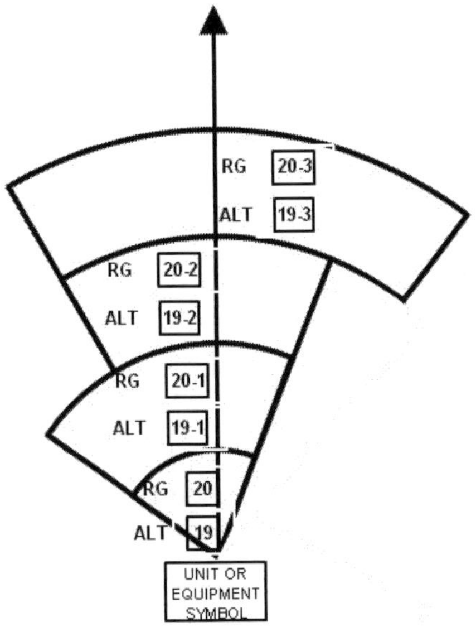 RG 20-3 ALT 19-3 RG 20-2 ALT 19-2 RG 20-1 ALT 19-1 RG 20 ALT 19 UNIT OR EQUIPMENT SYMBOL	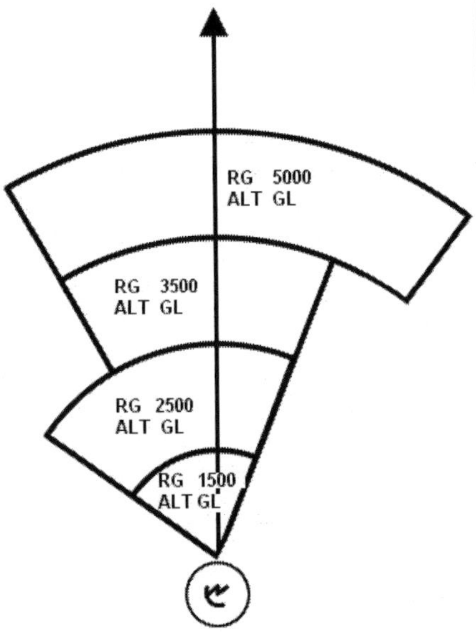 RG 5000 ALT GL RG 3500 ALT GL RG 2500 ALT GL RG 1500 ALT GL

PROTECTION

8-13. Table 8-4 (on pages 8-41 through 8-62) shows protection control measure symbols.

Table 8-4. Protection

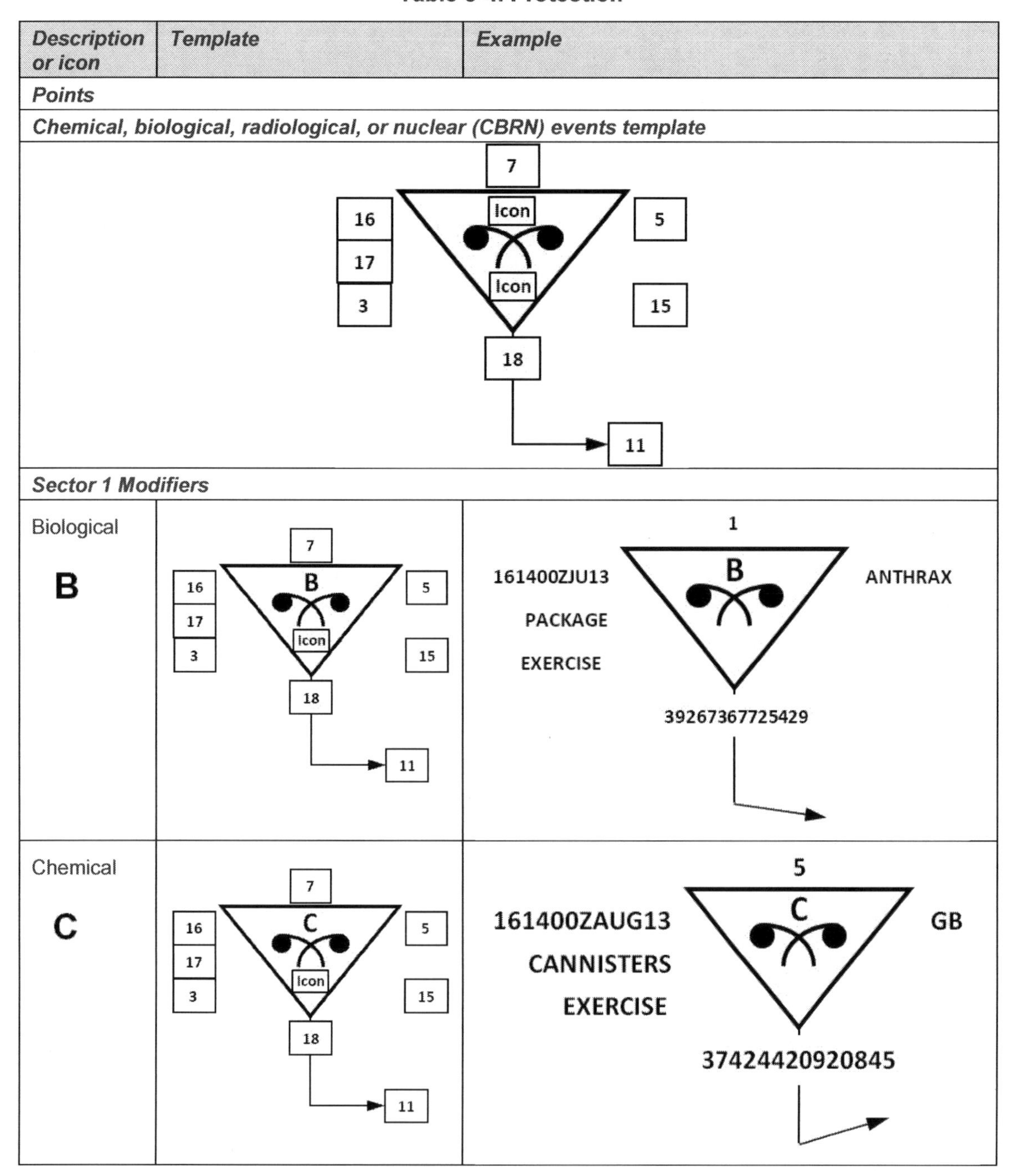

Description or icon	*Template*	*Example*
Points		
Chemical, biological, radiological, or nuclear (CBRN) events template		
	7 16 17 3 Icon Icon 5 15 18 11	
Sector 1 Modifiers		
Biological **B**	7 16 17 3 B Icon 5 15 18 11	1 161400ZJU13 B ANTHRAX PACKAGE EXERCISE 39267367725429
Chemical **C**	7 16 17 3 C Icon 5 15 18 11	5 161400ZAUG13 C GB CANNISTERS EXERCISE 37424420920845

Table 8-4. Protection (continued)

Description or icon	*Template*	*Example*
Nuclear and nuclear detonation for friendly ground zero **N**	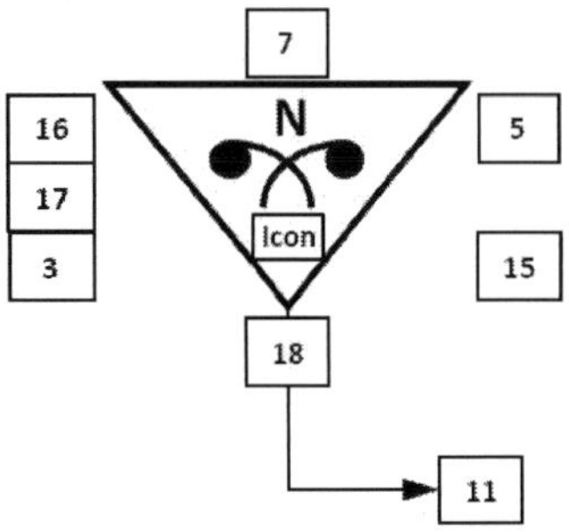 7, 16, 17, 3, N, Icon, 5, 15, 18, 11	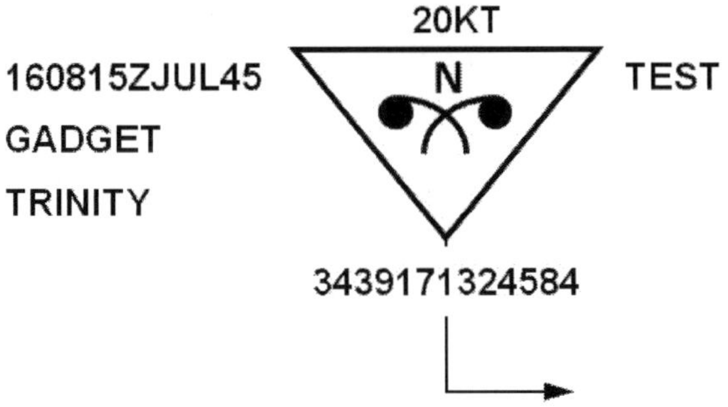 20KT 160815ZJUL45 GADGET TRINITY N TEST 3439171324584
Radiological **R**	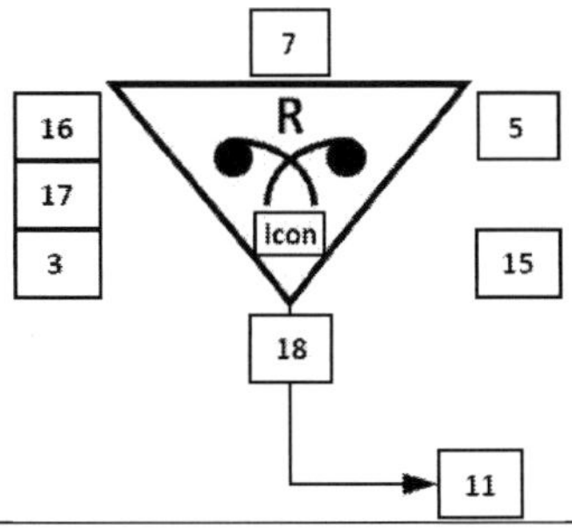 7, 16, 17, 3, R, Icon, 5, 15, 18, 11	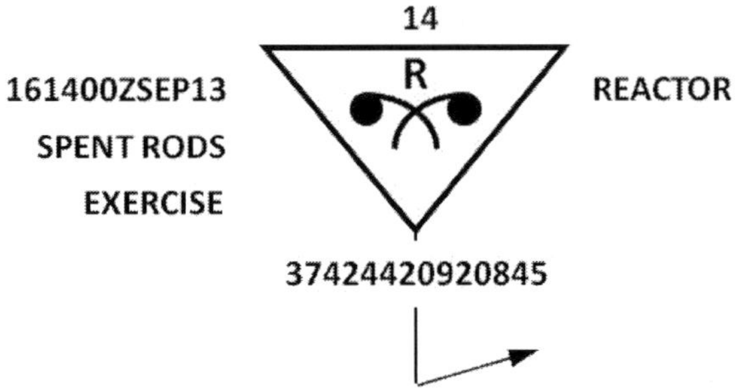 14 161400ZSEP13 SPENT RODS EXERCISE R REACTOR 37424420920845
Sector 2 Modifier		
Toxic industrial material **TIM**	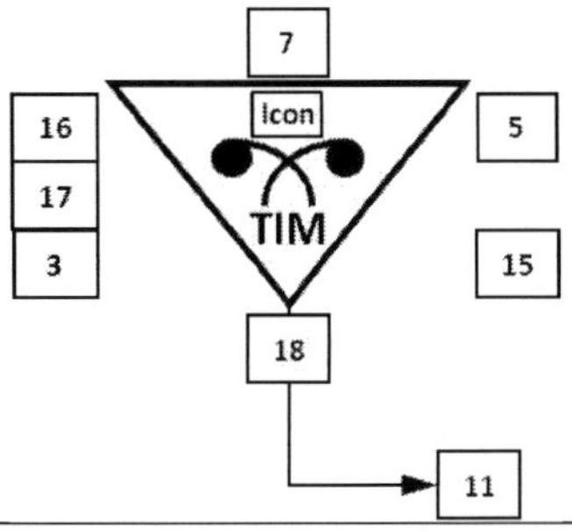 7, 16, 17, 3, Icon, TIM, 5, 15, 18, 11	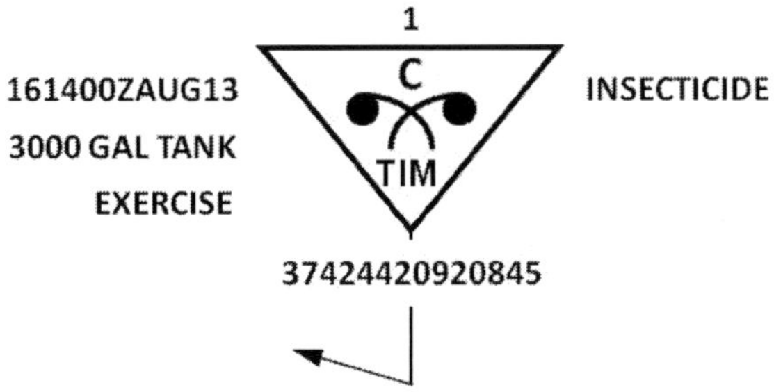 1 161400ZAUG13 3000 GAL TANK EXERCISE C TIM INSECTICIDE 37424420920845
Planned or on order or templated chemical, biological, radiological or nuclear events	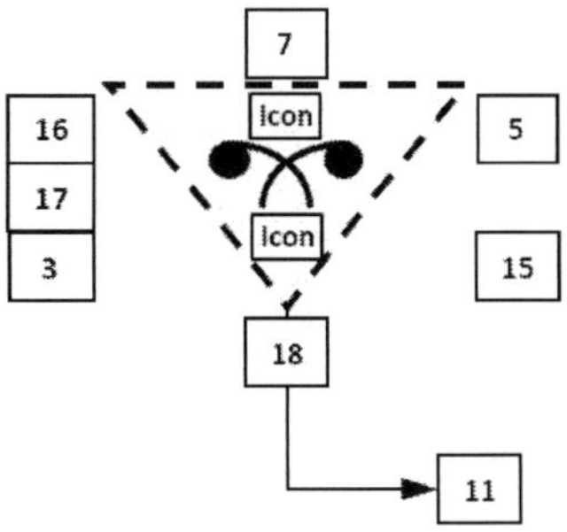 7, 16, 17, 3, Icon, Icon, 5, 15, 18, 11	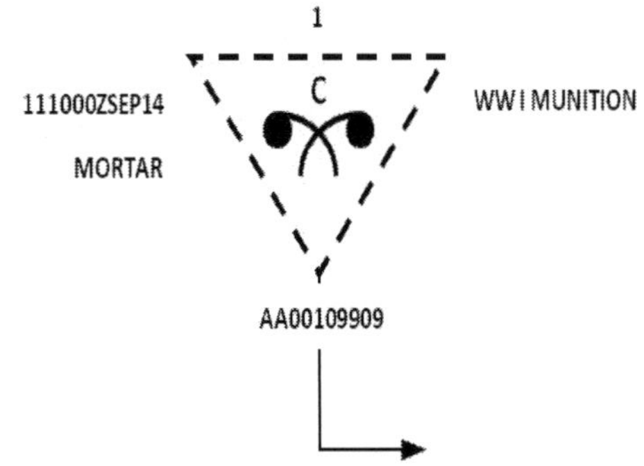 1 111000ZSEP14 MORTAR C WW I MUNITION AA00109909

Table 8-4. Protection (continued)

Description	*Template*	*Example*
Decontamination points		
General decontamination point	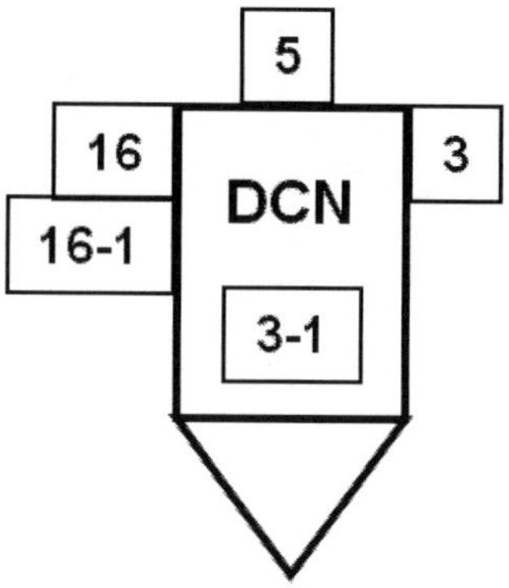 5 16 DCN 3 16-1 3-1	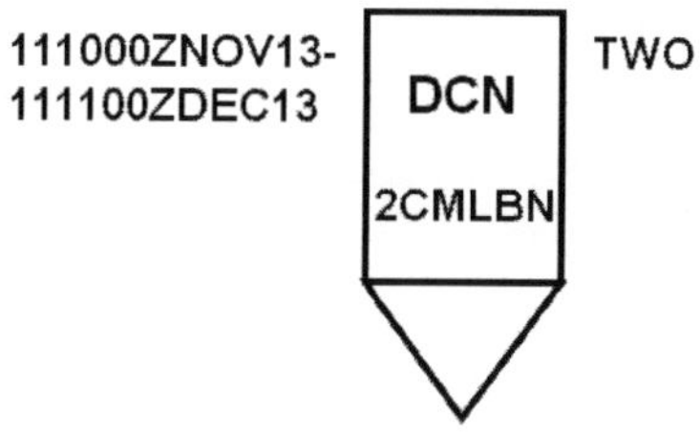 111000ZNOV13- 111100ZDEC13 DCN 2CMLBN TWO
Alternate decontamination point	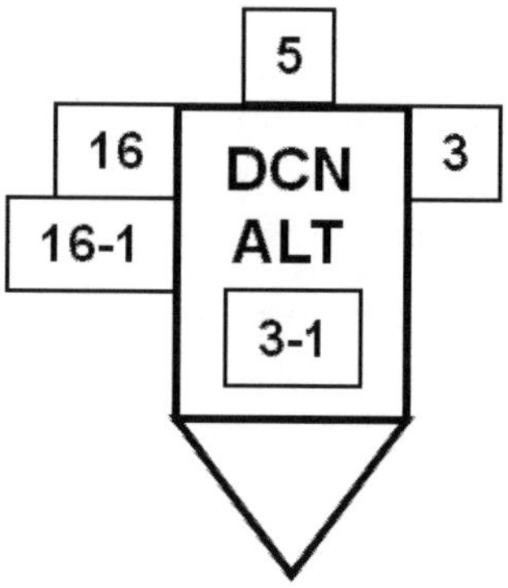 5 16 DCN ALT 3 16-1 3-1	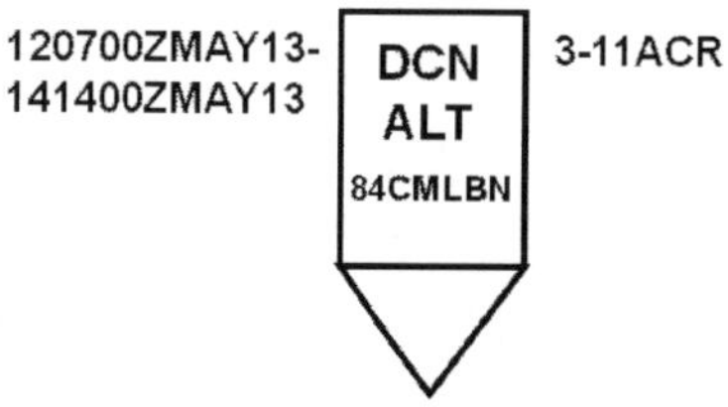 120700ZMAY13- 141400ZMAY13 DCN ALT 84CMLBN 3-11ACR
Equipment decontamination point	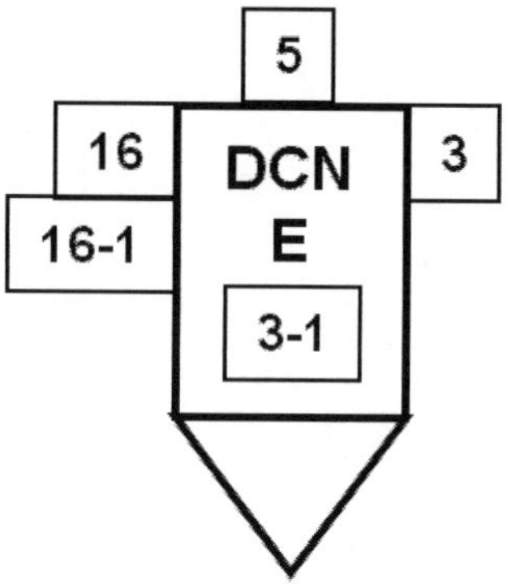 5 16 DCN E 3 16-1 3-1	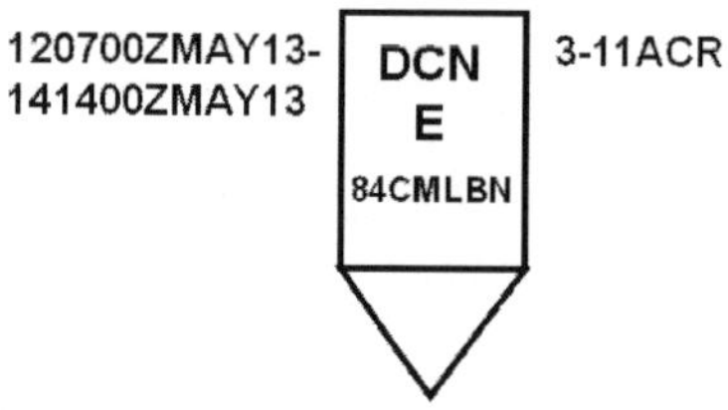 120700ZMAY13- 141400ZMAY13 DCN E 84CMLBN 3-11ACR

Table 8-4. Protection (continued)

Description	Template	Example
Equipment or troop decontamination point	5 16 16-1 DCN E/TP 3-1 3	CONTRACTOR OPERATED 210700ZAPR13-071800ZMAY13 DCN E/TP 2MEB V CORPS
Main equipment decontamination point	5 16 16-1 DCN ME 3-1 3	VEHICLES 240700ZAUG13-251800ZAUG13 DCN ME 1AD V CORPS
Operational decontamination point	5 16 16-1 DCN O 3-1 3	250700ZAPR13-291800ZAPR13 DCN O 6BCT 3
Through decontamination point	5 16 16-1 DCN TH 3-1 3	010900ZAUG13-021500ZAUG13 DCN TH 2ID 6-37FA

Table 8-4. Protection (continued)

Description	*Template*	*Example*
Troop decontamination point	5 16 DCN 3 16-1 TP 3-1	011900ZJUN13- 021500ZJUN13 DCN TP 10MTND HHC
Forward troop decontamination point	5 16 DCN 3 16-1 FTP 3-1	210700ZAUG13- 221800ZAUG13 DCN FTP 2BCT ALPHA
Wounded personnel decontamination point	5 16 DCN 3 16-1 W 3-1	14 FIELD HOSPITAL DCN W 22CMLBN 2ID

Table 8-4. Protection (continued)

Description	*Template*	*Example*
Displaced persons, refugees, and evacuees		
Type	*Icon*	*Example*
Civilian collection point	5 16 CIV 3 16-1 3-1	HOST NATION ONLY CIV NATO UN
Detainee collection point	5 16 DET 3 16-1 3-1	DET MND-S 10MPBN
Enemy prisoner of war collection point	5 16 EPW 3 16-1 3-1	160530ZJUL13 EPW V CORPS 11MPBN

Table 8-4. Protection (continued)

Description	*Template*	*Example*
Obstacles		
Abatis ***Note***: Abatis are normally trees large enough to block the road in a criss-cross pattern.		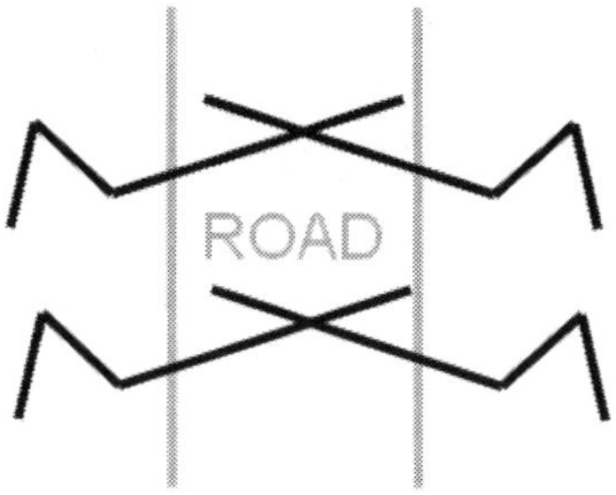 ROAD
Mines		
Antipersonnel mine		
Antipersonnel mine with dashed arrow showing direction of effects	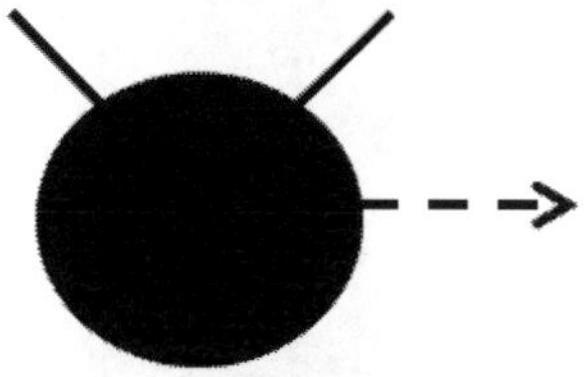	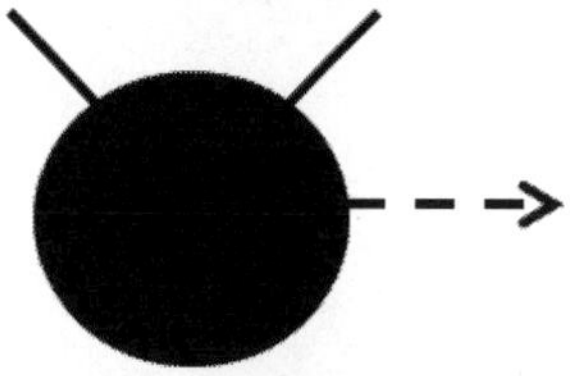
Antitank mine		
Antitank mine with anti-handling device		
Unspecified mine	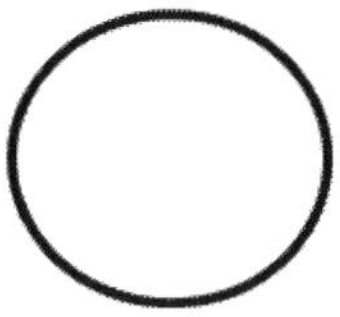	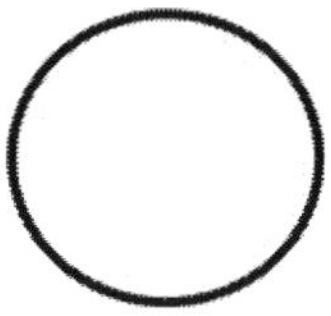

Table 8-4. Protection (continued)

Description	*Template*	*Example*
Wide area mine		
Booby trap		
Earthwork, small trench, or fortification		
Engineer regulating point	5 16 3 16-1 ERP 3-1	8 ERP 2ENBN
Fort		
Foxhole		

Table 8-4. Protection (continued)

Description	*Template*	*Example*
Antitank obstacle: tetrahedrons, dragon's teeth, and other similar obstacles		
Fixed and prefabricated antitank obstacle, tetrahedrons, dragon's teeth, and other similar obstacles		
Moveable antitank obstacle, tetrahedrons, dragon's teeth, and other similar obstacles		
Moveable and prefabricated antitank obstacle, tetrahedrons, dragon's teeth, and other similar obstacles		
Surface shelter		
Underground shelter		
Road Blocks		
Planned		
Explosive state of readiness 1 (safe)		

Table 8-4. Protection (continued)

Description	*Template*	*Example*
Explosive state of readiness 1 (armed but passable)	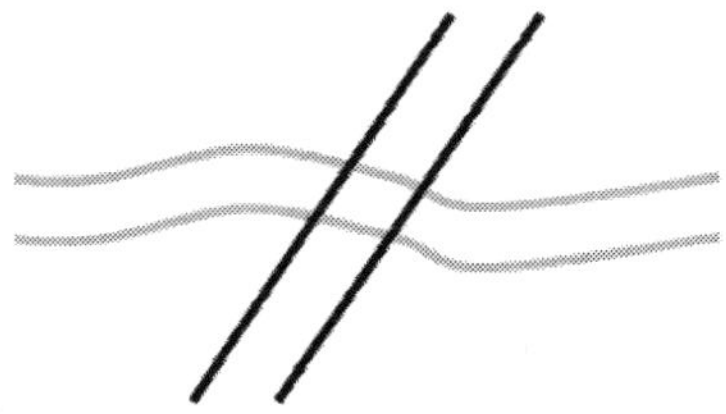	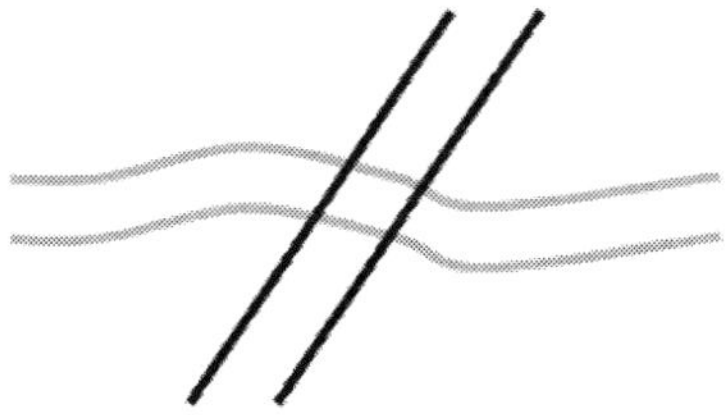
Road block complete	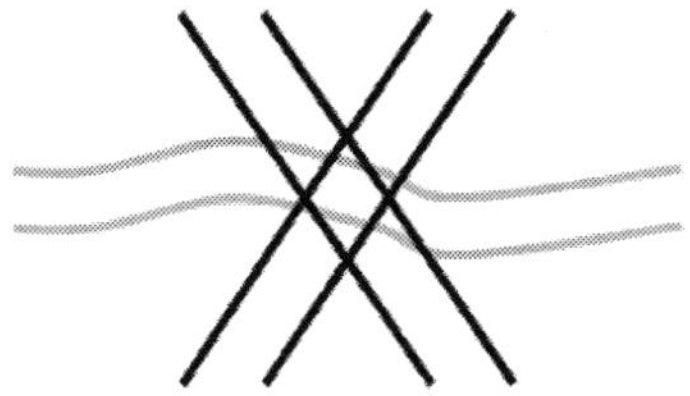	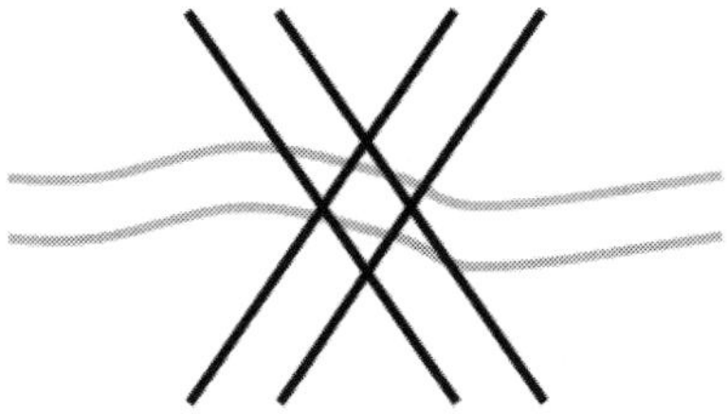
Obstacle bypass		
Obstacle bypass easy		
Obstacle bypass difficult	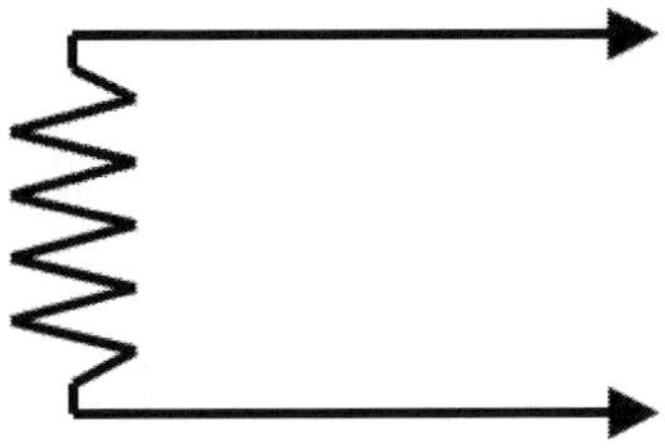	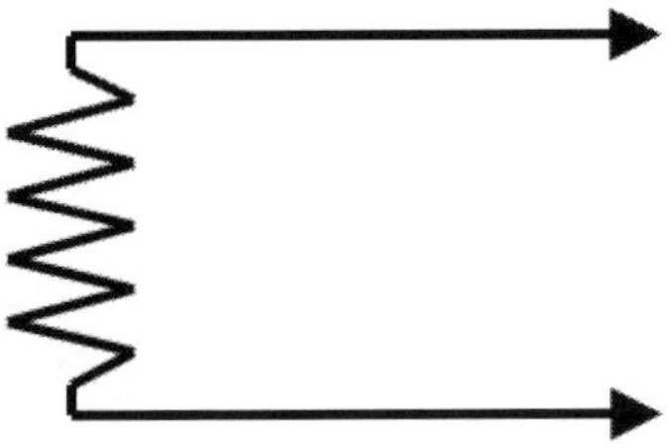
Obstacle bypass impossible	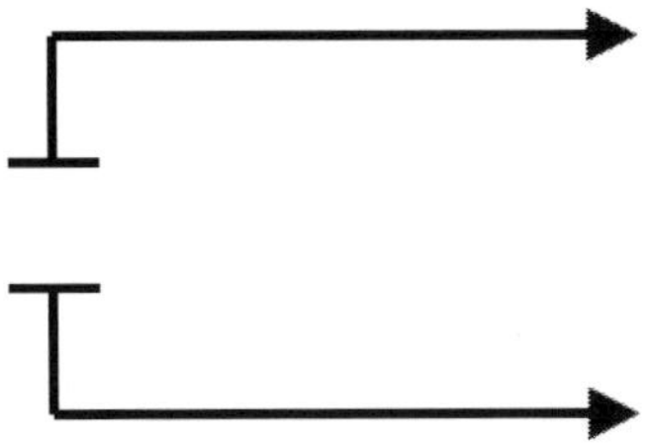	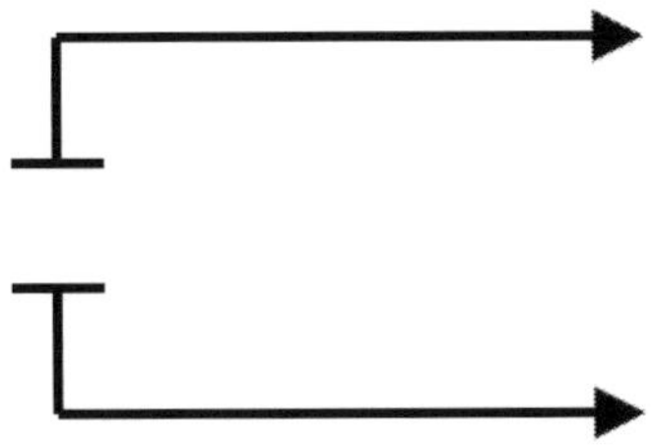

Table 8-4. Protection (continued)

Description	*Template*	*Example*
Obstacle effects		
Block	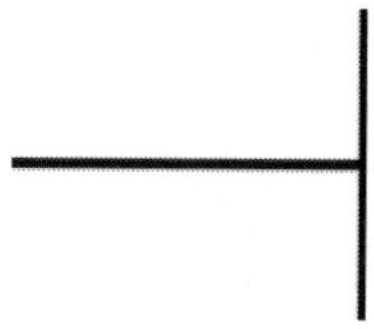	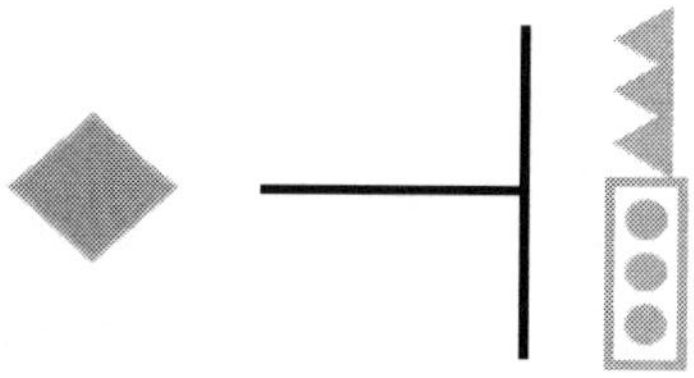
Disrupt	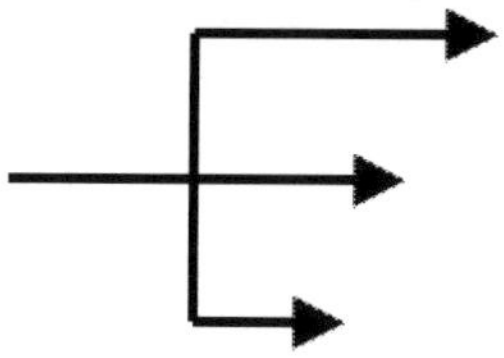	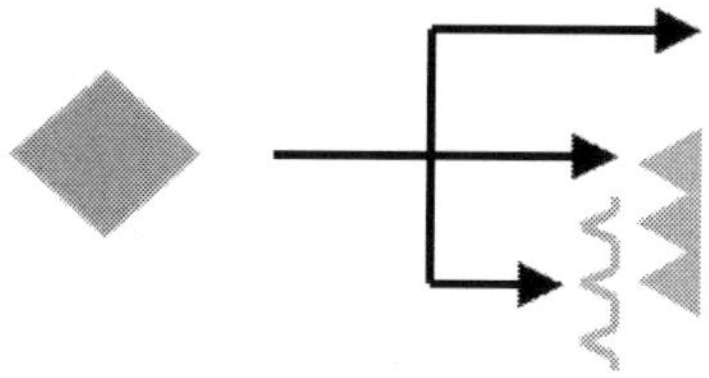
Fix		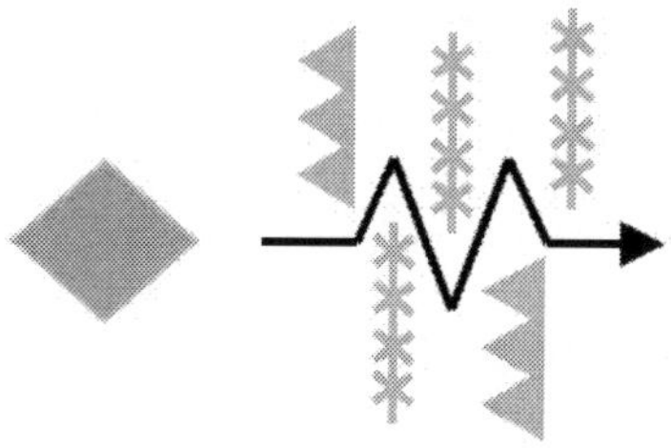
Turn		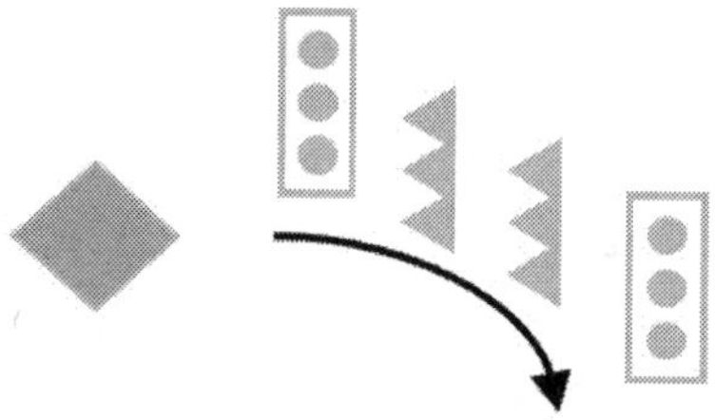

Table 8-4. Protection (continued)

Description	*Template*	*Example*
Lines		
Obstacles		
Antitank ditches ***Note***: The points of the antitank ditch point away from the enemy.		
Antitank ditch *completed*		
Antitank ditch *under construction*		
Antitank ditch *reinforced with antitank mines*		

Table 8-4. Protection (continued)

Description	*Template*	*Example*
Antitank wall – should point down		
Fortified line		
Mine cluster – may be an area		
Obstacle line	3	1-3 IN
Trip wire		
Wire obstacles		
Wire obstacle unspecified	X X X X X X X X	X X X X X X X X
Wire obstacle single fence		

Table 8-4. Protection (continued)

Description	*Template*	*Example*
Wire obstacle double fence		
Wire obstacle double apron fence		
Wire obstacle low wire fence		
Wire obstacle high wire fence		
Wire obstacle single strand concertina		
Wire obstacle double strand concertina		
Wire obstacle triple strand concertina		

Table 8-4. Protection (continued)

Description	*Template*	*Example*
Crossing sites or water crossings		
Assault crossing area	***Note***: This symbol is always parallel to the river.	
Bridge of gap crossing		
Ferry		
Ford easy		

Table 8-4. Protection (continued)

Description	*Template*	*Example*
Ford difficult		
Lane	16 16-1	120600ZFEB13- 121000ZFEB13
Raft site		
Areas		
Chemical, biological, radiological, or nuclear (CBRN) contaminated area		
Icon Icon ***Notes:*** Hatched lines are in yellow and outline is in black. Use sector 1 and sector 2 modifier icons and fields from CBRN events.		

Table 8-4. Protection (continued)

Description	*Template*	*Example*
	Sector 1 Modifiers	
Biological contaminated area	B Icon	B
Chemical contaminated area	C Icon	C
Nuclear contaminated area	N Icon	N
Radiological contaminated area	R Icon	R
	Sector 2	
Toxic industrial material contaminated area	Icon TIM	C TIM Chemical toxic industrial material contaminated area

Table 8-4. Protection (continued)

Description	*Template*	*Example*
Dose rate contour lines	cGy cGy cGy	30cGy 100cGy 300cGy
Minimum safe distance zones	1 2 3	1 2 3
Displaced persons, refugees, and evacuees		
Type	*Icon*	*Example*
Detainee holding area	DET	DET
Enemy prisoner of war holding area	EPW	EPW
Refugee holding	REF	REF

Table 8-4. Protection (continued)

Type	*Icon*	*Example*
Obstacles		
Limited access area	Mobility Icon	No tracked vehicles.
Obstacle belt	3	3-4CAV
Obstacle zone	3	1-7IN
Minefield completed (Unspecified)	***Note***: The type or types of mines are shown for the minefield.	

Table 8-4. Protection (continued)

Type	*Icon*	*Example*
Minefield planned (unspecified)	*Note*: The type or types of mines are shown for the minefield.	
Minefield antipersonnel		
Minefield antitank		
Minefield gap	16 16-1	S W O R D 030700ZOCT14- 031900ZOCT14
Scatterable mines minefield with self-destruct date-time group		
	S MINE SYMBOL 16	
Antipersonnel scatterable mines minefield with self-destruct date-time group	S 16	S 032200ZJUL13

Table 8-4. Protection (continued)

Type	*Icon*	*Example*
Antitank scatterable mines minefield with self-destruct date-time group	S 16	S 191000ZDEC12
Unspecified scatterable mines minefield with self-destruct date-time group	S 16	S 221200 ZDEC12
Mined area	M M M M	M M M M
Obstacle free area	FREE 3 16 16-1	FREE 2 EN BN 011730OCT12- 030900NOV12

Table 8-4. Protection (continued)

Type	*Icon*	*Example*
Obstacle restricted area	3 16 16-1	1AD (US) 210700ZMAY13- 250900ZMAY13
Unexploded ordnance area	UXO UXO	UXO UXO

Sustainment

8-14. Table 8-5 (on pages 8-63 through 8-73) shows sustainment control measure symbols.

Table 8-5. Sustainment

Type	*Template*	*Example*
	Points	
Ambulance control point	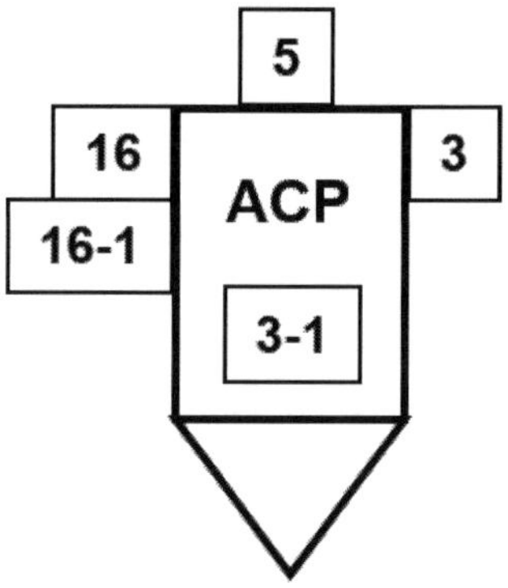 5 16 16-1 ACP 3 3-1	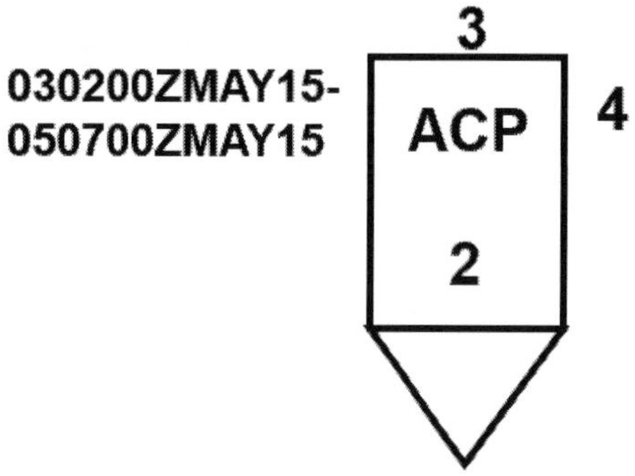 030200ZMAY15-050700ZMAY15 3 ACP 2 4
Ambulance exchange point	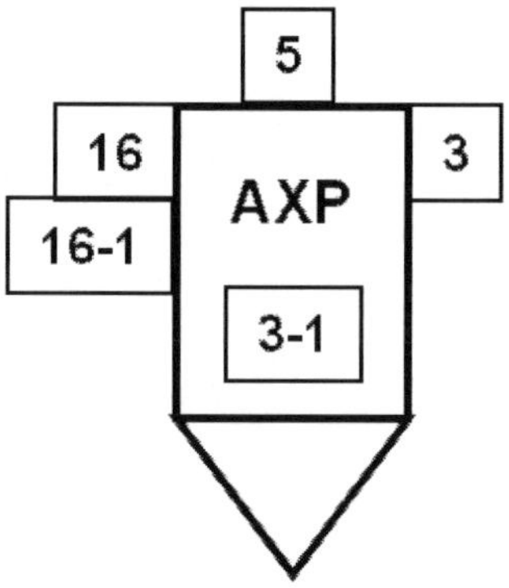 5 16 16-1 AXP 3 3-1	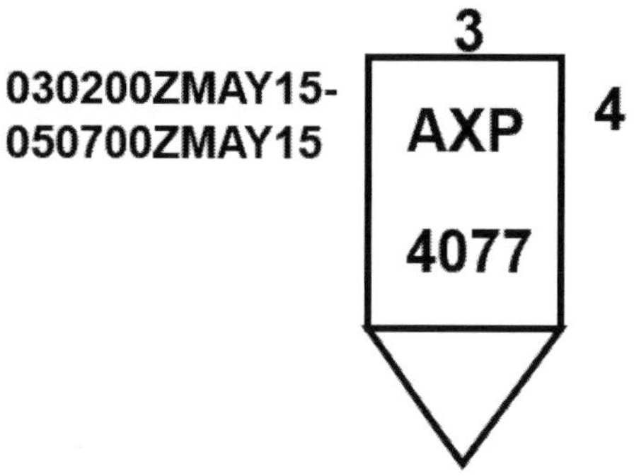 030200ZMAY15-050700ZMAY15 3 AXP 4077 4
Ambulance load point	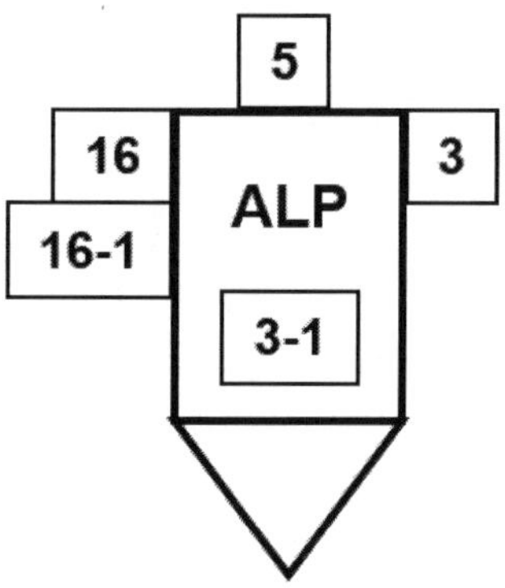 5 16 16-1 ALP 3 3-1	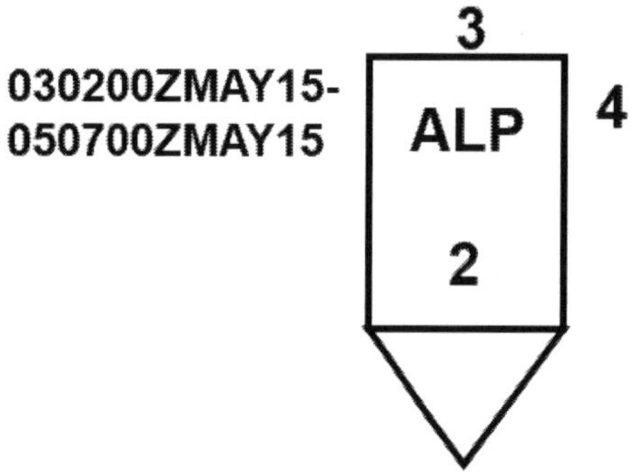 030200ZMAY15-050700ZMAY15 3 ALP 2 4

Table 8-5. Sustainment (continued)

Type	*Template*	*Example*
Ambulance relay point	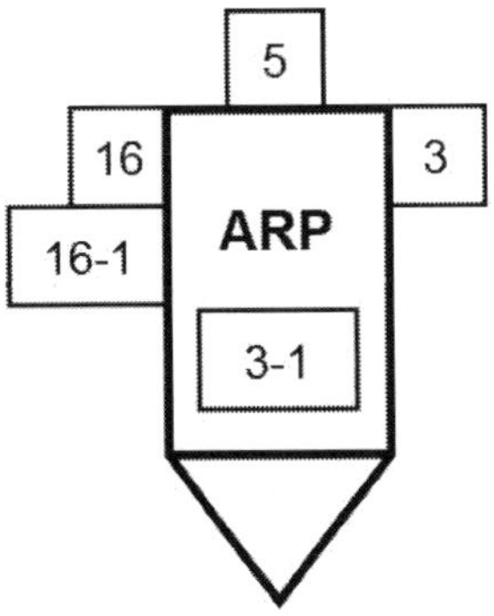 5 16 ARP 3 16-1 3-1	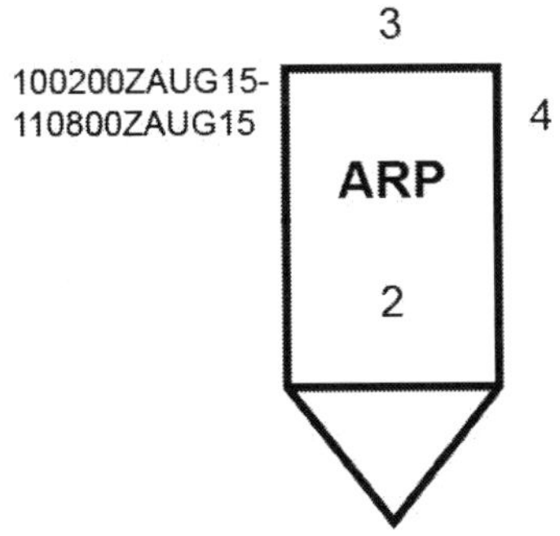 3 100200ZAUG15- 110800ZAUG15 ARP 4 2
Ammunition supply point	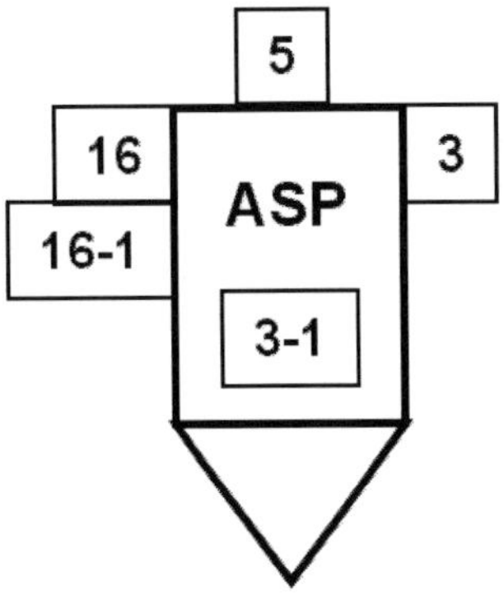 5 16 ASP 3 16-1 3-1	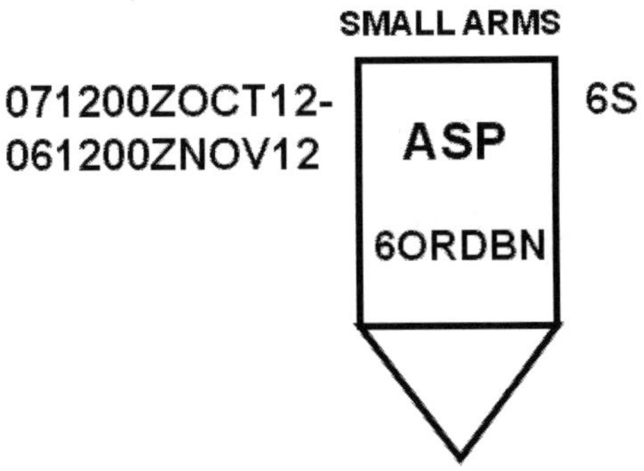 SMALL ARMS 071200ZOCT12- 061200ZNOV12 ASP 6S 6ORDBN
Ammunition transfer holding point	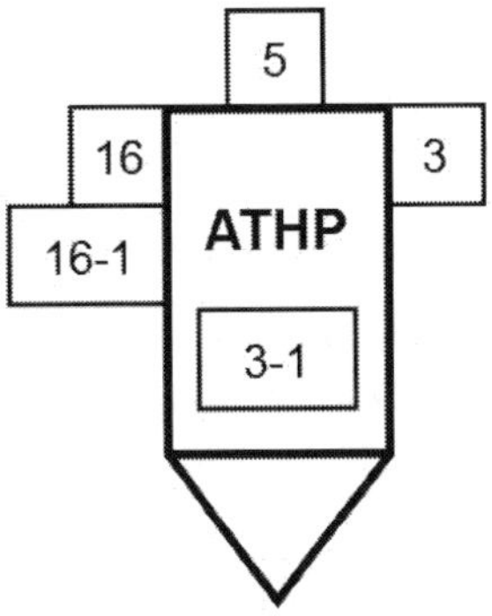 5 16 ATHP 3 16-1 3-1	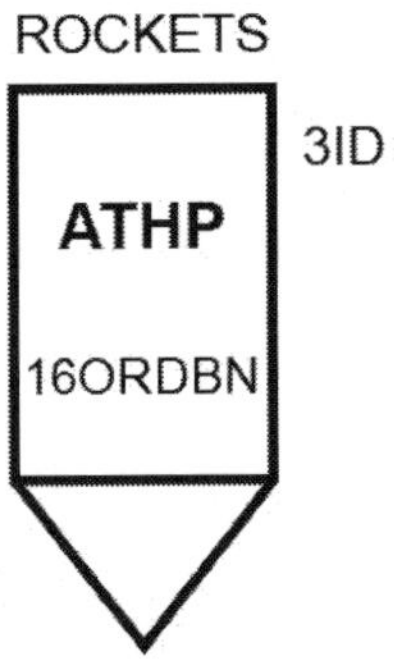 ROCKETS ATHP 3ID 16ORDBN

Table 8-5. Sustainment (continued)

Type	Template	Example
Cannibalization point	5 16 16-1 CAN 3 3-1	162100ZMAY13 CAN 10ORDBN 3SUSTBDE
Casualty collection point	5 16 16-1 CCP 3 3-1	CCP 3BDE RED
Logistics release point	5 16 16-1 LRP 3 3-1	091042ZSEP14-111400ZSEP14 LRP 143FSC 3-187IN
Maintenance collection point	5 16 16-1 MCP 3 3-1	VEHICLES MCP MND-S 2

Table 8-5. Sustainment (continued)

Type	*Template*	*Example*
Rearm, refuel, and resupply point	5 16 R3P 3 16-1 3-1	051200ZOCT12- 071800ZOCT12 R3P FOXTROT 2ACR
Refuel on the move	5 16 ROM 3 16-1 3-1	ROM H CO 2/3ACR
Traffic control post	5 16 TCP 3 16-1 3-1	TCP 2A 2MPBN
Trailer transfer point	5 16 TTP 3 16-1 3-1	1410000ZMAR13- 1914000ZMAR13 TTP CARLOS 51TPTCO

Table 8-5. Sustainment (continued)

Type	*Template*	*Example*
Unit maintenance collection point	5 16 UMCP 3 16-1 3-1	1710030ZMAR13- 1914000ZMAR13 UMCP 1 2-6IN
Supply Points		
United States classes of supply points		
Class I Subsistence, including health and welfare items.	5 16 3 16-1 3-1	I CORPS 107QMBN
Class II Clothing, individual equipment, tentage, tool sets and tool kits, hand tools, administrative, and housekeeping supplies and equipment (including maps). This includes items of equipment, other than major items, prescribed in authorization or allowance tables and items of supply (not including repair parts).	5 16 I I 3 16-1 3-1	**See North Atlantic Treaty Organization Class II example below.**

Table 8-5. Sustainment (continued)

Type	*Template*	*Example*
Class III Petroleum, oil, and lubricants (POL), petroleum and solid fuels, including bulk and packaged fuels, lubricating oils and lubricants, petroleum specialty products; solid fuels, coal, and related products.	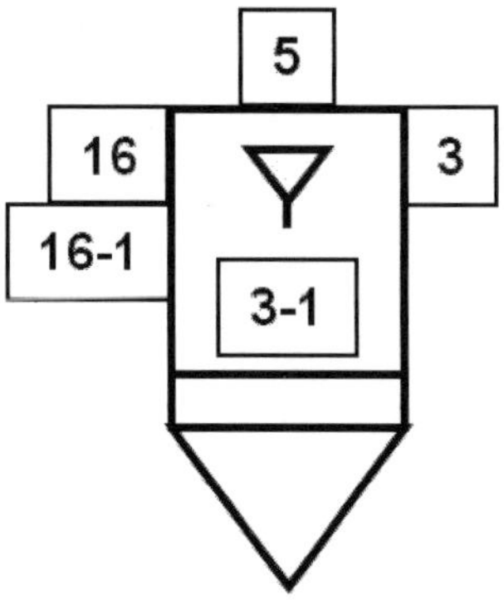 5 16 3 16-1 3-1	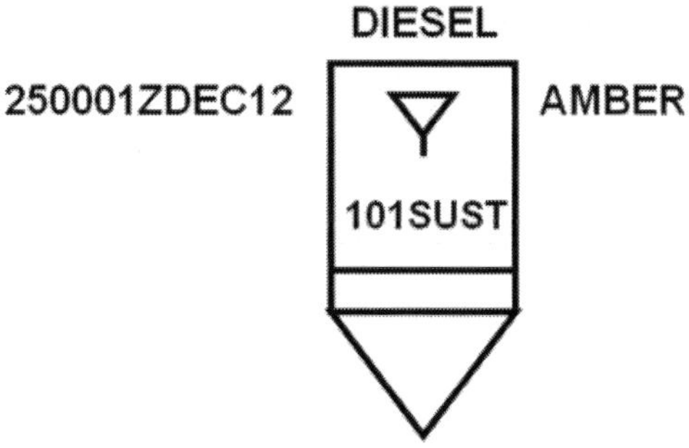 DIESEL 250001ZDEC12 AMBER 101SUST
Class IV Construction materials, to include installed equipment and all fortification and barrier materials.	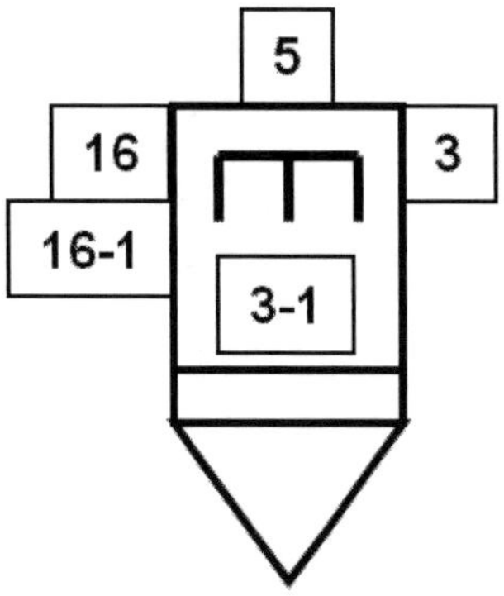 5 16 3 16-1 3-1	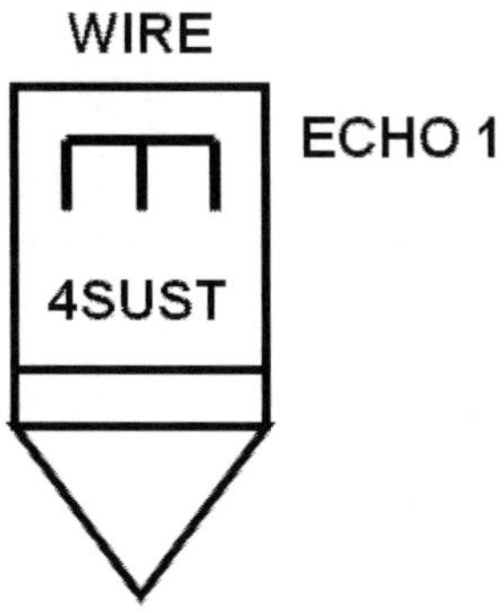 WIRE ECHO 1 4SUST
Class V Ammunition of all types (including chemical, radiological, and special weapons), bombs, explosives, mines, fuses, detonators, pyrotechnics, missiles, rockets, propellants, and other associated items.	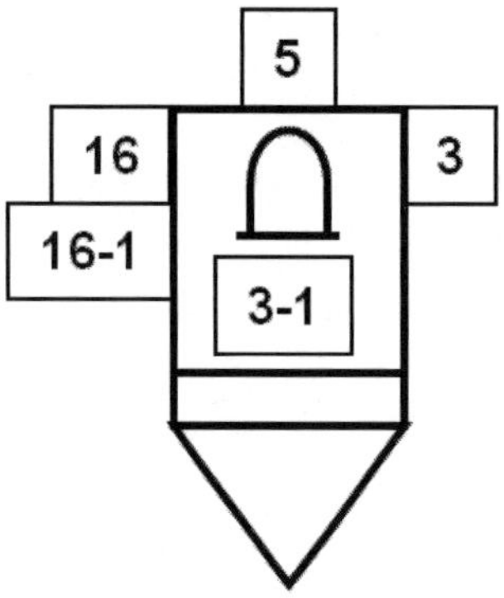 5 16 3 16-1 3-1	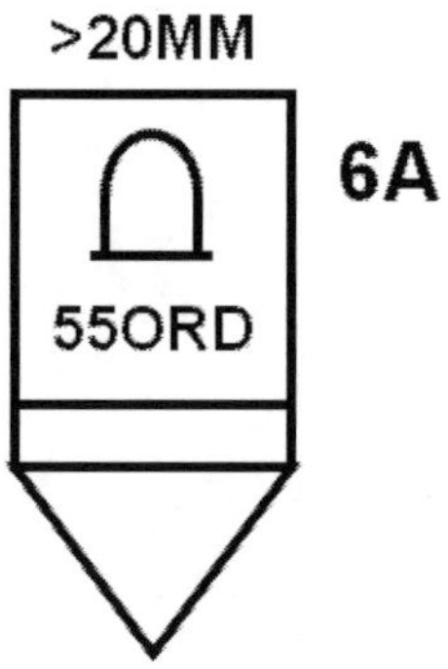 >20MM 6A 55ORD

Table 8-5. Sustainment (continued)

Type	*Template*	*Example*
Class VI Personal demand items (nonmilitary sales items)	5 16 3 16-1 3-1	EAGLE 1CD
Class VII Major items: A final combination of end products which is ready for its intended use: (principal item) for example, launchers, tanks, mobile machine shops, vehicles.	5 16 3 16-1 3-1	PUSAN 19TSC
Class VIII Medical materiel, including medical peculiar repair parts.	5 16 3 16-1 3-1	III CORPS 2 MED
Class IX Repair parts and components, including kits, assemblies and subassemblies, reparable and nonreparable, required for maintenance support of all equipment.	5 16 3 16-1 3-1	GENERATORS 1811030ZMAY13- 1914000ZMAY13 12B 6ORDBN

Table 8-5. Sustainment (continued)

Type	*Template*	*Example*
Class X Material to support nonmilitary programs; such as agricultural and economic development, not included in Class 1 through Class 9.	5 16 CA 3 16-1 3-1	AGRICULTURAL CA GREEN 7SUST
North Atlantic Treaty Organization (NATO) supply points ***Note***: The descriptions for the NATO classes of supply are from STANAG 2291, *Classes of Supply of NATO Forces*.		
Class I Those items which are consumed by personnel or animals at the approximately uniform rate, irrespective of local changes in combat or terrain conditions.	5 16 I 3 16-1 3-1	I 2 3SUST
Class II Supplies for which allowances are established by tables of organization and equipment.	5 16 I I 3 16-1 3-1	I I 5W MND-S

Table 8-5. Sustainment (continued)

Type	*Template*	*Example*
Class III Fuels and lubricants for all purposes, except for operating aircraft or for use in weapons such as flame throwers.	5 16 3 16-1 3-1	**See United States class III example above.**
Class IV Supplies for which initial issue allowances are not prescribed by approved issue tables. Normally such supplies include fortification and construction materials, as well as additional quantities of items identical to those authorized for initial issue (class II), such as additional vehicles.	5 16 IV 3 16-1 3-1	IV 7 412EN
Class V Ammunition, explosives, and chemical agents of all types.	5 16 3 16-1 3-1	**See US Class V example above.**

Table 8-5. Sustainment (continued)

Type	*Template*	*Example*
Lines		
Supply routes **Note:** The line of the route normally follows the path of a road.		
Type	***Template***	***Example***
Alternate supply route	ASR 3	ASR GREEN
Main supply route	MSR 3	MSR 1
Traffic flow of supply routes ***Note***: Traffic flow modifier is always used with either a main supply route (MSR) or an alternate supply route (ASR) control measure symbol.		
Alternating traffic	MSR 3 ALT	MSR 3 ALT
One-way traffic	ASR 3 ***Note***: Direction of traffic is always determined by the direction of the arrowhead.	ASR 9
Two-way traffic	MSR 3	MSR IRON

Table 8-5. Sustainment (continued)

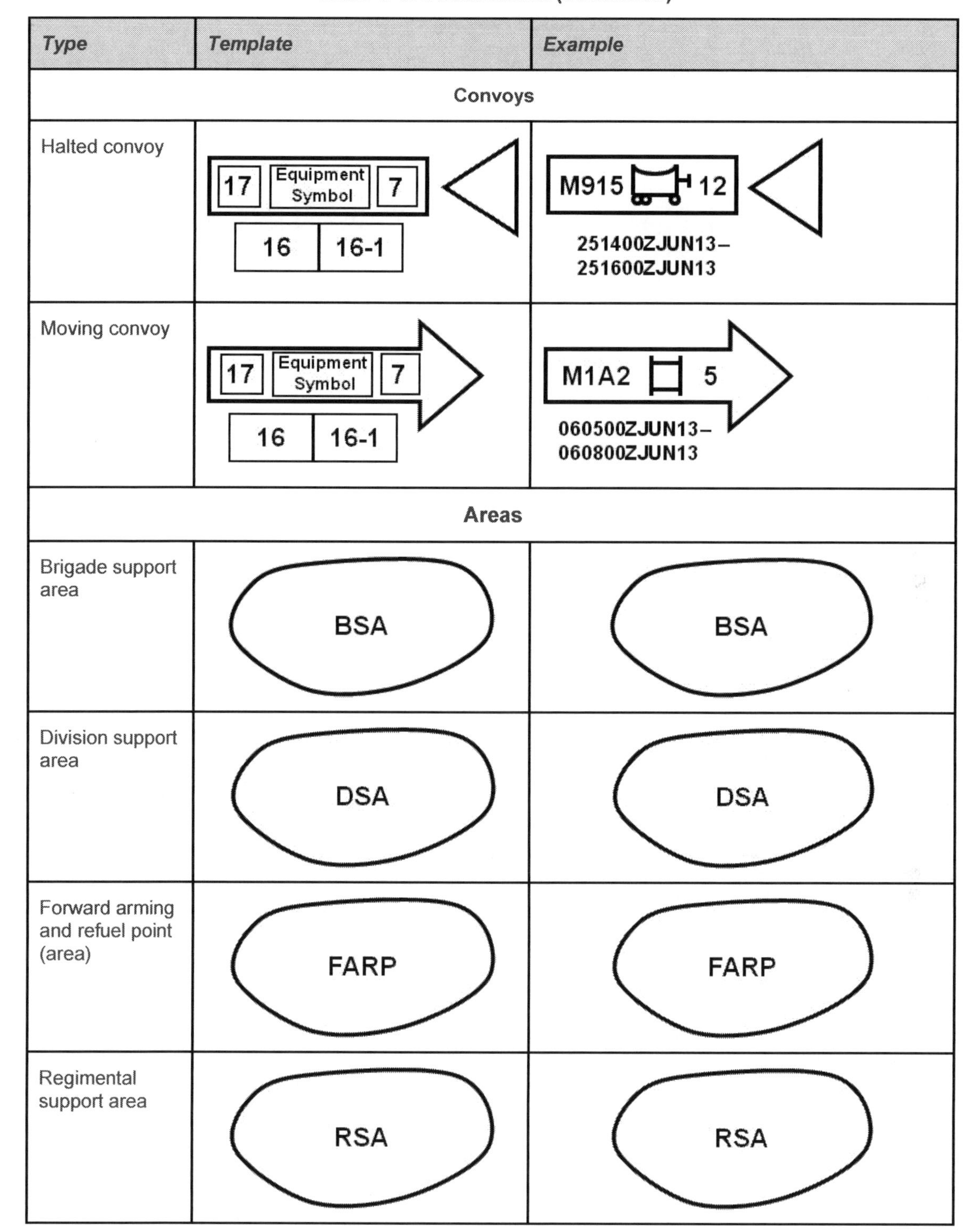

Type	*Template*	*Example*
	Convoys	
Halted convoy	17 Equipment Symbol 7 16 16-1	M915 12 251400ZJUN13– 251600ZJUN13
Moving convoy	17 Equipment Symbol 7 16 16-1	M1A2 5 060500ZJUN13– 060800ZJUN13
	Areas	
Brigade support area	BSA	BSA
Division support area	DSA	DSA
Forward arming and refuel point (area)	FARP	FARP
Regimental support area	RSA	RSA

INTELLIGENCE

8-15. Table 8-6 (on pages 8-74 through 8-75) shows intelligence control measure symbols.

Table 8-6. Intelligence

Type	*Template*	*Example*
Decoy, dummy, feint, or phoney	SYMBOL ***Note:*** The decoy, dummy, feint, or phoney icon is added to another symbol, such as equipment or installations and control measures to form the decoy, dummy, feint, or phoney symbol. This icon is normally used with the control measures axis of advance, direction of attack, or minefield.	
Dummy (equipment)	ICON	Dummy tank
Dummy (installation)	ICON	
Dummy (minefield)	MINE SYMBOL	Dummy antitank minefield

Table 8-6. Intelligence (continued)

Type	*Template*	*Example*
Feint (axis of advance)	UNIT SYMBOL, 3, 16, 16-1	See axis of advance.
Feint (direction of attack)	UNIT SYMBOL, 3, 16, 16-1	See direction of attack.
Areas		
Decoy mined area	M M M M	M M M M
Fenced decoy mine area	M M M M	M M M M

Airspace Control Measures

8-16. Table 8-7 (on pages 8-76 through 8-84) shows airspace control measure symbols.

Table 8-7. Airspace control

Type	*Template*	*Example*
	Airspace coordinating measures	
	Points	
Air control point	ACP 3	ACP 7
Communications checkpoint	CCP 3	CCP 1
Downed aircrew pickup point		
Pop-up point	PUP	PUP
	Lines	
Identification friend or foe off line	IFF OFF IFF OFF	IFF OFF IFF OFF
Identification friend or foe on line	IFF ON IFF ON	IFF ON IFF ON

Table 8-7. Airspace control (continued)

Type	*Template*	*Example*
	Corridors or routes (areas)	
General corridor route	NAME: 3 WIDTH: 22 MIN ALT: 19 MAX ALT: 19-1 DTG START: 16 DTG END: 16-1  ICON 3	
Air corridor	NAME: 3 WIDTH: 22 MIN ALT: 19 MAX ALT: 19-1 DTG START: 16 DTG END: 16-1 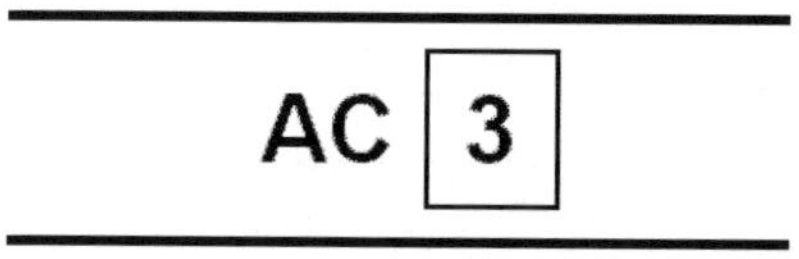 AC 3	NAME: GOLD WIDTH: 700M MIN ALT: 500M MAX ALT: 4000M DTG START: 240700ZMAY12 DTG END: 280700ZMAY12  ACP 1 AC GOLD ACP 2
Example of an air corridor with multiple legs	NAME: GOLD WIDTH: 400M MIN ALT: 500M MAX ALT: 4000M DTG START: 240700ZMAY12 DTG END: 290700ZMAY12 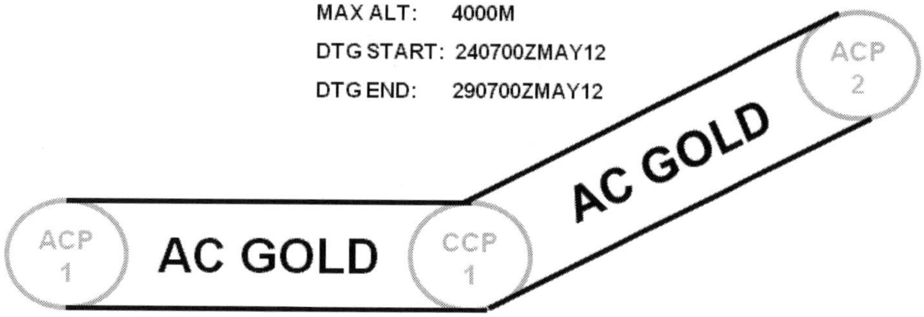 ACP 1 AC GOLD CCP 1 AC GOLD ACP 2	

Table 8-7. Airspace control (continued)

Type	*Template*	*Example*
Low-level transit route	NAME: 3 WIDTH: 22 MIN ALT: 19 MAX ALT: 19-1 DTG START: 16 DTG END: 16-1 LLTR 3	NAME: COBRA WIDTH: 100M MIN ALT: 50M MAX ALT: 1000M DTG START: 090700ZOCT12 DTG END: 091700ZOCT12 ACP 1 LLTR COBRA ACP 2
Minimum risk route	NAME: 3 WIDTH: 22 MIN ALT: 19 MAX ALT: 19-1 DTG START: 16 DTG END: 16-1 MRR 3	NAME: RED WIDTH: 500M MIN ALT: 1000M MAX ALT: 7000M DTG START: 110200ZSEP12 DTG END: 140300ZSEP12 ACP 1 MRR RED ACP 2
Safe lane (North Atlantic Treaty Organization)	NAME: 3 WIDTH: 22 MIN ALT: 19 MAX ALT: 19-1 DTG START: 16 DTG END: 16-1 SL 3	NAME: LION WIDTH: 200M MIN ALT: 500M MAX ALT: 1000M DTG START: 240700ZFEB12 DTG END: 240900ZFEB12 ACP 1 SL LION ACP 2

Table 8-7. Airspace control (continued)

Type	Template	Example
Special corridor	NAME: 3 WIDTH: 22 MIN ALT: 19 MAX ALT: 19-1 DTG START: 16 DTG END: 16-1 SC 3	NAME: OWL WIDTH: 500M MIN ALT: 100M MAX ALT: 12000M DTG START: 220700ZJUN13 DTG END: 300700ZJUN13 ACP 1 SC OWL ACP 2
Standard Army aircraft flight route	NAME: 3 WIDTH: 22 MIN ALT: 19 MAX ALT: 19-1 DTG START: 16 DTG END: 16-1 SAAFR 3	NAME: BLUE WIDTH: 200M MIN ALT: 50M MAX ALT: 1000M DTG START: 260930ZMAY12 DTG END: 280700ZMAY12 ACP 1 SAAFR BLUE ACP 2
Temporary minimum risk route	NAME: 3 WIDTH: 22 MIN ALT: 19 MAX ALT: 19-1 DTG START: 16 DTG END: 16-1 TMRR 3	NAME: 51 WIDTH: 500M MIN ALT: 1000M MAX ALT: 7000M DTG START: 120200ZSEP12 DTG END: 120300ZSEP12 ACP 1 TMRR 51 ACP 2

Table 8-7. Airspace control (continued)

Type	*Template*	*Example*
Transit corridor	NAME: 3 WIDTH: 22 MIN ALT: 19 MAX ALT: 19-1 DTG START: 16 DTG END: 16-1 TC 3	NAME: KING WIDTH: 300M MIN ALT: 700M MAX ALT: 2000M DTG START: 260700ZMAR13 DTG END: 280700ZMAR13 ACP 1 TC KING ACP 2
Transit route	NAME: 3 WIDTH: 22 MIN ALT: 19 MAX ALT: 19-1 DTG START: 16 DTG END: 16-1 TR 3	NAME: ABLE WIDTH: 600M MIN ALT: 1700M MAX ALT: 12000M DTG START: 220700ZMAR13 DTG END: 300700ZMAR13 ACP 1 TR ABLE ACP 2
Unmanned aerial vehicle or unmanned aircraft route	NAME: 3 WIDTH: 22 MIN ALT: 19 MAX ALT: 19-1 DTG START: 16 DTG END: 16-1 UAV 3	NAME: DRAGON WIDTH: 400M MIN ALT: 500M MAX ALT: 4000M DTG START: 200700ZMAY13 DTG END: 210700ZMAY13 ACP 1 UAV DRAGON ACP 2

Table 8-7. Airspace control (continued)

Type	*Template*	*Example*
Areas		
Engagement zones		
Air-to-air restricted operations zone (North Atlantic Treaty Organization)	AAROZ [3] MIN ALT: [19] MAX ALT: [19-1] TIME FROM: [16] TIME TO: [16-1]	AARROZ 101AAD MIN ALT: 100M MAX ALT: 27000M TIME FROM: 130001ZFEB13 TIME TO: 132400ZFEB13
Fighter engagement zone (North Atlantic Treaty Organization)	FEZ [3] MIN ALT: [19] MAX ALT: [19-1] TIME FROM: [16] TIME TO: [16-1]	FEZ CTF MIN ALT: 200M MAX ALT: 50000M TIME FROM: 160100ZJUN13 TIME TO: 150700ZJUL13
High-density airspace control zone	HIDACZ [3] MIN ALT: [19] MAX ALT: [19-1] TIME FROM: [16] TIME TO: [16-1]	HIDACZ 32AADC MIN ALT: 150000M MAX ALT: 37000M TIME FROM: 120700ZMAY13 TIME TO: 140630ZMAY13

Table 8-7. Airspace control (continued)

Type	*Template*	*Example*
Joint engagement zone	JEZ [3] MIN ALT: [19] MAX ALT: [19-1] TIME FROM: [16] TIME TO: [16-1]	JEZ JTF MIN ALT: 2000M MAX ALT: 415000M TIME FROM: 160900ZMAR13 TIME TO: 250100ZMAR13
Restricted operations zone	ROZ [3] MIN ALT: [19] MAX ALT: [19-1] TIME FROM: [16] TIME TO: [16-1]	ROZ 11ADA BDE MIN ALT: 900M MAX ALT: 7000M TIME FROM: 030001ZJUL08 TIME TO: 032400ZJUL08
Unmanned aircraft restricted operations zone (North Atlantic Treaty Organization)	UAVROZ [3] MIN ALT: [19] MAX ALT: [19-1] TIME FROM: [16] TIME TO: [16-1]	UAVROZ 1CD MIN ALT: 25M MAX ALT: 2000M TIME FROM: 030001ZOCT13 TIME TO: 032400ZOCT13

Table 8-7. Airspace control (continued)

Type	*Template*	*Example*
Air defense control measures		
Base defense zone	BDZ	BDZ
Missile engagement zones		
Missile engagement zone ***Note***: This is a change to MIL-STD-2525D.	MEZ [3] MIN ALT: [19] MAX ALT: [19-1] TIME FROM: [16] TIME TO: [16-1]	MEZ 2-4 ADA BN MIN ALT: 2000M MAX ALT: 15000M TIME FROM: 160100ZFEB13 TIME TO: 150100ZMAR13
High-altitude missile engagement zone ***Note***: This is a change to MIL-STD-2525D.	HIMEZ [3] MIN ALT: [19] MAX ALT: [19-1] TIME FROM: [16] TIME TO: [16-1]	HIMEZ 3-3 ADA BN MIN ALT: 12000M MAX ALT: 25000M TIME FROM: 010100ZSEP13 TIME TO: 150100ZOCT13

Table 8-7. Airspace control (continued)

Type	*Template*	*Example*
Low-altitude missile engagement zone This is a change to MIL-STD-2525D.	LOMEZ 3 MIN ALT: 19 MAX ALT: 19-1 TIME FROM: 16 TIME TO: 16-1	LOMEZ 1-7 ADA BN MIN ALT: 1000M MAX ALT: 5000M TIME FROM: 220100ZFEB13 TIME TO: 230100ZMAR13
Short-range air defense engagement zone	SHORADEZ 3 MIN ALT: 19 MAX ALT: 19-1 TIME FROM: 16 TIME TO: 16-1	SHORADEZ MNTF MIN ALT: 100M MAX ALT: 8000M TIME FROM: 130100ZAUG13 TIME TO: 132200ZAUG13
Weapons-free zone	WFZ 3 3-1 TIME FROM: 16 TIME TO: 16-1	WFZ V CORPS RED 1 TIME FROM: 070805ZDEC07 TIME TO: 210805ZDEC07

ABBREVIATIONS AND ACRONYMS FOR USE WITH CONTROL MEASURE SYMBOLS

8-17. Table 8-8 provides a list of abbreviations and acronyms for echelons and functional organizations to be used with boundaries.

Table 8-8. Abbreviations and acronyms for use with boundaries

Echelon	*Abbreviation or acronym*	*Examples* ***Note***: A 2- or 3-letter country code in parentheses may follow unit designation, such as 3 DIV (UK).
Army group	AG	1AG
Army	A	3A
Corps	CORPS	IICORPS ***Note:*** Corps uses Roman numerals.
Division	DIV	1DIV
Air assault division	AAD	101AAD
Airborne division	ABD	6ABD
Armoured division	AD	2AD
Cavalry division	CD	1CD
Infantry division	ID	52ID
Mechanized division	MD	4MD
Mountain division	MTND	10MTND
Multinational division	MND	1MND or MND(S) ***Note:*** Multinational divisions may use geographical references in parentheses.
Brigade	BDE	2BDE
Air assault brigade	AAB	8AAB
Airborne brigade	ABB	3ABB
Brigade combat team	BCT	4BCT
Fires brigade	FB	41FB
Multinational brigade	MNB	2MNB
Naval infantry brigade	NIB	4NIB
Separate armor brigade	SAB	194SAB
Separate infantry brigade	SIB	197SIB
Regiment	REGT	21REGT
Airborne regiment	ABR	901ABR
Group	GP	41GP
Battle group	BG	5BG
Battalion	BN	7BN
Company	CO[1]	ACO or 2CO
Platoon	PLT	2PLT
Team	TM	BTM

[1] North Atlantic Treaty Organization (NATO) uses COY

8-18. Table 8-9 (on page 8-86) provides a list of abbreviations and acronyms used in control measure symbols for unit functions.

Table 8-9. Abbreviation and acronyms used in control measure symbols for unit functions

Function	*Abbreviation or acronym*
Air defense	ADA ***Note:*** ADA used to prevent confusion with AD for armored division.
Armor	AR
Antitank or anti-armor	AT
Aviation	AVN
Cavalry	CAV
Chemical	CML
Chemical, biological, radiological, and nuclear (CBRN)	CB ***Note:*** CB used in lieu of CBRN.
Civil affairs	CA
Combined arms	CAR
Counterintelligence	CI
Electronic warfare	EW
Engineer	EN
Explosive ordnance disposal	EOD
Field artillery	FA
Infantry	IN
Logistics	LOG
Maintenance	MNT ***Note:*** MNT used in lieu of MAINT.
Medical	MED
Military intelligence	MI
Military police	MP
Naval	NAV
Ordnance	ORD
Quartermaster	QM
Reconnaissance	REC ***Note:*** REC used in lieu of RECON.
Signal	SIG
Special forces	SF
Special operations force	SOF
Surveillance	SUR ***Note:*** SUR used in lieu of SURVEIL.
Sustainment	SUST
Transportation	TPT ***Note:*** TPT used in lieu of TRANS.

Chapter 9

Tactical Mission Tasks

This chapter defines tactical mission tasks and provides symbols for them. This chapter also includes missions task verbs from STANAG 2019 (ED. 6)/APP 6(C).

TACTICAL MISSION TASKS DEFINED

9-1. A task is a clearly defined and measurable activity accomplished by individuals or organizations. A *tactical mission task* is a specific activity performed by a unit while executing a form of tactical operation or form of maneuver. A tactical mission task may be expressed as either an action by a friendly force or effects on an enemy force (FM 3-90-1). The tactical mission tasks describe the results or effects the commander wants to achieve.

SYMBOLS FOR TACTICAL MISSION TASKS

9-2. Table 9-1 (on pages 9-1 through 9-6) shows the tactical mission tasks that have symbols. Not all tactical mission tasks have symbols. Most of the tactical mission tasks shown in table 9-1 are defined in chapter 1. Some tactical mission task symbols will include unit symbols, and the tactical mission task "delay until a specified time" will use an amplifier. However, no modifiers are used with tactical mission task symbols. Tactical mission task symbols are used in course of action sketches, synchronization matrixes, and maneuver sketches. They do not replace any part of the operation order. Tactical mission task symbols are sized to accommodate the scale of the display or map being used. Where practical, the tactical mission task symbol connects with the task organization composition symbol centered of the left or right side of the symbol or at the center of the bottom of the symbol, depending on the orientation of the symbols. Figure 9-1 (on page 9-6) shows an example of a tactical mission task symbol connected to task organization composition symbol.

Table 9-1. Tactical mission task symbols

Task	*Symbol* ***Note:*** The friendly or hostile frame (gray) is not part of the symbol; it is for orientation only.
Ambush	
Attack by fire	
Block	B

Table 9-1. Tactical mission task symbols (continued)

Task	*Symbol* *Note:* The friendly or hostile frame (gray) is not part of the symbol; it is for orientation only.
Breach	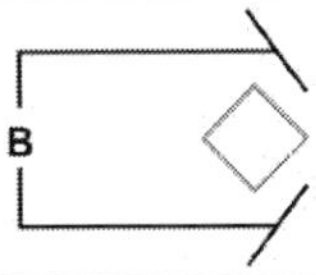 B
Bypass	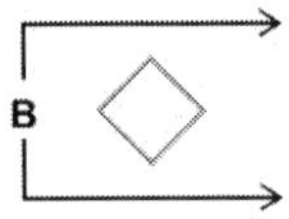 B
Canalize	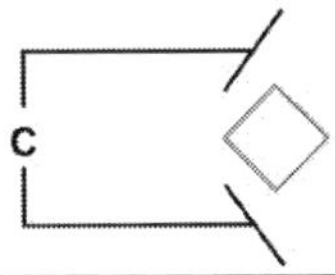 C
Clear	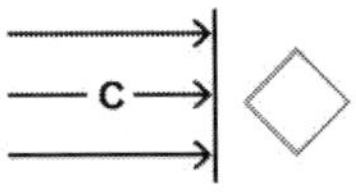 C
Contain	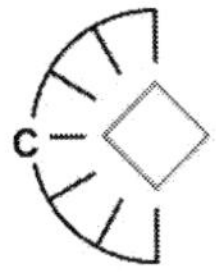 C
Control	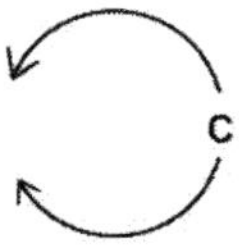 C
Cordon and knock	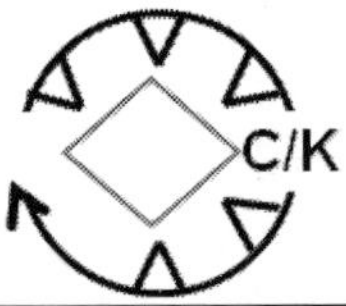 C/K
Cordon and search	 C/S
Counterattack	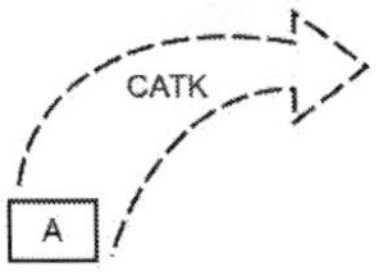 CATK A
Counterattack by fire	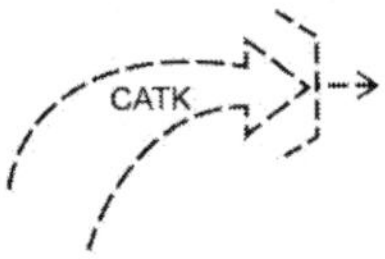 CATK

Table 9-1. Tactical mission task symbols (continued)

Task	*Symbol* *Note:* The friendly or hostile frame (gray) is not part of the symbol; it is for orientation only.
Delay or delay (until a specific time)	W D
Demonstration	DEM
Destroy	D
Disengage or disengagement	DIS
Disrupt	D
Envelopment	E
Exfiltrate	EX
Exploit	
Feint	
Fix	F
Follow and assume	A

Table 9-1. Tactical mission task symbols (continued)

Task	*Symbol* *Note:* The friendly or hostile frame (gray) is not part of the symbol; it is for orientation only.
Follow and support	A
Infiltration or infiltrate	IN
Interdict	I
Isolate	I
Neutralize	N
Occupy	O
Passage of lines (forward)	P(F)
Passage of lines (rearward)	P(R)
Penetration or penetrate	P
Relief in place	RIP

Table 9-1. Tactical mission task symbols (continued)

Task	***Symbol*** ***Note:*** The friendly or hostile frame (gray) is not part of the symbol; it is for orientation only.
Retain	R
Retirement	R
Secure	S
Security	
A A A ***Note:*** Unit to perform security is placed in the center of symbol.	
Type	***Icon***
Security (screen)	**S**
Security (cover)	**C**
Security (guard)	**G**
Seize	A S
Support by fire	
Suppress	S
Turn	T

Table 9-1. Tactical mission task symbols (continued)

Task	*Symbol* ***Note:*** The friendly or hostile frame (gray) is not part of the symbol; it is for orientation only.
Withdraw	W
Withdraw under pressure	WP

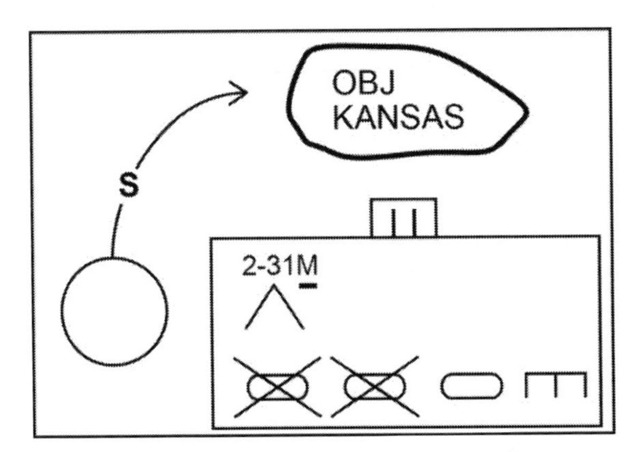

Figure 9-1. Example of tactical mission task symbol connected to task organization composition symbol

Chapter 10

Course of Action Sketch

This chapter discusses the purpose and makeup of a course of action sketch.

PURPOSE OF COURSE OF ACTION SKETCH

10-1. A *course of action* is a scheme developed to accomplish a mission (JP 5-0). It constitutes a broad potential solution to an identified problem. A course of action statement clearly portrays how the unit will accomplish the mission. The course of action statement should be a brief expression of how the combined arms concept will be conducted. The course of action sketch is the graphic portrayal of the course of action statement.

MAKEUP OF COURSE OF ACTION SKETCH

10-2. The course of action sketch provides a picture of the movement and maneuver aspects of the concept, including the positioning of forces. The course of action sketch becomes the basis for the operation overlay. At a minimum, the course of action sketch includes the array of generic forces and control measures, such as—

- Unit and subordinate unit boundaries.
- Unit movement formations (but not subordinate unit formations).
- Line of departure, or line of contact and phase lines, if used.
- Reconnaissance and security graphics.
- Ground and air axes of advance.
- Assembly areas, battle positions, strong points, engagement areas, and objectives.
- Obstacle control measures and tactical mission graphics.
- Fire support coordination and airspace control measures.
- Main effort.
- Location of command posts and critical information systems nodes.
- Enemy locations, known or templated.
- Population concentrations.

Most symbols for use on the course of action sketch are shown in chapters 4 through 9. However, the unit symbols do not provide decisionmakers with a quick and easy method of portraying detailed information relating to task organization composition or combat effectiveness. Task organization composition symbols portray detailed information for course of action sketches.

TASK ORGANIZATION COMPOSITION SYMBOLS

10-3. The task organization portion of the operation order specifies the resources available to the land maneuver commander in a detailed list. The headquarters and individual units of the task organization are portrayed graphically with unit symbols shown in chapter 4. These symbols provide a rapid and easily understood means—through situation maps, overlays, and annotated aerial photographs—to express an operation plan, concept, or friendly or hostile (enemy) situation. Figure 10-1 (on page 10-2) shows a comparison of a unit symbol and task organization composition symbols.

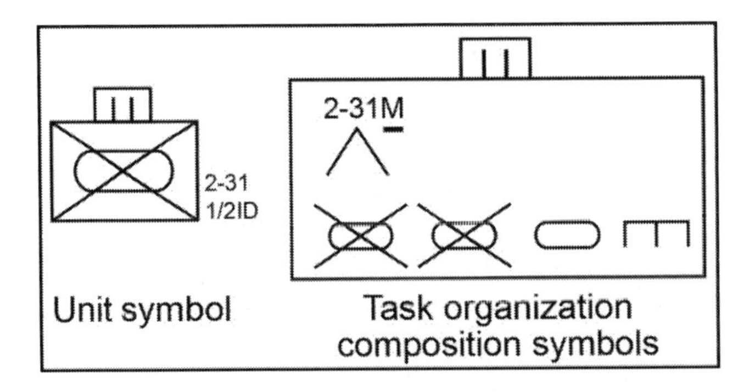

Figure 10-1. Comparison of sample unit and task organization composition symbols

10-4. A task organization composition symbol is built using a rectangular frame. Figure 10-2 shows icon and amplifier locations for task organization composition symbols (see chapter 3 for an explanation of icons and amplifiers). Centered in the middle on the top of the frame are the echelon (field 6) and task force (field 8), if required. Inside the top portion of the frame, on the left side, the unique designation (field 2) of the organization is shown. Inside the top portion of the frame, on the right side, the combat effectiveness (field 12) of the organization is shown. Inside the middle portion of the frame are symbols for any unit that is reinforced, reduced, or both (reinforced and reduced). Inside the bottom portion of the frame are symbols for remaining units. If no units are reinforced, reduced, or both (reinforced and reduced), then units are shown below the unique designation. If any unit is not one echelon lower than the designated unit on line 1, then the echelon indicator amplifier is used above the task organization unit icon (field 6). (See table 4-7 on page 4-33 for echelon amplifiers.) Figure 10-2 depicts sample unit and task organization composition symbols. While there are four icons shown in the sample, there is no limitation as to the number of icons in a row or number of rows.

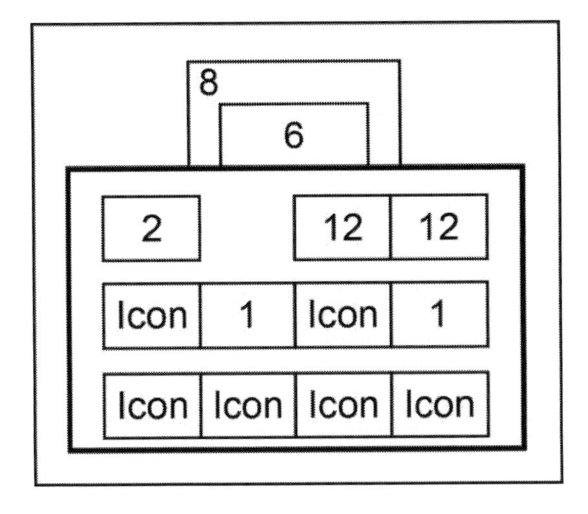

Figure 10-2. Task organization composition symbol example

Task Organization Icons

10-5. In all cases, task organization icons are the same as unit icons and amplifiers. Table 10-1 shows the most commonly used icons and modifiers in their appropriate configurations.

Table 10-1. Task organization icons

Function	*Symbol*	*Function*	*Symbol*
Air defense		Field artillery	
Armor		Infantry	
Armored reconnaissance (cavalry) ***Note***: Reconnaissance (cavalry) unit that is both armored and tracked)		Air assault infantry	
Antitank		Airborne infantry	
Attack helicopter	A	Light infantry	L
Air reconnaissance (cavalry)		Mechanized infantry ***Note:*** Infantry unit that is both armored and tracked.	
Assault or lift helicopter		Medium infantry (Stryker)	
Combined arms		Mountain infantry	
Engineer		Reconnaissance (cavalry or scout)	

Combat Effectiveness Icons

10-6. Combat effectiveness refers to the ability of a unit to perform its mission. Factors such as ammunition, personnel, status of fuel, and weapon systems availability are assessed and rated. The commander uses this information to provide a net assessment of the unit's ability to perform its mission. This assessment can then be expressed graphically using combat effectiveness icons. Table 10-2 (on page 10-4) shows two sets of combat effectiveness icons, which may be also used with task organization composition symbol.

10-7. Table 10-2 depicts combat effectiveness icons for the overall combat rating of the unit in the center column. Table 10-2 specifies combat effectiveness icons for the status of selected items of interest in the right column. The four selected items shown in the right column are ammunition; weapons; petroleum, oils, and lubricants (POL); and personnel. Standard operating procedures will specify the items of interest to be reported. The commander may add to this list for internal reporting and tracking.

Table 10-2. Combat effectiveness icons

Commander's assessment of unit's ability to perform its mission	*Effectiveness pie charts*	*Selected status pie chart* (Selected status pie chart: Personnel, Ammunition, POL, Weapons)
No problems in any area		
Some problems in personnel		
Major problems in weapon systems		
Cannot perform mission: personnel, ammunition, and weapons problems		
Legend: POL — petroleum, oils, and lubricants		

Example of a Task Organization Composition Symbol

10-8. See figure 10-3 for an example of a task organization symbol for a brigade combat team.

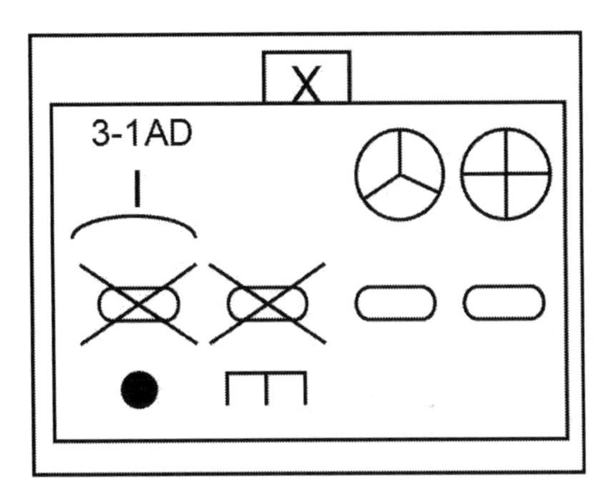

Figure 10-3. Brigade combat team example

References

All URLs accessed on 21 October 2015.

REQUIRED PUBLICATIONS

These documents must be available to intended users of this publication.

JP 1-02. *Department of Defense Dictionary of Military and Associated Terms*. 8 November 2010.

MIL-STD 2525D. *Joint Military Symbology*. 10 June 2014. This publication available at http://quicksearch.dla.mil/qsDocDetails.aspx?ident_number=114934.

RELATED PUBLICATIONS

These documents contain relevant supplemental information.

INTERNATIONAL STANDARDIZATION AGREEMENTS

Most North Atlantic Treaty Organization publications are available at http://www.nato.int.

AAP 15 (2014). *NATO Glossary of Abbreviations Used in NATO Documents and Publications (English and French)*. 5 May 2014.

STANAG 1059 (ED. 8). *Letter Codes for Geographical Entities*. 29 October 2007.

STANAG 1241 (ED. 5). *NATO Standard Identity Description Structure for Tactical Use*. 7 April 2005.

STANAG 2019 (ED. 6)/APP-6(C). *NATO Joint Military Symbology*. 24 May 2011.

STANAG 2991 (ED. 4)/APP-19 (D). *NATO Combat Engineer Glossary*. November 2003.

STANAG 3680 (ED. 5)/AAP-6 (2012) (2). *NATO Glossary of Terms and Definitions (English and French)*. 20 April 2014.

DEPARTMENT OF DEFENSE AND JOINT PUBLICATIONS

Most joint publications are available online at http://www.dtic.mil/doctrine/new_pubs/jointpub.htm.

DODD 2310.01E. *DoD Detainee Program*. 19 August 2014.

DODD 3025.18. *Defense Support of Civil Authorities (DSCA)*. 29 December 2010.

JP 1. *Doctrine for the Armed Forces of the United States*. 25 March 2013.

JP 1-0. *Joint Personnel Support*. 24 October 2011.

JP 1-04. *Legal Support to Military Operations*. 17 August 2011.

JP 1-05. *Religious Affairs in Joint Operations*. 20 November 2013.

JP 2-0. *Joint Intelligence*. 22 October 2013.

JP 2-01. *Joint and National Intelligence Support to Military Operations*. 5 January 2012.

JP 2-01.3. *Joint Intelligence Preparation of the Operational Environment*. 21 May 2014.

JP 2-03. *Geospatial Intelligence in Joint Operations*. 31 October 2012.

JP 3-0. *Joint Operations*. 11 August 2011.

JP 3-01. *Countering Air and Missile Threats*. 23 March 2012.

JP 3-02. *Amphibious Operations*. 18 July 2014.

JP 3-02.1. *Amphibious Embarkation and Debarkation*. 25 November 2014.

JP 3-03. *Joint Interdiction*. 14 October 2011.

JP 3-04. *Joint Shipboard Helicopter and Tiltrotor Aircraft Operations*. 6 December 2012.

JP 3-05. *Special Operations*. 16 July 2014.

JP 3-07. *Stability Operations*. 29 September 2011.

JP 3-07.2. *Antiterrorism*. 14 March 2014.

JP 3-07.3. *Peace Operations.* 1 August 2012.

JP 3-07.4. *Counterdrug Operations.* 14 August 2013.

JP 3-08. *Interorganizational Coordination During Joint Operations.* 24 June 2011.

JP 3-09. *Joint Fire Support.* 12 December 2014.

JP 3-09.3. *Close Air Support.* 25 November 2014.

JP 3-10. *Joint Security Operations in Theater.* 13 November 2014.

JP 3-11. *Operations in Chemical, Biological, Radiological, and Nuclear Environments.* 4 October 2013.

JP 3-12 (R). *Cyberspace Operations*. 5 February 2013.

JP 3-13. *Information Operations.* 27 November 2012.

JP 3-13.1. *Electronic Warfare.* 8 February 2012.

JP 3-13.2. *Military Information Support Operations.* 21 November 2014.

JP 3-13.3. *Operations Security.* 4 January 2012.

JP 3-13.4. *Military Deception.* 26 January 2012.

JP 3-14. *Space Operations.* 29 May 2013.

JP 3-15. *Barriers, Obstacles, and Mine Warfare for Joint Operations.* 17 June 2011.

JP 3-15.1. *Counter-Improvised Explosive Device Operations*. 9 January 2012.

JP 3-16. *Multinational Operations.* 16 July 2013.

JP 3-17. *Air Mobility Operations.* 30 September 2013.

JP 3-18. *Joint Forcible Entry Operations.* 27 November 2012.

JP 3-22. *Foreign Internal Defense.* 12 July 2010.

JP 3-24. *Counterinsurgency.* 22 November 2013.

JP 3-27. *Homeland Defense.* 29 July 2013.

JP 3-28. *Defense Support of Civil Authorities.* 31 July 2013.

JP 3-29. *Foreign Humanitarian Assistance.* 3 January 2014.

JP 3-30. *Command and Control of Joint Air Operations.* 10 February 2014.

JP 3-31. *Command and Control for Joint Land Operations.* 24 February 2014.

JP 3-33. *Joint Task Force Headquarters*. 30 July 2012.

JP 3-34. *Joint Engineer Operations.* 30 June 2011.

JP 3-35. *Deployment and Redeployment Operations.* 31 January 2013.

JP 3-40. *Countering Weapons of Mass Destruction.* 31 October 2014.

JP 3-41. *Chemical, Biological, Radiological, and Nuclear Consequence Management.* 21 June 2012.

JP 3-50. *Personnel Recovery.* 2 October 2015.

JP 3-52. *Joint Airspace Control.* 13 November 2014.

JP 3-57. *Civil-Military Operations.* 11 September 2013.

JP 3-60. *Joint Targeting.* 31 January 2013.

JP 3-61. *Public Affairs.* 25 August 2010.

JP 3-63. *Detainee Operations.* 13 November 2014.

JP 3-68. *Noncombatant Evacuation Operations.* 23 December 2010.

JP 4-0. *Joint Logistics*. 16 October 2013.

JP 4-01. T*he Defense Transportation System.* 6 June 2013.

JP 4-01.2. *Sealift Support to Joint Operations.* 22 June 2012.

JP 4-01.5. *Joint Terminal Operations.* 6 April 2012.

JP 4-01.6. *Joint Logistics Over-the-Shore*. 27 November 2012.

JP 4-02. *Health Service Support.* 26 July 2012.

JP 4-03. *Joint Bulk Petroleum and Water Doctrine*. 9 December 2010.
JP 4-05. *Joint Mobilization Planning.* 21 February 2014.
JP 4-06. *Mortuary Affairs*. 12 October 2011.
JP 4-08. *Logistics in Support of Multinational Operations*. 21 February 2013.
JP 4-09. *Distribution Operations.* 19 December 2013.
JP 4-10. *Operational Contract Support.* 16 July 2014.
JP 5-0. *Joint Operation Planning.* 11 August 2011.
JP 6-0. *Joint Communications System.* 10 June 2015.
JP 6-01. *Joint Electromagnetic Spectrum Management Operations*. 20 March 2012.

ARMY PUBLICATIONS

Most Army doctrinal publications are available online at http://www.apd.army.mil/.
ADP 1-01. *Doctrine Primer*. 2 September 2014.
ADP 1-02. *Operational Terms and Military Symbols.* 31 August 2012.
ADP 3-0. *Unified Land Operations*. 10 October 2011.
ADP 3-05. *Special Operations*. 31 August 2012.
ADP 3-09. *Fires*. 31 August 2012.
ADP 3-28. *Defense Support of Civil Authorities.* 26 July 2012.
ADP 3-37. *Protection*. 31 August 2012.
ADP 3-90. *Offense and Defense*. 31 August 2012.
ADP 4-0. *Sustainment.* 31 July 2012.
ADP 5-0. *The Operations Process*. 17 May 2012.
ADP 6-0. *Mission Command.* 17 May 2012.
ADP 6-22. *Army Leadership.* 1 August 2012.
ADP 7-0. *Training Units and Developing Leaders.* 23 August 2012.
ADRP 1. *The Army Profession.* 14 June 2015.
ADRP 2-0. *Intelligence.* 31 August 2012.
ADRP 3-0. *Unified Land Operations*. 16 May 2012.
ADRP 3-05. *Special Operations.* 31 August 2012.
ADRP 3-07. *Stability.* 31 August 2012.
ADRP 3-09. *Fires.* 31 August 2012.
ADRP 3-28. *Defense Support of Civil Authorities*. 14 June 2013.
ADRP 3-37. *Protection.* 31 August 2012.
ADRP 3-90. *Offense and Defense*. 31 August 2012.
ADRP 4-0. *Sustainment.* 31 July 2012.
ADRP 5-0. *The Operations Process*. 17 May 2012.
ADRP 6-0. *Mission Command.* 17 May 2012.
ADRP 6-22. *Army Leadership.* 1 August 2012.
ADRP 7-0. *Training Units and Developing Leaders.* 23 August 2012.
AR 15-6. *Procedures for Investigating Officers and Boards of Officers.* 2 October 2006.
AR 600-100. *Army Leadership*. 8 March 2007.
ATP 1-0.2. *Theater-Level Human Resources Support.* 4 January 2013.
ATP 1-05.03. *Religious Support and External Advisement.* 3 May 2013.
ATP 1-06.2. *Commanders' Emergency Response Program (CERP).* 5 April 2013.
ATP 1-06.3. *Banking Operations*. 23 January 2015.

ATP 1-19. *Army Music*. 13 February 2015.

ATP 1-20. *Military History Operations*. 9 June 2014.

ATP 2-01. *Plan Requirements and Assess Collection*. 19 August 2014.

ATP 2-01.3. *Intelligence Preparation of the Battlefield/Battlespace.* 10 November 2014.

ATP 2-19.3. *Corps and Division Intelligence Techniques*. 26 March 2015.

ATP 2-19.4. *Brigade Combat Team Intelligence Techniques.* 10 February 2015.

ATP 2-22.4. *Technical Intelligence.* 4 November 2013.

ATP 2-22.7. *Geospatial Intelligence.* 26 March 2015.

ATP 2-22.9. *Open-Source Intelligence.* 10 July 2012.

ATP 2-33.4. *Intelligence Analysis*. 18 August 2014.

ATP 2-91.7. *Intelligence Support to Defense Support of Civil Authorities*. 29 June 2015

ATP 2-91.8. *Techniques for Document and Media Exploitation.* 5 May 2015.

ATP 3-01.15. *IADS, Multi-Service Tactics, Techniques, and Procedures for an Integrated Air Defense System*. 9 September 2014.

ATP 3-01.18. *Stinger Techniques*. 8 December 2014.

ATP 3-01.91. *Terminal High Altitude Area Defense (THADD) Techniques*. 26 August 2013.

ATP 3-04.64. *Multi-Service Tactics, Techniques, and Procedures for the Tactical Employment of Unmanned Aircraft Systems.* 22 January 2015.

ATP 3-05.1. *Unconventional Warfare.* 6 September 2013.

ATP 3-05.2. *Foreign Internal Defense*. 18 August 2015.

ATP 3-05.11. *Special Operations Chemical, Biological, Radiological, and Nuclear Operations.* 30 April 2014.

ATP 3-05.20. *Special Operations Intelligence.* 3 May 2013.

ATP 3-05.68. *Special Operations Noncombatant Evacuation Operations*. 30 September 2014.

ATP 3-06.1. *Aviation Urban Operations Multi-Service Tactics, Techniques, and Procedures for Aviation Urban Operations.* 19 April 2013.

ATP 3-07.5. *Stability Techniques.* 31 August 2012.

ATP 3-07.10. *Advising Multi-Service Tactics, Techniques, and Procedures for Advising Foreign Security Forces*. 1 November 2014.

ATP 3-07.20. *IMSO Multi-Service Tactics, Techniques, and Procedures for Integrated Monetary Shaping Operations.* 26 April 2013.

ATP 3-07.31. *Peace Ops Multi-Service Tactics, Techniques, and Procedures for Peace Operations*. 1 November 2014.

ATP 3-09.12. *Field Artillery Target Acquisition*. 24 July 2015.

ATP 3-09.13. *The Battlefield Coordination Detachment*. 24 July 2015.

ATP 3-09.23. *Field Artillery Cannon Battalion*. 24 September 2015.

ATP 3-09.24. *Techniques for the Fires Brigade.* 21 November 2012.

ATP 3-09.30. *Techniques for Observed Fire*. 2 August 2013.

ATP 3-09.34. *Kill Box Multi-Service Tactics, Techniques, and Procedures for Kill Box Planning and Employment*. 16 April 2014.

ATP 3-09.50. *The Field Artillery Cannon Battery*. 7 July 2015.

ATP 3-11.23. *Multi-Service Tactics, Techniques, and Procedures for Weapons Of Mass Destruction Elimination Operations*. 1 November 2013.

ATP 3-11.37. *Multi-Service Tactics, Techniques, and Procedures for Chemical, Biological, Radiological, and Nuclear Reconnaissance and Surveillance*. 25 March 2013.

ATP 3-11.50. *Battlefield Obscuration.* 15 May 2014.

ATP 3-13.10. *EW Reprogramming Multi-Service Tactics, Techniques, and Procedures for Reprogramming Electronic Warfare (EW) Systems*. 17 June 2014.

ATP 3-14.5. *Army Joint Tactical Ground Station (JTAGS) Operations*. 15 October 2014.

ATP 3-17.2. *Airfield Opening Multi-Service Tactics, Techniques, and Procedures for Airfield Opening*. 18 June 2015.

ATP 3-18.4. *Special Forces Special Reconnaissance*. 18 August 2015.

ATP 3-18.11. *Special Forces Military Free-Fall Operations*. 24 October 2014.

ATP 3-18.14. *Special Forces Vehicle-Mounted Operations Tactics, Techniques, and Procedures*. 12 September 2014.

ATP 3-20.15. *Tank Platoon.* 13 December 2012.

ATP 3-22.40. *NLW Multi-Service Tactics, Techniques, and Procedures for the Employment of Nonlethal Weapons.* 13 February 2015.

ATP 3-27.5. *AN/TPY-2 Forward Based Mode Radar Operations*. 13 April 2015.

ATP 3-28.1. *DSCA Multi-Service Tactics, Techniques, and Procedures for Defense Support of Civil Authorities (DSCA)*. 25 September 2015.

ATP 3-34.5. *Environmental Considerations*. 10 August 2015.

ATP 3-34.80. *Geospatial Engineering*. 23 June 2014.

ATP 3-34.84. *MDO Multi-Service Tactics, Techniques, and Procedures for Military Diving Operations.* 13 February 2015.

ATP 3-35 (FM 3-35). *Army Deployment and Redeployment.* 23 March 2015.

ATP 3-37.10. *Base Camps.* 26 April 2013.

ATP 3-37.34. *Survivability Operations.* 28 June 2013.

ATP 3-39.10. *Police Operations*. 26 January 2015.

ATP 3-39.32. *Physical Security.* 30 April 2014.

ATP 3-50.3. *Survival, Evasion, and Recovery Multi-Service Tactics, Techniques, and Procedures for Survival, Evasion, and Recovery.*11 September 2012.

ATP 3-52.1. *Airspace Control Multi-Service Tactics, Techniques, and Procedures for Airspace Control.* 9 April 2015.

ATP 3-52.2. *TAGS Multi-Service Tactics, Techniques, and Procedures for the Theater Air-Ground System*. 30 June 2014.

ATP 3-52.3. *JATC Multi-Service Tactics, Techniques, and Procedures for Joint Air Traffic Control.* 18 April 2014.

ATP 3-53.1. *Military Information in Special Operations*. 23 April 2015.

ATP 3-53.2. *Military Information in Conventional Operations*. 7 August 2015.

ATP 3-55.3. *ISR Optimization Multi-Service Tactics, Techniques, and Procedures for Intelligence, Surveillance, and Reconnaissance Optimization.* 14 April 2015.

ATP 3-55.6. *ATCARS Multi-Service Tactics, Techniques, and Procedures for Airborne Target Coordination and Attack Radar Systems.*22 October 2012.

ATP 3-55.12. *Combat Camera Multi-Service Tactics, Techniques, and Procedures for Combat Camera (COMCAM) Operations*. 12 April 2013.

ATP 3-57.10. *Civil Affairs Support to Populace and Resources Control*. 6 August 2013.

ATP 3-57.20. *Multi-Service Techniques for Civil Affairs Support to Foreign Humanitarian Assistance.* 15 February 2013.

ATP 3-57.30. *Civil Affairs Support to Nation Assistance.* 1 May 2014.

ATP 3-57.60. *Civil Affairs Planning.* 27 April 2014.

ATP 3-57.70. *Civil-Military Operations Center.* 5 May 2014.

ATP 3-57.80. *Civil-Military Engagement.* 31 October 2013.

ATP 3-60. *Targeting*. 7 May 2015.

ATP 3-60.1. *Dynamic Targeting Multi-Service Tactics, Techniques, and Procedures for Dynamic Targeting.* 10 September 2015.

ATP 3-60.2. *SCAR Multi-Service Tactics, Techniques, and Procedures for Strike Coordination and Reconnaissance.* 10 January 2014.

ATP 3-75. *Ranger Operations*. 26 June 2015.

ATP 3-76. *Special Operations Aviation*. 15 October 2015.

ATP 3-90.8. *Combined Arms Countermobility Operations*. 17 September 2014.

ATP 3-90.15. *Site Exploitation*. 28 July 2015.

ATP 3-90.90. *Army Tactical Standard Operating Procedures.* 1 November 2011.

ATP 3-91. *Division Operations*. 17 October 2014.

ATP 3-91.1. *The Joint Air Ground Integration Center*. 18 June 2014.

ATP 3-93. *Theater Army Operations*. 26 November 2014.

ATP 4-0.1. *Army Theater Distribution*. 29 October 2014.

ATP 4-01.45. *TCO Multi-Service Tactics, Techniques, and Procedures for Tactical Convoy Operations.* 18 April 2014.

ATP 4-02.2. *Medical Evacuation*. 12 August 2014.

ATP 4-02.3. *Army Health System Support to Maneuver Forces*. 9 June 2014.

ATP 4-02.55. *Army Health System Support Planning*. 16 September 2015.

ATP 4-02.84. *Multi-Service Tactics, Techniques, and Procedures for Treatment of Biological Warfare Agent Casualties.* 25 March 2013.

ATP 4-11. *Army Motor Transport Operations.* 5 July 2013.

ATP 4-12. *Army Container Operations.* 10 May 2013.

ATP 4-13. *Army Expeditionary Intermodal Operations.* 16 April 2014.

ATP 4-14. *Expeditionary Railway Center Operations.* 29 May 2014.

ATP 4-15. *Army Watercraft Operations*. 3 April 2015.

ATP 4-32. *Explosive Ordnance Disposal (EOD) Operations*. 30 September 2013.

ATP 4-32.2. *Explosive Ordnance Multi-Service Tactics, Techniques, and Procedures for Explosive Ordnance.* 15 July 2015.

ATP 4-32.16. *EOD Multi-Service Tactics, Techniques, and Procedures for Explosive Ordnance Disposal.* 6 May 2015.

ATP 4-35. *Munitions Operations and Distribution Techniques*. 5 September 2014.

ATP 4-42.2. *Supply Support Activity Operations*. 9 June 2014.

ATP 4-43. *Petroleum Supply Operations*. 6 August 2015.

ATP 4-44. *Water Support Operations.* 2 October 2015.

ATP 4-48. *Aerial Delivery*. 23 June 2014.

ATP 4-90. *Brigade Support Battalion*. 2 April 2014.

ATP 4-92. *Contracting Support to Unified Land Operations*. 15 October 2014.

ATP 5-0.1. *Army Design Methodology*. 1 July 2015.

ATP 5-19. *Risk Management*. 14 April 2014.

ATP 6-01.1. *Techniques for Effective Knowledge Management*. 6 March 2015.

ATP 6-02.40. *Techniques for Visual Information Operations*. 27 October 2014.

ATP 6-02.75. *Techniques for Communications Security (COMSEC) Operations*. 17 August 2015.

ATTP 3-90.4. *Combined Arms Mobility Operations*. 10 August 2011.

ATTP 4-10. *Operational Contract Support Tactics, Techniques, and Procedures*. 20 June 2011.

FM 1-04. *Legal Support to the Operational Army.* 18 March 2013.

FM 1-06. *Financial Management Operations.* 15 April 2014.

FM 2-0. *Intelligence Operations*. 15 April 2014.

FM 2-22.2. *Counterintelligence*. 21 October 2009.

FM 2-22.3. *Human Intelligence Collector Operations*. 6 September 2006.

FM 2-91.6. *Soldier Surveillance and Reconnaissance: Fundamentals of Tactical Information Collection.* 10 October 2007.

FM 3-01. *U.S. Army Air and Missile Defense Operations.* 15 April 2014.

FM 3-01.7. *Air Defense Artillery Brigade Operations.* 11 February 2010.

FM 3-04. *Army Aviation.* 29 July 2015.

FM 3-05. *Army Special Operations*. 9 January 2014.

FM 3-05.231. *Special Forces Personnel Recovery*. 13 June 2003.

FM 3-06. *Urban Operations*. 26 October 2006.

FM 3-07. *Stability.* 2 June 2014.

FM 3-09. *Field Artillery Operations and Fire Support*. 4 April 2014.

FM 3-11. *Multi-Service Doctrine for Chemical, Biological, Radiological, and Nuclear Operations.* 1 July 2011.

FM 3-13. *Inform and Influence Activities.* 25 January 2013.

FM 3-14. *Army Space Operations*. 19 August 2014.

FM 3-16. *The Army in Multinational Operations*. 8 April 2014.

FM 3-18. *Special Forces Operations*. 28 May 2014.

FM 3-21.10. *The Infantry Rifle Company*. 27 July 2006.

FM 3-22. *Army Support to Security Cooperation*. 22 January 2013.

FM 3-24. *Insurgencies and Countering Insurgencies*. 13 May 2014.

FM 3-27. *Army Global Ballistic Missile Defense Operations*. 31 March 2014.

FM 3-34. *Engineer Operations*. 2 April 2014.

FM 3-38. *Cyber Electromagnetic Activities*. 12 February 2014.

FM 3-39. *Military Police Operations*. 26 August 2013.

FM 3-50. *Army Personnel Recovery*. 2 September 2014.

FM 3-52. *Airspace Control*. 8 February 2013.

FM 3-53. *Military Information Support Operations*. 4 January 2013.

FM 3-55. *Information Collection*. 3 May 2013.

FM 3-57. *Civil Affairs Operations*. 31 October 2011.

FM 3-81. *Maneuver Enhancement Brigade.* 21 April 2014.

FM 3-90-1. *Offense and Defense, Volume 1*. 22 March 2013.

FM 3-90-2. *Reconnaissance, Security, and Tactical Enabling Tasks, Volume 2.* 22 March 2013.

FM 3-94. *Theater Army, Corps, and Division Operations*. 21 April 2014.

FM 3-96. *Brigade Combat Team*. 8 October 2015.

FM 3-98. *Reconnaissance and Security Operations*. 1 July 2015.

FM 3-99. *Airborne and Air Assault Operations*. 6 March 2015.

FM 4-01. *Army Transportation Operations*. 3 April 2014.

FM 4-02. *Army Health System*. 26 August 2013.

FM 4-02.7. *Multiservice Tactics, Techniques, and Procedures for Health Service Support in a Chemical, Biological, Radiological, and Nuclear Environment.* 15 July 2009.

FM 4-30. *Ordnance Operations*. 1 April 2014.

FM 4-40. *Quartermaster Operations*. 22 October 2013.

FM 4-95. *Logistics Operations*. 1 April 2014.

FM 6-0. *Commander and Staff Organization and Operations*. 5 May 2014.

FM 6-02. *Signal Support to Operations*. 22 January 2014.

FM 6-02.53. *Tactical Radio Operations*. 5 August 2009.

FM 6-05. *CF-SOF, Multi-Service Tactics, Techniques, and Procedures for Conventional Forces and Special Operations Forces Integration, Interoperability, and Interdependence*. 13 March 2014.

FM 6-22. *Leader Development*. 30 June 2015.

FM 27-10. *The Law of Land Warfare*. 18 July 1956.

FM 100-30. *Nuclear Operations*. 29 October 1996.

UNITED STATES LAW

Most acts and public laws are available at http://thomas.loc.gov/home/thomas.php.

Title 10. United States Code. *Armed Forces*.

Title 22. United States Code. *Foreign Relations and Intercourse*.

Title 29. United States Code. *Labor*.

Title 32. United States Code. *National Guard*.

GENEVA CONVENTION

1949 Geneva Convention relative to the Treatment of Prisoners of War. Available at https://www.icrc.org/ihl/INTRO/375.

PRESCRIBED FORMS

None

REFERENCED FORMS

Unless otherwise indicated, DA forms are available on the Army Publishing Directorate (APD) Web site at http://www.apd.army.mil/. DD forms are available on the Office of the Secretary of Defense Web site at www.dtic.mil/whs/directives/infomgt/forms/formsprogram.htm. Standard and optional forms are available on the U.S. General Services Administration Web site at www.gsa.gov.

DA Form 2028. *Recommended Changes to Publications and Blank Forms*.

DD Form 1833-Test (V2). *Isolated Personnel Report (ISOPREP) Instructions*.